Manual of Engineering Drawing

Third edition

Colin H Simmons

I.Eng, FIED, Mem ASME.
Engineering Standards Consultant

Member of BS. & ISO Committees dealing with
Technical Product Documentation specifications
Formerly Standards Engineer, Lucas CAV

Dennis E Maguire

CEng. MIMechE, Mem ASME, R.Eng.Des, MIED

Design Consultant
Formerly Senior Lecturer, Mechanical and
Production Engineering Department, Southall College
of Technology
City & Guilds International Chief Examiner in
Engineering Drawing

Neil Phelps

IEng MIED, MIET
Practicing mechanical design engineer.

Member of BSI and ISO Committees dealing with
Technical Product Specification and Documentation

AMSTERDAM • BOSTON • HEIDELBERG • LONDON • NEW YORK • OXFORD
PARIS • SAN DIEGO • SAN FRANCISCO • SINGAPORE • SYDNEY • TOKYO

Newnes is an imprint of Elsevier

Newnes is an imprint of Elsevier
Linacre House, Jordan Hill, Oxford OX2 8DP, UK
30 Corporate Drive, Suite 400, Burlington, MA 01803, USA

First edition by Arnold 1995
Reprinted by Butterworth Heinemann 2001, 2002
Second edition 2004
Reprinted 2004, 2005, 2006

British Library Cataloguing in Publication Data
Simmons, C. H. (Colin H.)
  Manual of engineering drawing : a guide to ISO and ASME standards. - 3rd ed.
  1. Engineering drawings 2. Mechanical drawing - Standards
  I. Title II. Maguire, D. E. (Dennis E.)
  604.2

Library of Congress Cataloging-in-Publication Data
Library of Congress Catalog Number: 2008938627

ISBN–13: 978-0-7506-8985-4

Printed and bound in Great Britain

09 10 11 12 13   10 9 8 7 6 5 4 3 2 1

Contents

Preface

I received the request to prepare a third edition of *Manual of Engineering Drawing* with mixed feelings. It was not that I did not want to do a revision, in fact I was keen to do so, being very conscious that some of the contents was in need of updating to reflect the latest developments, made by the ever-changing world of technology, and by ISO/BS Standardization, and I also saw it as an opportunity to enhance the book's content, by introducing new chapters on topical subject matters. But, I was aware that my dear friend and co-author over the past thirty years would be unable to play a part in this revision by virtue of ill-health.

I discussed my dilemma with Neil Phelps, a practising mechanical design engineer and fellow colleague, on various ISO and British Standards committees, with whom I have worked closely for many years, and was delighted when Neil expressed his desire and willingness to assist in the revision and become a co-author.

I welcome Neil on board and feel assured that with his valued expertise, input and acumen this *Manual of Engineering Drawing* will enjoy continued success in the future, as it as proven to be over the past decades.

This latest edition of the *Manual of Engineering Drawing* has been revised to include and explain latest developments in the fields of Technical Product Specification and Geometric Product Specification, in line with the latest published ISO, European, and British Standards, including BS 8888:2008–'Technical Product Specification', which in 2000 replaced BS 308 'Recommendations for Engineering Drawing Practice'.

Included in the revised updates are:
The importance and advantages that may be obtained, by having an effective Configuration Management and Control, within a Management system, whether the system be of a highly sophisticated CAD type or that of a manual type.

Computing developments and the impacts on industry and commerce in relation to CAD Organization and various applications.

In line with 'Standardization being a continuous process medium,' this revision also includes the addition of the following four new comprehensive chapters:

'3 D Annotation' which deals with the Digital Product Definition Data Practices, aligned to ISO 16792 Technical Product Documentation: Management and presentation of digital product definition data and ASME Y 14.41.

'The Duality Principle' this being a Geometric Product specification approach based on the concept that any given workpiece exists in several different "worlds" or as several different versions at the same time, as considered by the Designer and the Verification Engineer. It is the essential link between the Design Intent and the Verification of the end product.

'The Differences between the American ASME Y 14.5 M Geometric Dimensioning and Tolerancing (GD & T) and ISO/BS 8888 Geometric Dimensioning and Tolerancing Standards', the subtle differences of common terms, and the effects of them being interpreted in different ways are identified and analysed …

'Surface Texture' The Composition and Application of the Related Graphical Symbology
The mandatory positions for the indication of surface texture requirements are illustrated and, explained in line with BS/EN/ISO 1302 compliances.

The text that follows covers the basic aspects of engineering drawing practice required by college and university students, and also professional drawing office personnel. Applications show how regularly used standards should be applied and interpreted.

Geometrical constructions are a necessary part of engineering design and analysis and examples of two-and three-dimensional geometry is provided. Practice is invaluable, not only as a mean of understanding principles, but in developing the ability to visualize shape and form in three dimensions with a high degree of fluency. It is sometimes forgotten that not only does a draughtsman produce original drawings but is also required to read and absorb the content of drawings he receives without ambiguity.

The section on engineering diagrams is included to stimulate and broaden technological interest, further study, and be of value to students engaged on project work. Readers are invited to redraw a selection of the examples give for experience, also to appreciate the necessity for the insertion and meaning of every line.

Please accept our apologies for continuing to use the term 'draughtsmen', which is the generally understood collective noun for drawing office personnel, but implies equality in status.

In conclusion, may we wish all readers every success in their studies and careers. We hope they will obtain much satisfaction from employment in the absorbing activities related to creative design and considerable pleasure from the construction and presentation of accurately defined engineering drawings incapable of misinterpretation.

Colin Simmons

Acknowledgements

The authors express their special thanks to the British Standards Institution Chiswick High Road, London, W4 4AL for kind permission to reprint extracts from their publications.

We are also grateful to the International Organization for Standardization, Genève 20, Switzerland, for granting us permission to use extracts from their publications.

We very much appreciate the encouragement and friendly assistance given to us by:-

Ford Motor Company Ltd
SKF (UK) Ltd
KGB Micros Ltd
Norgren Martonair Ltd
Loctite Holdings Ltd
Staefa Control System Ltd
Autodesk Ltd

Barber and Colman Ltd
Bauer Springs Ltd
Delphi Diesel Systems
GKN Screws and Fasteners Ltd
Glacier Vandervell Ltd
GGB Bearing Technologies
F S Ratcliffe Ltd
Salterfix Ltd

Matthew Deans and his staff at Elsevier:
Jonathan Simpson, Melanie Benson, and Lyndsey Dixon.

To, Brian and Ray for sheet metal and machine shop examples, models.

Our final thanks go to Colin's and Dennis' patient spouses for their understanding and encouragement since work was started in 1973 on the first edition of *Manual of Engineering Drawing*.

Chapter 1

Drawing office management and organization

Every article used in our day-to-day lives will probably have been produced as a result of solutions to a sequence of operations and considerations, namely:

1. Conception
2. Design and analysis
3. Manufacture
4. Verification
5. In-service (maintenance)
6. Disposal.

The initial stage will commence when an original marketable idea is seen to have a possible course of development. The concept will probably be viewed from an artistic and a technological perspective.

The appearance and visual aspects of a product are very important in creating an acceptable good first impression.

The technologist faces the problem of producing a sound, practical, safe design, which complies with the initial specification and can be produced at an economical cost.

During every stage of development there are many progress records to be maintained and kept up to date so that reference to the complete history is available to responsible employees and regulatory bodies.

For many years various types of drawings, sketches and paintings have been used to convey ideas and information. In the last decade 3D models and rapid prototypes have also become a common way of conveying design intent. However, a good recognizable picture will often remove ambiguity when discussing a project and assist in overcoming a possible language barrier.

British Standards are listed in the British Standards Catalogue and the earliest relevant Engineering Standards date back to 1903. Standards were developed to establish suitable dimensions for a range of sizes of metal bars, sheets, nuts, bolts, flanges, etc. following the Industrial Revolution and used by the Engineering Industry. The first British Standard for Engineering Drawing Office Practice published in September 1927 only contained 14 clauses as follows:

1. Sizes of drawings and tracings, and widths of tracing cloth and paper
2. Position of drawing number, date and name
3. Indication of scale
4. Method of projection
5. Types of line and writing
6. Colour of lines
7. Dimension figures
8. Relative importance of dimensions
9. Indication of materials on drawings
10. Various degrees of finish
11. Screw threads
12. Flats and squares
13. Tapers
14. Abbreviations for drawings.

There were also five figures illustrating:

1. Method of projection
2. Types of line
3. Views and sections
4. Screw threads
5. Tapers.

First angle projection was used for the illustrations and the publication was printed on A5 sheets of paper.

During the early days of the Industrial Revolution manufacturers simply compared and copied component dimensions to match those used on the prototype. However, with the introduction of quantity production where components were required to be made at different factory sites, measurement by more precise means was essential. Individual manufacturers developed their own standard methods. Clearly, for the benefit of industry in general a National Standard was vital. Later the more comprehensive British Standard of Limits and Fits was introduced. There are two clear aspects, which are necessary to be considered in the specification of component drawings:

1. The drawing shows the dimensions for the component in three planes. Dimensions of the manufactured component need to be verified because some variation of size in each of the three planes (length, breadth, and thickness) will be unavoidable. The designer's contribution is to provide a Characteristics Specification, which in current jargon is defined as the 'Design Intent Measurand'.
2. The metrologist produces a 'Characteristics Evaluation' which is simply the Measured Value.

The drawing office is generally regarded as the heart of any manufacturing organization. Products, components, ideas, layouts, or schemes which may be presented by a designer in the form of rough freehand sketches, may be developed stage by stage into working drawings and annotated 3D models by the draughtsman. There is generally very little constructive work which can be done by other departments within the firm without an approved drawing of some form being available. The drawing is the universal means of communication.

Drawings are made to an accepted standard, and in the United Kingdom, it is BS 8888, containing normative and informative references to international standards. These standards are acknowledged and accepted throughout the world.

The contents of the drawing (and annotated 3D models) are themselves, where applicable, in agreement with separate standards relating to materials, dimensions, processes, etc. Larger organizations employ standards engineers who ensure that products conform to British and also international standards where necessary. Good design is often the product of teamwork where detailed consideration is given to the aesthetic, economic, ergonomic and technical aspects of a given problem. It is therefore necessary to impose the appropriate standards at the design stage, since all manufacturing instructions originate from this point.

A perfect drawing communicates an exact requirement, or specification, which cannot be misinterpreted and which may form part of a legal contract between supplier and user.

Engineering drawings can be produced to a good professional standard if the following points are observed:

a. the types of lines used must be of uniform thickness and density;
b. eliminate fancy printing, shading and associated artistry;
c. include on the drawing only the information which is required to ensure accurate clear communication;
d. use only standard symbols and where no other method of specification exist, appropriate abbreviations;
e. ensure that the drawing is correctly dimensioned (adequately but not over-dimensioned) with no unnecessary details.

Remember that care and consideration given to small details make a big contribution towards perfection, but that perfection itself is no small thing. An accurate, well-delineated engineering drawing can give the draughtsman a responsible considerable pride and job satisfaction.

The field of activity of the draughtsman may involve the use, or an appreciation, of the following topics.

1. *Company communications* Most companies have their own systems which have been developed over a period of time for the following:
   (a) internal paperwork;
   (b) numbering of drawings and contracts;
   (c) coding of parts and assemblies;
   (d) production planning for component manufacture;
   (e) quality control and inspection;
   (f) updating, modification, and reissuing of drawings.
2. *Company standards* Many drawing offices use their own standard methods which arise from satisfactory past experience of a particular product or process. Also, particular styles may be retained for easy identification, e.g., certain prestige cars can be recognized easily since some individual details, in principle, are common to all models.
3. *Standards for dimensioning* Interchangeability and quality are controlled by the application of practical limits, fits and geometrical tolerances.
4. *Material standards* Physical and chemical properties and non-destructive testing methods must be borne in mind. Note must also be taken of preferred sizes, stock sizes, and availability of rod, bar, tube, plate, sheet, nuts, bolts, rivets, etc., and other bought-out items.
5. *Draughting standards and codes of practice* Drawings must conform to accepted standards, but components are sometimes required which in addition must conform to certain local requirements or specific regulations, for example relating to safety when operating in certain environments or conditions. Assemblies may be required to be flameproof, gastight, waterproof, or resistant to corrosive attack, and detailed specifications from the user may be applicable.
6. *Standard parts* are sometimes manufactured in quantity by a company, and are used in several different assemblies. The use of standard parts reduces an unnecessary variety of materials and basically similar components.
7. *Standards for costs* The draughtsman is often required to compare costs where different methods of manufacture are available. A component could possibly be made by forging, by casting, or by fabricating and welding, and a decision as to which method to use must be made. The draughtsman must obviously be well aware of the manufacturing facilities and capacity offered by his own company, the costs involved when different techniques of production are employed, and also an idea of the likely costs when work is sub-contracted to specialist manufacturers, since this alternative often proves an economic proposition.
8. *Data sheets* Tables of sizes, performance graphs, and conversion charts are of considerable assistance to the design draughtsman.

Figure 1.1 shows the main sources of work flowing into a typical industrial drawing office. The drawing office provides a service to each of these sources of supply, and the work involved can be classified as follows.

1. *Engineering* The engineering departments are engaged in:
   (a) current production;
   (b) development;
   (c) research;

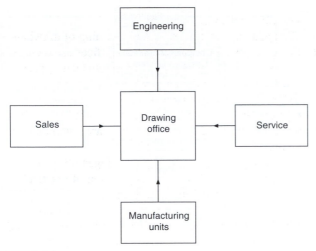

FIGURE 1.1

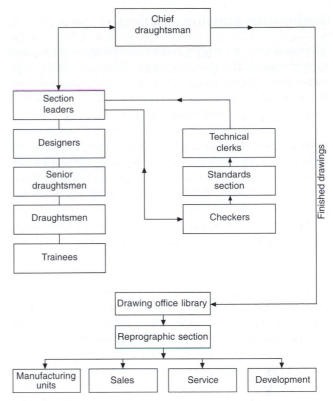

FIGURE 1.2

(d) manufacturing techniques, which may include a study of metallurgy, heat-treatment, strength of materials and manufacturing processes;

(e) advanced project planning;

(f) field testing of products.

2. *Sales* This department covers all aspects of marketing existing products and market research for future products. The drawing office may receive work in connection with: (a) general arrangement and outline drawings for prospective customers; (b) illustrations, charts and graphs for technical publications; (c) modifications to production units to suit customers' particular requirements; (d) application and installation diagrams; (e) feasibility investigations.

3. *Service* The service department provides a reliable, prompt and efficient after-sales service to the customer. The drawing office receives work associated with (a) maintenance tools and equipment; (b) service kits for overhauls; (c) modifications to production parts resulting from field experience; (d) service manuals.

4. *Manufacturing units* Briefly, these cover all departments involved in producing the finished end-product. The drawing office must supply charts, drawings, schedules, etc. as follows:

   a. working drawings of all the company's products;
   b. drawings of jigs and fixtures associated with manufacture;
   c. plant-layout and maintenance drawings;
   d. modification drawings required to aid production;
   e. reissued drawings for updated equipment;
   f. drawings resulting from value analysis and works' suggestions.

Figure 1.2 shows the organization in a typical drawing office. The function of the chief draughtsman is to take overall control of the services provided by the office. The chief draughtsman receives all work coming into the drawing office, which he examines and distributes to the appropriate section leader. The section leader is responsible for a team of draughtsmen of various grades. When work is completed, the section leader then passes the drawings to the checking section. The standards section scrutinizes the drawings to ensure that the appropriate standards have been incorporated. All schedules, equipment lists, and routine clerical work are normally performed by technical clerks. Completed work for approval by the chief draughtsman is returned via the section leader.

Since drawings may be produced manually, or by electronic methods, suitable storage, retrieval and duplication arrangements are necessary. Systems in common use include:

(a) filing by hand into cabinets the original master drawings, in numerical order, for individual components or contracts;

(b) microfilming and the production of microfiche;

(c) computer storage.

The preservation and security of original documents is of paramount importance in industry. It is not normal practice to permit originals to leave the drawing office. A drawing may take a draughtsman several weeks to develop and complete and therefore has considerable value. The reprographic staff will distribute copies which are relatively inexpensive for further planning, production and other uses. A library section will maintain and operate whatever archive arrangements are in operation. A large amount of drawing office work comes from continuous product development and modification so easy access to past designs and rapid information retrieval is essential.

Engineering drawing practices

The comments so far refer to drawing offices in general and typical organizational arrangements which are likely to be found within the engineering industry. Good communication by the use of drawings of quality relies on ensuring that they conform to established standards.

BS 5070, Parts 1, 3, and 4 dealing with engineering diagram drawing practice, is a companion standard to BS 8888 and caters for the same industries; it provides recommendations on a wide variety of engineering diagrams. Commonly, as a diagram can be called a 'drawing' and a drawing can be called a 'diagram', it is useful to summarize the difference in the scopes of these standards. BS 8888 covers what are commonly accepted to be drawings that define shape, size and form. BS 5070 Parts 1, 3, and 4 covers diagrams that are normally associated with flow of some sort, and which relate components (usually indicated by symbols) functionally one to another by the use of lines, but do not depict their shape, size or form; neither may they in general indicate actual connections or locations.

Therefore, any drawing or diagram, whether produced manually or on computer aided draughting equipment, must conform to established standards and will then be of a satisfactory quality for commercial understanding, use and transmission by electronic and microfilming techniques. All of the examples which follow conform to the appropriate standards.

Drawing practice and the computer (CAD: Computer aided draughting and design)

The computer has made a far bigger impact on drawing office practices than just being able to mimic the traditional manual drawing board and tee square technique. However, it depends on drawing office requirements and if only single, small, two dimensional drawings and sketches are occasionally required and storage of originals is not an issue, then a manual drawing system may still be appropriate. CAD can however perform a much more effective role in the design process and many examples of its ability follow – but it will not do the work on its own. The input by the draughtsman needs to follow the same standards applied in the manual method and this fact is often not understood by managers hoping to purchase CAD and obtain immediate answers to design enquiries. The draughtsman needs the same technical appreciation as before plus additional computing skills to use the varied software programs which can be purchased.

To introduce CAD, an organization must set out clear objectives which are appropriate to their present and future requirements and Fig. 1.3 includes aspects of policy which could appear in such plans. The following need consideration:

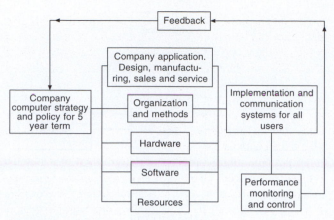

FIGURE 1.3 General computer policy relationships.

(a) CAD management roles;
(b) creation, training and maintenance of capable CAD operators;
(c) CAD awareness of design project team members in addition to their leaders;
(d) the flow of work through the system and the selecting of suitable types of project;
(e) associated documentation;
(f) possible changes to production methods;
(g) needs involving the customer and sub-contractor;
(h) system needs relating to planning, security and upgrading;
(i) CAD library and database (storage of drawings, symbols, etc.) and archive procedures;
(j) Configuration management.

When introducing or replacing a CAD system it is very important to take time to consider points (f) to (i) as these can have a major impact on through life costs. If you ask yourself the question 'In 5, 10, or 15+ years time, will I still need to maintain the electronic product information'. If the answer is 'yes or possibly' then the following needs careful consideration.

The need to consider possible down stream applications such as Computer aided Machining (CAM), Inspection (CAI) and Analysis, i.e., finite element analysis (FEA) and compatibility with existing systems (or the requirement to replace them) being very important.

Compatibility with potential customer and sub-contractor systems should also be investigated because if this is a controlling requirement then selection of a suitable system could be limited.

Longevity of the manufactured products will also play a significant role in system selection. In many areas of industry regulatory or contract requirements require retention of design records for a period of time after the product is withdrawn from service. This period of time, when added to the length of service could be considerable, i.e., in excess of 25 years. As a result, costs associated with hardware and software upgrades and system maintenance need to be considered. In some cases, the costs associated with upgrades can be offset against new products, but nevertheless, ongoing maintenance of legacy databases can be costly. It is interesting to note that this was a problem that was often not foreseen in the early days of CAD. The rush by industry to move from the drawing board to CAD has in some cases, led to companies changing CAD systems and struggling to maintain legacy drawings without costly redraws. Even today, with some very good CAD file translators on the market, 100% data transfer can't always be guaranteed without some remedial work.

A possible solution for consideration could be maintaining the drawings of a stable product on microfilm or possibly an enduring electronic format such as Adobe Acrobat files (.PDF). Some companies are already future proofing their design information by retaining drawings and models in their native CAD format and also in .PDF format and on microfilm.

Configuration control is also equally important to CAD systems as it is to manual systems and extra care needs to be taken to ensure unauthorised access to master files.

Many similar aspects will be appropriate in particular applications but good intentions are not sufficient. It is necessary to quantify objectives and provide dates, deadlines, numbers, individual responsibilities and budgets which are achievable if people are to be stretched and given incentive after full consultation. Present lines of communication will probably need to be modified to accommodate CAD, and planning integration is vital. A possible approach here is the appointment of a CAD Director with the ultimate responsibility for CAD technology assisted by a Systems Manager and an Applications Manager.

A CAD Director has the task of setting and implementing objectives and needs to be in a position to define binding policy and direct financial resources. He will monitor progress. A Systems Manager has the role of managing the computer hardware, the software and the associated data. Company records and designs are its most valuable asset. All aspects of security are the responsibility of the Systems Manager. Security details are dealt with in the next chapter. The Applications Manager is responsible for day-to-day operations on the CAD system and the steady flow of work through the equipment. He will probably organize training for operators in the necessary computer skills. Both of these managers need to liaise with the design project leaders to provide and maintain a draughting facility which is capable of increasing productivity to a considerable degree.

Figure 1.4 shows the probable position of the CAD Director in the management structure. His department will be providers of computer services to all other computer users within the company.

Why introduce BS 8888 and withdraw BS 308?

For 73 years, BS 308 was a highly regarded drawing office practice document. Why the change and what was behind the decision to withdraw BS 308 and replace it with BS 8888?

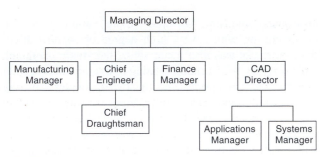

FIGURE 1.4

A drawing standard

From time immemorial, drawings have been the medium used to convey ideas and intentions. Hence the adage that 'a picture is worth a thousand words'. No need for language, the picture tells it all. In recent years there has, unfortunately, developed another opinion since CAD appeared on the scene, that there is no need for a draughtsman now as the computer does it all. The truth of the matter is that the computer is able to extend the range of work undertaken by the draughtsman and is really a very willing slave, the quality of the product produced from a drawing is solely down to the competence of the designer or draughtsman. The evolution of the Industrial Revolution required the 'pictures' to be more detailed. In the pre-mass-production era, manufacture was based on 'matched fits', with the assistance of verbal communication. The advent of mass production however, demanded more specific and precise specifications.

A national form of draughting presentation was needed to promote a common understanding of the objectives and in September 1927, BS 308 came to fruition, as the recognized National Code of Practice for Engineering Drawing.

The initial issue was A5-size and contained only 14 clauses. Dimensioning was covered in four paragraphs and tolerancing in only one. The recommendations were based on just two example drawings. The recommended projection was first angle.

Revisions

The life span of BS 308 was 73 years and five revisions were made. The first in December 1943, followed by others in 1953, 1964, 1972, and 1985. The 1972 revision was a major one, with the introduction of three separate parts replacing the single document:

The fifth (1985) revision replaced the Imperial standard with a Metric edition.

BS 308 was finally withdrawn and replaced by BS 8888 in 2000. The revisions were necessary to keep abreast of technological innovations.

As manufactured products became more sophisticated and complex, the progress and development of manufacturing and verification techniques accelerated. Advances in the electronics industry ensured more applications in manufacturing with a very high degree of sophistication. Much progress was also made since that single paragraph in the original 1927 version relating to tolerancing, together with the four paragraphs and the two examples covering dimensioning. Geometrical tolerancing was not referred to at all in early versions. The subject gained prominence during the 1960s, especially when it was realized that a symbolic characterization would assist in the understanding of the subject by users and replace the use of lengthy notes relating to geometric controls.

This activity was addressed by the major revision in 1972 with the publication of Part 3, devoted entirely to the dimensioning of geometric tolerancing.

The replacement of BS 308

Formerly, the Chief Designer and the drawing office set, and were responsible for, company manufacturing standards and procedures, for other disciples to follow. This practice gradually eroded away because of the advancement of progressive and sophisticated techniques in the manufacturing and verification fields.

Increasing commercial pressure for Design for Manufacture and Design for Inspection, created the demand for equal status. During the period separate standards were gradually developed for design, manufacture and measurement. Each discipline utilized similar terms but often with slightly different interpretations despite their apparent commonality.

An urgent need to harmonize the meaning of these terms was recognized by ISO. An international meeting in 1989 formed a Joint Harmonization Group.

The Danish Standards Association funded a project to bring all design, measurement, and metrology standards together using definitions common to all, but with appendices for each discipline.

A full ISO committee (ISO/TC 213) was formed, with the Danish being responsible for the secretariat. The task allocated to this very vibrant committee progressed considerably, with many new international standards being published.

A major happening that would affect the future of BS 308 was the UK's agreement in 1993 with the European Standards Authority (CEN), whereby BSI would withdraw standards relating to technical drawing in favour of the implemented ISO standards covering the same subject. Initially, BSI systematically withdrew various clauses of BS 308 as the relevant ISO Standards were introduced.

PD 308 was introduced in June 1996 as a guidance document to assist the transition from BS 308 to the implementation of ISO drawing standards. In 1999, as was the case in 1927, major decisions were deemed necessary, and the following were made:

- To transfer the United Kingdom totally to the ISO Standards base.

- To prepare an applications standard to serve as both a specification for specifying and graphically representing products, and as a route map to the ISO Standards.
- To withdraw BS 308.

From this positive commitment, BS 8888 was created and published on 15 August 2000.

The complete comprehensive title of BS 8888 is:

BS 8888. Technical product specification (TPS). Specification.

Basic differences

The fundamental differences between BS 308 and BS 8888 are:

- The title: Technical product documentation (TPD) Specification for defining, specifying and graphically representing products.
- Confirmation of the conventional use of the comma as the decimal marker.
- BS 308 was a Code of Practice, a guidance document. BS 8888 is essentially an applications specification, providing a route map to over 150 ISO standards. The operative word is 'specification'. BS 8888 carried forward and contains a significant number of valuable clauses contained in BS 308, which, at present, is not in any ISO documentation.
- BS 8888 is capable of accommodating significant technical changes, known to be in development, plus the facility to accommodate future additions and changes.
- With over 150 related ISO standards, BS 8888 has a much broader field of application than its predecessor and its 30 related ISO standards.
- BS 8888 provides common understanding, and acceptance between the designer and the metrologist of 'uncertainty'. These are caused by differences between the Design Intent Measurand (Characteristics Specification) and the Measured Value (Characteristics Evaluation) of the actual manufactured part.
- BS 8888 is a uniform source of reference and will be regularly updated to keep abreast of developments as new international standards are finalized and implemented.
- It will capture any fundamental changes and will reflect moves towards an integrated system for definition, manufacture and verification.
- BS 8888 links each standard to the appropriate stage of the design process and lays the foundations for future development.

BS 8888 has been revised every two years, however, this period may be extended to three or five years in the future. BS 8888 is available as a hard copy, and interactive CD with all referenced standards and in the future as a Web Based facility.

Chapter 2

Product development and computer aided design

Work undertaken by a drawing office will vary considerably with different branches of industry. Generally, work of a 'design and make' nature will follow a plan which sets out stages in development from the time a potential client makes an enquiry until the completed product is delivered. The function of the product will dictate many of the associated activities.

A vehicle manufacturer will not design and make all of the parts used but subcontract components from specialists. The engine incorporates electrical and mechanical components and these need to conform to agreed upon specifications. They must also be designed for installation in specified areas and be suitable for operation in well-defined conditions. Component manufacturers strive to improve quality and performance in conjunction with the end user.

The stages in design and development for components in this category are shown typically, step-by-step in Fig. 2.1.

Step 1 A client requiring a certain product is often not completely familiar with specific details and needs the experience and advice from a specialist producer to clarify initial ideas. When a range of viable alternatives is presented, opinions can be focused and firm decisions made.

Step 2 The Chief Engineer or Design Authority in a company has the responsibility of producing the company specifications for a product including all applicable legislations, which the product will need to comply with and the levels of configuration control required through the product life cycle. He will no doubt seek advice where aspects of the total design are outside his range of experience, and where design is involved on the fringes of technology. However, a top executive plan needs to be carefully prepared because at the outset the company must know whether or not it wishes to entertain, or get involved with design proposals to satisfy the client. For example, while rewards may be great, the firm may not be able to cope with the scale of financial and labour demands and delivery requirements in view of current work. They simply may not wish to take the risk and, in view of available production capacity, the firm may prefer not to tender for a possible order.

Step 3 Drawings at this stage should be regarded only as provisional. The exercise is needed as an aid to think around the problem, with contributions being made by specialists within the firm to ensure feasibility.CAD (computer aided design) has many virtues at this stage of primary design. All information, defined in mathematical terms, can be stored in the system and manipulated on the display. After the basic geometry is established, design variations can be kept, and in redrawing alternatives sections of the previous proposals, which were found to be acceptable can be used repeatedly. At any point in development, the designer can take a printout, so that suggestions and comments can be made by other technical staff. Consideration of the level of configuration control to be applied at this stage is important. In lots of cases the decisions as to why and why not a decision was taken (Optioneering) needs to be formally recorded in some format.

It is essential that the company should appreciate the extent of their commitment if a firm order is accepted at a later date. This commitment includes not only the technical ability to complete the design and manufacture a satisfactory product but also the financial issues relating to its introduction on the factory production line.

Step 4 With the completion of preliminary design work an agreed design concept will be established, but it is necessary to obtain customer approval before work continues. If our product is to be used in conjunction with others in a large assembly, then, for example, expected overall dimensions and operational parameters need to be confirmed with the client before money is spent on further development.

Step 5 If all is well, working drawings will be prepared. These are not production drawings – at this stage, we as a company have only ensured that our proposals are in line with the requirements and that hopefully we shall be able to deliver. The object now is to prepare working drawings to formulate construction methods.

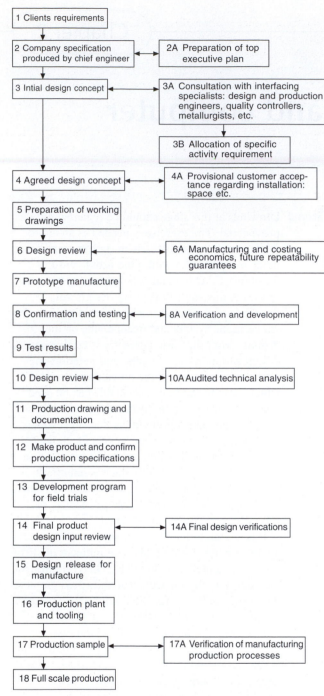

FIGURE 2.1

Step 6 A design review is necessary to check the feasibility of manufacturing to ensure that all aspects of design requirements have been incorporated in an economic manner and to guarantee future supplies.

Step 7 A prototype or a small batch may now be manufactured. The ultimate production methods of manufacture will not be employed here. For example, components which may be moulded could be machined from solid to eliminate casting costs.

Step 8 Prototypes are used for testing to make certain that operational requirements of the specification can be achieved. As a result design changes may be necessary. Product tests cover all areas where the component will be expected to function without failure, and these could include use in extremes of temperature and humidity, and also when subjected to shock, vibration and fatigue.

Step 9 Proven test results are vital to confirm the validity of these tests.

Step 10 A design review and analysis ensure that progress at this point will be acceptable in every technical aspect and to each responsible member of the team.

Step 11 Production drawing can commence now when the performance targets from the prototype have been confirmed. Drawings of the prototype will be reviewed and modifications made to use full-scale production processes during manufacture. For plant to be used efficiently, plans need to be prepared for loading and progressing work through the factory. The necessary documentation now commences.

Step 12 Manufacture of the final product following production of the prototype has involved modifications and different manufacturing processes. It is therefore prudent to check that the specifications can still be kept.

Step 13 Following trials where the equipment is used in its operational environment and its performance exhaustively checked, the design details can be released for full-scale production.

Step 14 Production involves not only the use of machines, but many jigs, fixtures, tools, gauges, inspection procedures need to be planned, and auxiliary equipment designed to move materials on and off production lines.

Step 15 Inevitably teething troubles occur and samples are taken to verify that all plant and equipment operates as planned. Economic production requires that down time is eliminated before full-scale production commences.

Computer aided draughting and design

CAD is much more than drawing lines and geometry by electronic means. Similarly by the purchase of a CAD system, a design does not emerge at the push of a button. 'Buy a computer and you do not need a draughtsman' is also very different from reality (see Chapter 1). The engineering designer is very much responsible for decisions taken at all technical stages between conception and production. The computer is an aid and performs as it is directed with rapidity and accuracy. The following notes are included to indicate areas of useful activity to assist the draughtsman.

The preparation of two- and three-dimensional drawings and models and the projection of associated views is the 'bread and butter' work in the drawing office. Service manuals use exploded views so that people with no technical training can follow assembly sequences. Children stick together model kits with guidance using pictorial diagrams.

CAD programs are available where a three-dimensional model can be produced automatically given two-dimensional views. From the dimensions of the component, the computer will calculate surface areas, volumes, weights for different materials, centres of gravity, moments of inertia, and radii of gyration; it can also use the applicable values for stress and other calculations, which are a necessary part of design. Computer models permit a study of special relationships and applications given in the chapter which follows. Models can be manipulated into pleasing forms for artistic approval or for the basis for Rapid Prototyping or Rapid Machining before production work follows. Previous techniques included modelling with plasticine and plaster, and applications ranged from ornaments to boat hulls and car bodies. CAD has revolutionized modelling capabilities.

Sales departments utilize 3D illustrations in brochures and literature for promotional applications. Desktop publishing from within the company can very simply use illustrations generated as part of the manufacturing process. The scanning of photographs into a CAD system is also an asset especially as photographic work can be retouched, manipulated and animated. Multimedia applications with video and slide presentations form a large part of selling and advertising.

Structural design requires a thorough knowledge of engineering material properties. Calculations of stress, strain and deflection are essential to determine proportions and dimensions in structural applications. Computers now have the ability to perform millions of calculations per second and with the availability of powerful desktop models, finite element analysis has developed as a principal method. One advantage of finite element analysis is that design engineers can produce better designs and eliminate dubious options during the conceptual design phase. CAD systems permit the rapid generation of models of proposed designs as wire frames. The component can be defined as a collection of small loaded elements. The computer memory stores details of all the geometric data to define each part of the frame. Numerical analysis will then verify whether or not the suggested design will be capable of supporting the expected loads. Formerly, stress calculations were time consuming and in the early days of computing, although the calculation time was considerably shorter, computer time was relatively expensive. This is not the case now and for this type of design work CAD is an essential tool in the drawing office. However, it is advisable to ensure the designer using these tools has a good understanding of traditional methods of calculation in order to have confidence in the electronic output.

CAD is very suitable for repetitive and fast documentation where a product is one in a range of sizes. Assume that we manufacture a range of motor driven pumps operating at different pressures. Many parts will be used in different combinations in the range and the computer database documentation is programmed accordingly. Company standard designs will be offered when enquiries are received. A computerized tender can be sent with the appropriate specification and technical details. On receipt of an order, all of the documentation relating to manufacture, testing, despatch, and invoicing will be available. An obvious advantage is the speed of response to the customer's enquiry.

CAD will be linked to CAM (computer aided manufacture) whenever possible. Documentation will include parts lists, materials details of parts to be manufactured or bought out, stock levels, computerized instructions for numerical controlled machine tools, instructions for automated assemblies, welding equipment, etc. Printed circuit boards can be designed on CAD and manufactured by CAM.

Production tooling requires the design of many jigs and fixtures. A jig is a device which holds the component or is held on to the component, locating the component securely and accurately. Its function is to guide the cutting tool into the component or for marking off or positioning. A fixture is similar to a jig but it does not guide the tool. Generally a fixture will be of heavier construction and clamped to the machine tool table where the operation will be performed. Jigs are used frequently in drilling and boring operations. Fixtures are a necessary part of tooling for milling, shaping, grinding, planning and broaching operations. The use of jigs and fixtures enables production to proceed with accuracy, and hence interchangeability due to the maintenance of tolerances (see Chapter 19) and especially by the use of unskilled or semiskilled labour and robotics.

The traditional method of jig and tool draughting was to draw the component in red on the drawing board. The jig or fixture would then be designed around the component. This process ensures that the part is located and clamped correctly, can be loaded and unloaded freely, and that the machining operation can be performed without hindrance.

With a CAD system, the component drawing can be shown in colour on one of the 'layers' (see Chapter 3) and design work undertaken on the other layers. If designing uses three-dimensional CAD systems, the tooling can be designed around the model of the product again using appropriate layering.

Machining operations need to be checked to ensure that tools and cutters do not foul any other equipment in the vicinity. The path taken by the tool into its cutting position should be the most direct and the shortest in time. The actual cutting operation will take a different time and the tool may traverse the component several times, cutting away more material on each occasion. Machining sequences can be simulated on the screen and when the optimum method has been obtained, the numerical program is prepared. All relevant data for the machining operation is converted into coded instructions for continuous production.

Programs are available for the economic use of metallic and non-metallic materials. Many engineering components are manufactured by flame, laser, wire or water jet cutting intricate shapes from plate or sheet and these need to be positioned to minimize scrap. The cutting head is guided by computer using the X and Y coordinates at each point along the curve. Other applications use a variety of cutters and saws to shape materials singly or heaped into a pile, such as foams in upholstery or dress fabrics.

The tool draughtsman, for example, will use many standardized components in tooling and designing associated handling equipment for production. If a range of parts is similar, it is common practice to produce a single drawing with dimensions in a table of the separate features. A typical example is given in Fig. 7.2 and is the normal manual draughting procedure. CAD can however use a parametric technique where the component drawing is dimensioned by algebraic expressions understood by the computer. Each separate size of component will be given its own part number. When a particular part is required and called up, the computer calculates sizes and draws the part to the correct scale for the draughtsman to position where required on the assembly drawing. This is a very useful facility and only available through the introduction of CAD.

CAD always produces drawings finished to the same high standard, and of a uniform quality and style. All tracing costs are saved.

It will be seen from the above notes that CAD fits in with many of the separate procedures necessary for design and production, but it is vital that, before its introduction, software must be available with proven ability. Likewise, staff must receive training to extract the maximum advantages and benefits.

Technical product documentation

Individual companies generally develop their own systems largely depending on the type of work involved and the size of the undertaking, e.g., original designs, drawing revisions, modifications, repairs, new contracts, enquiries, and proposals.

These notes provide guidelines for new business routines where both manual and computer-based systems are used. They refer to internal communication within companies and between other organizations.

There are five short Standards dealing with the handling of computer-based technical information during the design process.

Part 1: BS EN ISO 11442–1. Security requirements.

This document details advice and precautions regarding the system installation, power supply, ventilation and cooling, magnetism and electrostatic environment, and also computer access.

Notes regarding service and maintenance, stand-by equipment and back-up copies are given. Useful comments relate to document authorization and copyright.

Part 2: BS EN ISO 11442–2. Original documentation.

Definitions are provided for various types of document used by industry in the Drawing Office.

Part 3: BS EN ISO 11442–3. Phases in the product design process.

Distribution of documents during each phase is detailed.

Part 4: BS EN ISO 11442–4. Document management and retrieval systems.

This section deals with activities in the design process and the handling of associated documents, e.g., identification and classification of administrative and technical documents. Provides helpful advice in the management of documentation in parallel with the phases of product development. Assistance also given for drawing revisions, document handling, classification and retrieval of data.

Ready-made 'Turnkey' data-processing systems are available and can be adapted by specialist suppliers.

Part 5: BS EN ISO 11442–5. Documentation in the conceptual design stage of the development phase.

Part 5 deals with documentation in the preparation of a design specification, design proposals and solutions.

Problems can arise from power cuts of short and extended time periods, from spikes, or fluctuations of power due to other electrical equipment being switched on. Stormy weather can cause surges and static build ups. A reliable power source with a stable supply is essential. Consideration should be given to the provision of a backup supply, if in doubt. Service and maintenance arrangements may require the issue of external contracts, as computer down time resulting in lost production can prove expensive.

Computers generate heat, and wide variations in environmental temperatures should be avoided. Air conditioning in the complex may be necessary if cooling is required and clean air cannot otherwise be guaranteed. Part of the computer complex may need to be out of bounds except to authorized personnel to maintain an acceptable environment. Care should be exercised in the selection of floor coverings and furniture to protect equipment from static electricity. Similarly tapes and discs need to be shielded from stray magnetic fields. Ensure that the CAD complex is kept locked and secure when not in use at night and weekends.

An organization must develop a routine for storing data on which company fortunes may depend. In the event of power failure, work in progress may be lost. It could also be lost due to operator error or computer malfunction, fire, flood, vandalism, etc. Backup routines must cover personal responsibility aspects, together with frequency of copying, storage medium and designated places of safety. Backup copies should not be stored in the same buildings as the originals.

Programs used for operating and applying CAD systems need to be checked at regular intervals to ensure that intended methods are being kept in practice. Computer aided designs and production information could easily be copied and some countries do not have legislation prohibiting unauthorized use. Documents should therefore include a clause

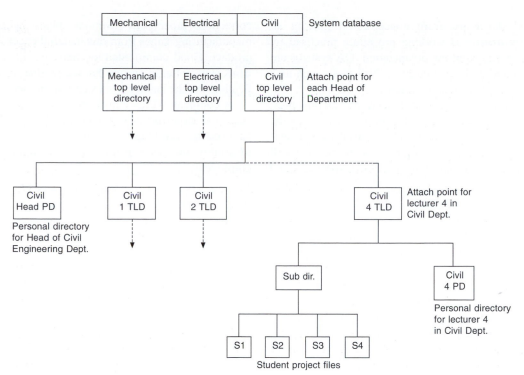

FIGURE 2.2   Directory tree for controlled access to database.

relating to copyright where design information is transmitted, it is recommended that the clause should appear before the text and again at the end.

Many grades of staff are involved in the design process; senior designers, detailers, checkers and technical clerks all make a positive contribution. Each member's duties must be carefully defined with rules applied, an authority given, so that each can only operate within his or her agreed sphere of activity. By means of passwords it is possible to access design information at appropriate levels. Revision procedures will ensure that modifications are only made at the correct point by authorized designated staff. Quality assurance systems require strict application of these methods.

Access into the computer network

Every CAD installation requires access responsibilities to be defined for the operating staff and the following example relates to an educational establishment.

A typical College of Technology may consist of three separate departments, each requiring to use a common computer facility where a central processing unit is installed. Each department is serviced using a tree and branch system leading to the desks of staff holding different levels of responsibility, and to student outlets in classrooms, drawing offices and laboratories. All members of staff and students need to gain access to the computer freely, and in their own time, and be able to store their work safely.

A Head of Department, however, may need to gain access to the students' work to monitor progress.

All members of the college staff would wish to have a personal file and keep confidential records. A lecturer must be free to allocate space to students in several classes, so he or she will open subdirectories as necessary and possibly delete work at the completion of a course.

Figure 2.2 shows a directory structure where access can only be made into the system provided the keyboard operator logs in a personal identity number. Each member of staff will be assigned two directories:

(a) a top level directory (TLD);
(b) a personal directory (PD).

The TLD is the attach point for the user into the system. The lecturer is free to open subdirectories for students' work and each student's file will be protected from the rest of the class. The Head of Department has access to a lecturer's TLD and through to a student's file.

The above system can be adapted for any graded organization where controlled access and protection for records is desirable.

Quality assurance

BS EN ISO 9000 series relates to quality systems and is produced in several sections. The principles of quality assurance (QA) embrace all activities and functions concerned with the attainment of quality. *BSI Quality Management Handbook QMH 100* is essential for reading.

Having purchased quality CAD equipment, the products which the company propose to manufacture need to be designed and developed from conception following an agreed *quality assurance* working procedure practised by all employees throughout the organization. QA systems are usually accredited and certified by a third party such as a professional institution or association.

An organization should be able to show that all drawings, documentation and necessary calculations relating to the design, are vigorously checked and approved by management. The stage by stage development of the product will follow an agreed work plan with checks, inspections and correction procedures. Similar plans will cover the manufacturing stages from raw material checks to the tested product. Good communication between all of the participants is essential to ensure that the product meets its specification and the customer's exact requirements.

A company which can demonstrate superior technical skill and expertise has a considerable asset which can be used to advantage in marketing. Proven excellence invariably increases pride and well-being in company employees.

Chapter 3

CAD organization and applications

Computing developments have made a rapid and immense impact on industry and commerce and as the degree of complexity has also increased, then training facilities have expanded accordingly. As a source of information and communication, the Technical Press and the Internet play a very important part. Journals from professional institutions offer impartial news, advice and guidance, opinions, and new product details. Manufacturers and the larger suppliers of CAD (computer aided design) equipment have set up centres around the country where exhibitions and demonstrations are organized. Higher education establishments, private organizations, and dealerships also give specialist courses for the benefit of students and users.

The mainstream engineering software programs have been written and developed in the United States and the United Kingdom. To perform complex tasks, additional programming may need to be seamlessly integrated so that they work in harmony as a unit.

There are literally hundreds of specialist applications available. Banks, building societies, airlines, all have their own systems and via the Internet, can freely communicate with each other. This fact has also given rise to another branch of industrial development, i.e. security.

Screen sizes have increased in size and the availability of the flat screen has reduced the size of workspace required by users.

The provision of multi-layers provides a very useful method of working on CAD. Imagine transparent sheets placed on top of each other, which may be shuffled and rearranged so that you can draw on the top. Each of the layers underneath the pile can be turned on or off, they may be given identification colours and selected parts of drawings moved from layer to layer if required. Assume that we want to draw plans for a house. Layer 1 could be used to draw a plan view of the building plot. Layout work is often easier if graph paper is used. On layer 2 we make our own construction grid, which is transparent graph paper with squares to any convenient scale of our choice. Using this grid under layer 3 we design a suitable ground floor layout. Copying the position of the outside walls from layer 3 and modified as required could start layer 4 showing the first floor layout. When all of the required plans and elevations are constructed, they can be repositioned on a drawing arrangement. If necessary, the site layout reduced to a smaller scale. When completed, the construction grid may be deleted. Tracing facilities and the ability to print layers together or apart are valuable draughting assets.

The physical equipment components of a computer system are known as the *hardware*. The programs and data used on the computer are defined as the *software*.

Another advantage of CAD is its ability to store line systems and other entities, which are frequently used on drawings. For example, software containing symbols to British, European and other International Standards is freely available for most engineering applications. The draughtsman can also create libraries of regularly used parts.

For repetitive use on a drawing, a typical item may be retrieved and positioned in seconds, also oriented at any angle to suit particular circumstances.

As a drawing aid, every CAD program must provide basic geometric features, permitting the operator to blend lines and arcs, etc. It is necessary in engineering drawing to be able to determine points of tangency between straight lines and curves and between curves of different radii.

Productivity is much improved by a program enabling you to easily draw polygons, ellipses, multiple parallel lines and multiple parallel curves. The speed of machine drawing is increased by the use of automatic fillets and chamfers. Layout work benefits when use is made of construction grids and the computer's ability to 'snap' automatically to particular geometric points and features, will speed the accurate positioning of line work. Copy, rotate and mirror facilities give assistance when drawing symmetrical parts. Automatic crosshatching within closed boundaries is useful in the construction of sectional views and when indicating adjacent parts and different materials. Many changes of hatch patterns are supplied with CAD programs. Filling areas in various colours is a requirement in artwork.

The ability to zoom in and out is an asset when drawing to scale. It is possible to work on fine detail in an assembly and then zoom out to observe the result in context.

CAD information is stored in digital form and hence, irrespective of the size of the final printed drawing; it is possible to accurately dimension components automatically.

Different 'type-set' and alternative style fonts are always supplied with CAD programs. If a special font is required to match an existing style then specialist vendors can supply. Alphabets in different languages present no problem. Quite clearly the physically largest affordable screen has many advantages. If the draughtsman is also involved with desktop publishing, it is ideal to be able to work on a screen that displays two A4 paper sheets side by side so that 'what you see is what you get'. The screen should give high resolution, necessary to provide an image that is flicker free. The quality of the display will have a big contribution to make in the avoidance of fatigue and eyestrain. Firsthand practical experience and a demonstration is important here for an ideal solution.

Plotting and printing equipment will vary according to drawing office requirements. It is true, however, that many CAD installations are judged by the quality of their plotted drawings. It is necessary to also have a demonstration and this will ensure that an excellent CAD system will have an output to do it justice.

A wide variety of plotters are available for reproductions from A4 to A0 in size, and in a quality suitable for production work or the most prestigious presentations.

Probably the best-known software in the drawing office is that from AutoCAD, which build products that conform to the most widely used DWG format permitting the transfer of information between networks.

In the 1970s, 2D drawing packages were introduced with the ability to slowly draw lines, circles and text. Rapid developments have taken place since with a vast increase in computing power. The computer industry has expanded, progressed and now produces software for an ever increasing number of engineering applications. Computing power is vital for the operation of highly sophisticated research projects, advanced design and modelling programs. Communication developments have had a profound effect regarding the methods that we use for our current solutions. We have the capability to transmit files of drawings and notes from the computer screen for use by collaborative partners, and the Internet can transmit information around the world in seconds.

Solid models suitably animated can also be viewed in 3D to clarify detail and this can be a considerable asset where perhaps there is a change of language. User manuals for domestic equipment are commonly drawn in solid modelling programs to illustrate sequences of assembly and improve clarity for non-technical customers.

A very important part of work in the drawing office is dealing and handling revisions and modifications. Modifications use quite a large proportion of drawing office time. It is possible to link drawings so that if you update the master, linked drawings are updated automatically.

Immediate transmission to all members of an associated group has considerable advantages. Examples here are recall notices for car owners and faulty items in domestic appliances.

There are many examples where various component parts are manufactured in different countries and brought together for assembly and testing. The aircraft industry is a typical case.

Drawings are reproduced in many sizes and small items present little difficulty with zoom facilities. Views drawn to different scales and a variety of orientations can be arranged on the same drawing print as an aid to comprehension. Windows giving an overall view of your drawing for fast zooming and panning are also of value.

Computer and software purchase

It is strongly recommended that before any purchases are made, the client seeks advice from a recognized and authorized dealer, as they would be able to check that the equipment can perform the tasks you expect in your style of working. Practical demonstrations are very necessary before issuing orders. CAD equipment is a tool and there are possibly many ways of doing the same job. In this computer age it may well be that an experienced dealer can indicate a better and more productive way.

Your supplier would also give you a written specification for computers and software indicating any other relevant equipment required for protection and safe operation. See Chapter 1 for additional organizational considerations when purchasing computer hardware and software.

Project development

The reader will appreciate that the design of, for example, a large construction project from its conception, will involve technical input from architects and engineering designers in a wide variety of associated disciplines. It is vital that all contributors to the overall scheme talk the same language and that only compatible computer software packages are in use for the separate areas of work. In addition, the management contractor must have access to the designs as work is in progress. Before the age of CAD, it was the practice to have countless meetings in order to co-ordinate progress.

Design obviously continues in steps, and in planning and construction work problems arise, and designers need to be in a position to make modifications to overcome them before progressing to the next phase.

A typical case study illustrating the activity associated with this type of work is the construction of the Civil Aviation Authority 'en-route' centre, built at Southampton. This prestige building and installation controls all the air traffic passing through Britain's airspace and houses, controllers operating banks of electronic and computer equipment where only an efficiency of 100% is acceptable. The building services engineer must ensure that the environment to keep both controllers and equipment comfortable is maintained 24 h a day, seven days a week.

Due to the extensive use of computers at the centre, a huge amount of electrical, heating, ventilating and air conditioning plant needed to be installed. Different specialist contractors were responsible for these services under the stewardship of the management contractor.

The fast track nature of the design and construction required an extensive application of CAD, where individual contractors responsible for electrical, mechanical, and ducting work were 'net-worked' on site, and could refer to CAD data from each other.

At this development, it was accepted by contractors that for some drawings it was practical to work in three dimensions to make it easier, for example, to ensure clearances between piping and ductwork in the more cramped areas. Layout drawings in 3D permitted engineers to demonstrate clearly to other parties where, for example, electrical cables and conduits were likely to plough straight through heating and ventilation ducts. Potential problems were solved on screen rather than emerging during construction. In addition, adequate access for maintenance purposes and replacement of equipment could be confirmed. The draughtsman can check designs by altering the angles from which arrangements are viewed on screen.

In the design of many heavy engineering plant layouts, it is often the practice to build a scale model of the plant as design work progresses. The function of the model is to keep a running check on the feasibility of the installation. Obvious improvements can then be incorporated.

Constructions of chemical plants and oil refineries are typical examples. After completion of the project, models may be used for publicity purposes and to assist in the education of technicians who operate and service the equipment. Three-dimensional modelling has many other applications in the film and entertainment industry and drawings in 3D can materially assist in comprehension.

When many workstations have to be installed for a design team, it is vital to agree on working methods. Recommendations for useful Standards in Construction Drawing Practice are detailed in Chapter 31.

Agreement is necessary on the organization of many aspects of work and in CAD, these include the use of layers, the groupings of the various sections of construction designs, use of colours so that for example, similar ductwork appears on the screen in the same shade, procedures for the transfer of data between several drawing offices, methods of structuring data for archiving, and to help future retrieval. In the light that no national or international standards exist, most organizations have produced their own specific CAD input standards to meet their business requirements. The quality of all drawing work needs to be uniform and conform to BS 8888 for a complete understanding and to avoid ambiguity. It is essential that all contributors work as a team and in harmony if planning deadlines are to be kept, as obviously, delays in one area of construction can hold-up another contractor's work, and may result in financial loss.

Size of computer

As a rough guide to selection, the larger the drawing and degree of complexity, the more important is the performance and power of the computer and its operator.

If a drawing contains large areas which are crosshatched or shaded, for example, it is important to be able to redraw illustrations quickly to prevent time wasting. Equally, when designing using 3D software, the assembly of complex models requires high performance computers to enable real time manipulation of the assembly.

It is easy to obtain demonstrations of computer power and this is recommended. When selecting software products required to operate with each other, it is necessary to check compatibility; your dealer should advise.

You will appreciate from the applications mentioned above that associated specialist software is being developed all the time both here and in the US. The one certain aspect is that future trends will use applications needing greater amounts of computer memory, so the chosen system must be expandable. Consideration must also be given to the question of storing drawings, filing systems and information retrieval.

Given the rapid progress and changes in the drawing office during the last 10 years the only prediction one can make is that the role of the draughtsman, far from diminishing, is more important than ever.

Parametric design

It is a common drawing office practice, where a range of parts are similar, to produce a single drawing with a table of dimensions for the features of each separate component. The user will then need to sort out the appropriate sizes of each detail relating to the part required. The drawing itself being representative of a number of similar parts cannot be drawn true to scale for them all.

A study of Fig. 3.1 will show a special screw, which has a family of parts. It is defined on a single drawing where the main dimensions are expressed algebraically as ratios of the shank diameter of the screw and other relevant parametric values. For a given thread size and screw length, the CAD system is able to produce a true-to-size drawing of any individual screw listed. This drawing may then be used as part of an assembly drawing, or fully dimensioned and suitable for manufacturing purposes. Four typical screws are indicated at the right-hand side of the illustration. It is always a positive advantage in design work to appreciate true sizes and use them in layouts.

Components such as bolts, nuts, washers, fasteners, spindles, seals, etc., fall naturally into families where similar geometric features are present. The parametric capability of the CAD system can be used to considerably improve productivity in this area of drawing office work.

It is not an uncommon practice in product development to modify existing standard components if possible and use them as the basis for new ones. Notice the visible connection between the features of the four components illustrated in Fig. 3.2. This is a further example of parametrication where the principles of variational geometry have been applied.

FIGURE 3.1

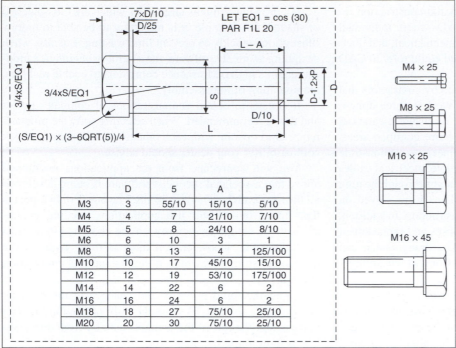

	D	5	A	P
M3	3	55/10	15/10	5/10
M4	4	7	21/10	7/10
M5	5	8	24/10	8/10
M6	6	10	3	1
M8	8	13	4	125/100
M10	10	17	45/10	15/10
M12	12	19	53/10	175/100
M14	14	22	6	2
M16	16	24	6	2
M18	18	27	75/10	25/10
M20	20	30	75/10	25/10

The family of parts is constructed from a large and small cylinder with different diameters, lengths and central bore sizes. A chamfer at the left-hand end, a vertical hole extending into a slot and a flat surface at the top are added details.

Parametric systems handle the full range of linear and angular dimensions including degrees and minutes. The computer will also calculate maximum and minimum limits of size from specified tolerance values. Dimensions can be defined numerically or as algebraic expressions. You can avoid the need to dimension every fillet radius for example by setting a default value for radii. This means that unless a specific value is stated for a particular radius on a part that it will automatically be drawn at a previously agreed size. Where many radii are present, as in the case of casting work, this feature is a considerable aid to drawing office productivity. A number of such defaults can be entered, to cover a variety of applications.

Areas of detail within a drawing, which are not required to be parametricated can be excluded by enclosing them in a group line and this avoids the need to dimension every detail.

FIGURE 3.2

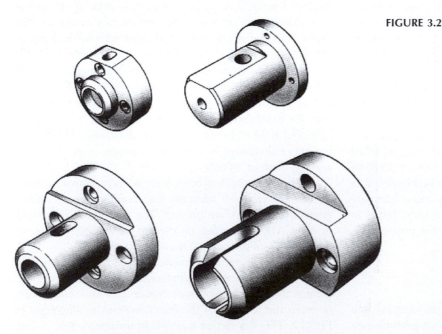

The geometry contained in the enclosed group may remain static or magnified when the part is parametricated.

A further advantage of expressing dimensional values in algebraic form allows the designer to simulate the movement of mechanisms and produce loci drawings of specific points. It is essential in the design of mechanisms to appreciate the path taken by every point, which moves.

Sheet metalwork application

The design of components to be manufactured from folded sheet metal is a field in which CAD systems can offer great assistance.

In the case of the bracket shown in Fig. 3.3 it would first be necessary to establish the overall dimensions of the part.

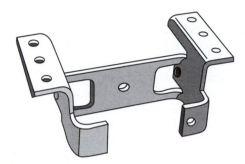

FIGURE 3.3

The second step would be to imagine that the bracket is folded back gradually as indicated in Fig. 3.4 into the flat sheet form. This shape would then be stamped from metal strip in a power press.

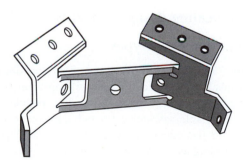

FIGURE 3.4

The dimensions of the flat pattern have to make allowance for the bend radius, the metal thickness and the type of metal used. Metals behave quite differently when bent and the CAD system can be programmed to calculate an appropriate bend allowance. After stamping the bracket can be refolded with suitably radiused bends.

In this particular case, the dimensions of the stamping are also needed for the design of the press tool set.

The design can be checked for material accuracy, weight, volume, and so on, before being committed to manufacture.

Computerized programs can be produced to operate lathes, mills, flame cutting machines, etc., and many other items of equipment in the manufacturing process.

Models may be constructed in several different ways, including: geometric modelling, meshed surfaces, sweeps, volumes of revolution and ruled surfaces. Each of these is summarized below.

Geometric modellers build models from geometric solids, which have the attribute that mathematical formulae exactly define any point in 3D space occupied by these solids. Shapes include planes, cylinders, spheres, cones, toroids, etc. These shapes are combined using Boolean operations to produce the component. The Boolean operations produce a 3D model by a combination of the following methods:

(a) resulting from the union of any two 3D objects or shapes;
(b) resulting from the difference between any two 3D objects or shapes;
(c) resulting from the volume that is common to any two 3D objects or shapes.

This approach is very successful for modelling machined components but cannot handle anything that might be described as having a freeform shape.

Meshed surfaces X, Y and *Z* co-ordinates are either calculated, transferred from 2D drawing views, or measured to provide basic modelling input. The modeller will then generate a 3D meshed surface joining up all the specified points. In order to build up a well-defined surface, the modeller interpolates between points defined in the user input in order to develop a fine enough mesh to show a smooth change in cross-section. This method can be used to produce the freeform shapes used in, for example, styling household appliances.

Sweeps where a 2D outline is defined graphically and then lofted or swept, by the modeller to give the outline a uniform thickness, as the third dimension. This produces objects of any shape in terms of the *x* and *y* dimensions, but a constant value for the *z* dimension. Sweeps can model all of those components that look like they are extruded, or have been cut from steel plate. For a model of a pipe a circular cross-section is swept or moved along a centreline of any shape.

Volumes of revolution for objects the shape of which is symmetrical about a central axis. The wheel is a simple example of this type of 3D object. The input is a half outline, or a cross-section through the object, which is rotated about the axis by the modeller, to produce a 3D illustration.

Ruled surfaces is a simple form of modelling, where any two sections or profiles can be joined at all points by straight lines. An airfoil, or a turbine blade is a typical example where this method can be applied.

Examples of various methods of CAD modelling are shown in Fig. 3.5.

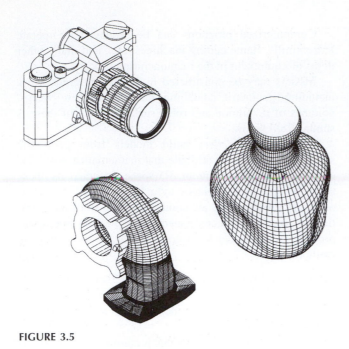

FIGURE 3.5

Pipework systems

There are many aspects of pipework draughtsmanship where the computer can considerably improve productivity. In many cases, by using 3D modelling software the design can be partly automated.

Having received an order to construct the plant, pipework systems basically require two types of drawings: flow charts and services drawings. Flow charts are functional diagrams showing the scheme and will include major items of plant. This diagrammatic arrangement is not to scale but shows the relative positions of main items and the connections between them. The diagram illustrates the feasibility of the system.

Equipment may be fixed at various levels. Assuming that a factory is to be built, then separate areas will be allocated to individual teams of draughtsmen who prepare layouts and services drawings for structural work, manufacturing areas, heating, ventilation, air conditioning, compressed air, and electrical services, etc. It is a standard practice to aid installation and to eliminate a clash of services, coordinated services drawings are produced. These drawings combine all relative services which are dimensioned relative to fixed datum's (see below). Ground site surveys are undertaken and various datum levels established to act as benchmarks for reference measurements. Steelwork can now be designed for the factory and manufactured to suit the site contours.

A 3D scale drawing could be constructed showing separate levels on which the items of plant are mounted. Straight lines representing the centrelines of interconnecting pipework are added. Pipes are sized to ensure adequate flow of liquids or gases and to withstand the pressure exerted by the contents. Realistic pipework can now be added. Suitable

bends, elbows and other fittings may be directly 'dragged and dropped' at the various corners where pipes change directions and levels.

Software is available with libraries of ready-made standard fittings. Note carefully, however, which Standards are applicable. ISO and US Standards are regularly used and specifications need to be checked. The drawing office will be responsible for preparing lists and schedules of equipment required for fabrication and the following are typical:

- Pipe lists quoting sizes and lengths taking into account bend radii. During erection, pipes are cut to length then welded into the pipelines.
- Lists of similar standard bends and elbows.
- Lists of similar welded joints and processes.
- Lists of unions joining pipes together for non-welded constructions.
- Valves of all different types, sizes and connections, i.e. screwed, bolted and welded.
- Hangers to support pipework and expansion devices to permit movement.
- Pumps and associated fittings.
- Instrumentation devices, pressure gauges, temperature measuring devices and flow meters, and filters.
- Equipment will be ordered from manufacturers using these records and costs calculated.

Another vital task that the computer can determine is to check clearances where pipes cross to ensure that there is an adequate space to allow erection and operation.

The above are typical process tasks that can be handled by piping software.

Communicating design concepts

Mockups and prototypes

Mockups and prototypes show how products and mechanisms look and perform but building them is a time-consuming process. A 3D model is life-like, popular and can be of considerable assistance for publicity purposes especially where the client has limited technical experience. Recent developments are easy to use and an economical method of demonstrating engineering design concepts.

Drawings can be communicated by email and have the advantage that they can be viewed by anyone who has a Windows PC. Products can be rotated through 360° to show how they appear from any angle so that movement through their cycle of operations can be demonstrated. Simulation may be sufficient to reduce the need for expensive prototypes.

Maximized sales and marketing opportunities may result from presenting new and novel product designs more effectively to customers and business prospects.

Models can shorten development cycles and assist in fast product design changes.

Animated drawings give you the opportunity to explode or collapse an assembly to demonstrate how the components fit together.

Rapid prototyping has become widely used to produce physical models to aid conveying design intent. Rapid prototyped models are invaluable at design reviews or for use in market testing of products. Derived from 3D models methods such as 3D Printing (using a starch based powder and inkjet technology) are relatively inexpensive and are ideal for producing good quality, 'touchy-feely' models which, although at times fragile, are an excellent aid to any design review. Stereolithography (liquid resin cured by laser), Selective Laser Sintering (thermoplastic powder fused by laser) and Fused Deposition (hot extruded thermoplastic) methods are more expensive but produce models which are more robust and can be used to make functional prototypes without the costs of expensive tooling. Rapid prototyping has advanced a long way in the last 10 years and the development of new modelling materials has almost elevated the technique to a full production capability.

Confidence in a particular project also results from confirmation that it is acceptable and suitable in the market place.

Production

Many products require a considerable amount of testing. Safety is always vital and must be the top priority.

It needs to be remembered that all products must be designed so that the production department can economically manufacture them. The design must also be suitable for easy assembly and repair. Financial constraints should never be forgotten, hence meticulous care is taken in pre-production phases to reduce the time-to-market and eliminate modifications to the product, once mass production begins.

Rendering controls

Presentations and proposals using photo-realistic images add excitement and visual impact. Before applying rendering features to a model, the background and lighting conditions should be adjusted to simulate mood, time and scene composition. It can also apply lighting, shadows and ray tracing for reflective and transparent materials, and if required, background scenery.

Smoothing areas of high contrast to improve appearance can enhance image quality. Accuracy is improved through fully associative design. Mating constraints are preserved and the relationships between parts and assemblies. Drawings update automatically. Errors may be prevented and designs optimized by using Collision Detection to observe in real time how moving parts interact. Visualization may be improved with enhanced graphical control of lights. Photo-realistic effects may also be created by means of sophisti-

cated ray-traced lighting. It is also possible to analyse the complete history of the design project and document an automated design process.

The rendering mode quickly displays a shaded image of the model with materials attached. It is often convenient to save alternative production quality images of your design for comparison purposes and use in other associated design projects. Alternative views can be a valuable and visible asset during training programs.

Materials options

Visual effects from libraries of life-like textures and materials can be added to 3D models easily, using commands available on the toolbar menu. Libraries are available with a wide selection of plastics, metals, woods, stones, and other textured materials which can be applied to entire parts, features, or individual faces. Realistic changes can also be made to suggest surface reflectance, roughness, transparency, and an irregular or indented appearance.

Typical CAD drawings and 3D models

The following examples are meant to convey to the reader the extensive range of draughting facilities available from software associated with basic programs. Obviously there is a certain amount of overlap in the scope of programs and often alternative ways of performing similar operations.

Figures 3.6 and 3.7 show pictorial drawings of an engine development.

A pictorial view can easily be generated after drawing orthographic views, which give the dimensions in three planes at right angles to each other. Figure 3.8 shows a drawing of a cycle. Figure 3.9 demonstrates the realistic effect of rendering. The viewing point and orientation is adjustable.

Architectural drawings for the design of a shopping mall are reproduced in Figs. 3.10 and 3.11. They show the outline of the development and how the completed construction could appear.

Figures 3.12 and 3.13 illustrate an architectural drawing from two different viewpoints. Alternative simulations may be used to assist the client in the choice of colour for the finished building.

Design concepts, which are rendered clearly and convincingly, certainly aid at the stage where decisions need to be made to finalize aspects of shape, form and finish. The presentation of alternative solutions using the same master drawing is also an added bonus.

Engineered components are often designed for clients without a technical background. To be able to observe the final product in three dimensions with its approved finish and in an ideal situation will reduce design time. Many people have difficulty in reading drawings, but with a presentation of an internal building detail, which perhaps shows a slate

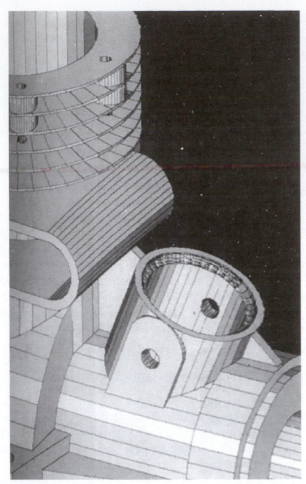

FIGURE 3.6

FIGURE 3.8

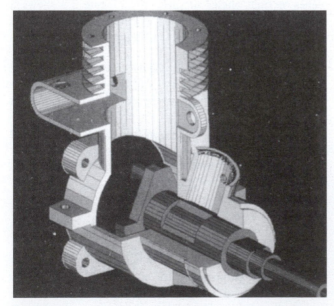

FIGURE 3.7

FIGURE 3.9 Illustrates the variety of subtle textures available within the materials library (Page No.: 22).

FIGURE 3.10

FIGURE 3.11

FIGURE 3.12

FIGURE 3.13

tracing giving shadowing, reflection, and refraction effects. A comprehensive library of materials and textures can be used to create a variety of surfaces, such as wood, glass, marble, granite, etc.

A wide selection of illumination tools and compatibility with associated software, allows the draughtsman to make walk-throughs, fly-throughs and animated product-assembly presentations.

Lighting studies are easy and accurate. You can produce a variety of artificial, natural, and mixed lighting effects. It is possible to arrange directional lights in various combinations and locations and control such characteristics as colour, intensity, attenuation, and shadowing.

In addition, a Sun Locator lets you work easily with sunlight effects. You can position the sun to a specific time of day and year to create realistic sun–shadow combinations. The feature allows architects to calculate, for example, whether a living room will receive enough sunlight at mid-day in late December. On a larger scale, in the design of shopping malls for example, the position of the sun in relation to a particular area can materially affect heating, lighting, and cooling loads.

You will appreciate that these programs can help to confirm design decisions and prevent misunderstandings while they are still easy and inexpensive to remedy.

A perspective drawing of an internal part of a building in Fig. 3.14 indicates the style and character of a finished construction.

Examples of animated presentations are given in Figs. 3.15 and 3.16. The impact and appeal of sales literature is often enhanced by the use of theatrical effects.

BS 4006 gives the specification for hand operated square drive socket wrenches and accessories. The tools are manufactured from chrome vanadium steel and Figs. 3.17 and 3.18 show a presentation for a sales catalogue.

Figure 3.19 illustrates exploded three-dimensional views of a turbocharger for an automobile.

Figure 3.20 shows an application where part of an assembly drawing has been copied into a word-processor and used to prepare a production-engineering document.

Figure 3.21 shows an assembly drawing of a fuel injector for a diesel engine. Drawn to BS and ISO standards, this is a typical professional CAD drawing which could be produced using most CAD software on the market. In industry, it is now common practice to include a small axonometric representation of the product, placed in a corner of the drawing, to aid its reading.

floor, and coloured textured walls, then the client can understand exactly how the structure will look. The drawing bridges the communication gap.

Creating renderings is fast and menus and dialogue boxes are used. The program features include shading and ray

FIGURE 3.14

FIGURE 3.15

FIGURE 3.16

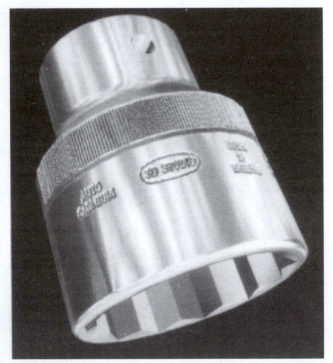

FIGURE 3.18

FIGURE 3.17 BS 4006 gives the dimensions, testing and design requirements for hand-operated square drive socket wrenches and accessories. The
tools are manufactured from chrome vanadium steel and these illustrations
show typical production drawings and presentations for sales brochures.

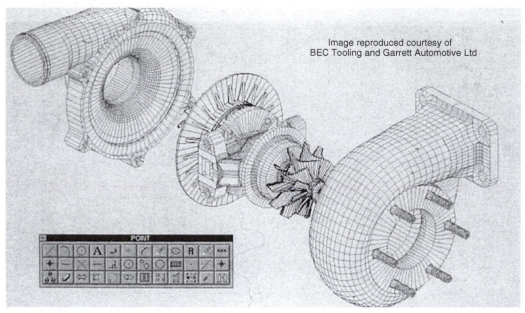

FIGURE 3.19   Illustrates exploded three-dimensional views of a turbocharger for an automobile.

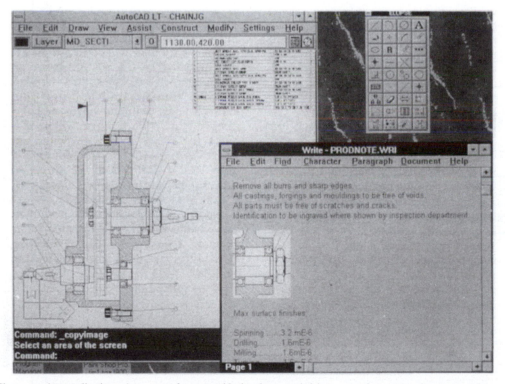

FIGURE 3.20   Shows another application where part of an assembly has been copied into a word-processor document and to prepare production engineering information.

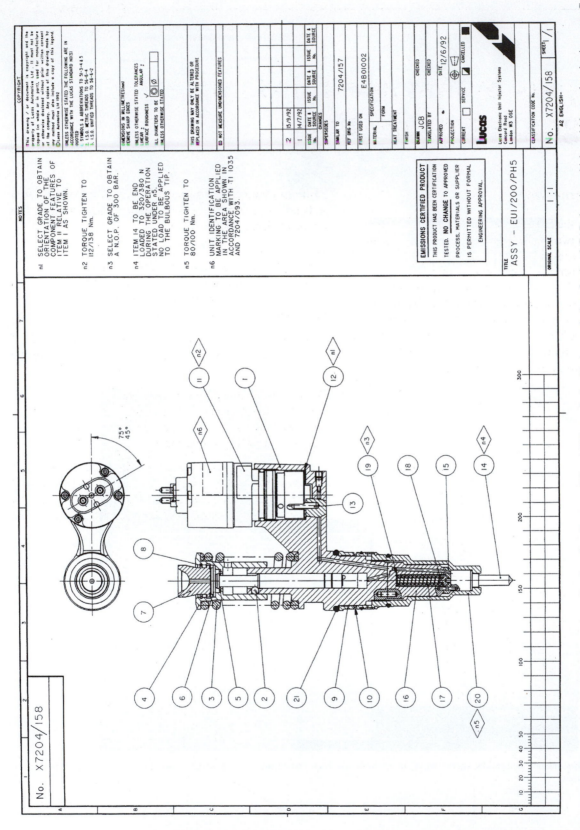

FIGURE 3.21 An assembly drawing of a fuel injector for a diesel engine. Drawn to BS and ISO standards, this is a typical professional CAD drawing which could be produced using most CAD software on the market.

Chapter 4

Principles of first and third angle orthographic projection

First angle projection

Assume that a small block is made 35 mm × 30 mm × 20 mm and that two of the corners are cut away as shown below in three stages (Fig. 4.1).

Figure 4.2 illustrates a pictorial view of the block and this has been arranged in an arbitrary way because none of the faces are more important than the others. In order to describe the orthographic views, we need to select a principal view and in this case we have chosen the view in direction of arrow A to be the view from the front.

The five arrows point to different surfaces of the block and five views will result. The arrows themselves are positioned square to the surfaces, that is at 90° to the surfaces and they are also at 90°, or multiples of 90° to each other. The views are designated as follows:

View in direction A is the view from the front,
View in direction B is the view from the left,
View in direction C is the view from the right,
View in direction D is the view from above,
View in direction E is the view from below.

In first angle projection the views in the directions of arrows B, C, D, and E are arranged with reference to the front view as follows:

The view from B is placed on the right,
The view from C is placed on the left,
The view from D is placed underneath,
The view from E is placed above.

The experienced draughtsman will commit the above rules to memory. It is customary to state the projection used on orthographic drawings to remove all doubt, or use the distinguishing symbol which is shown on the arrangement in Fig. 4.3.

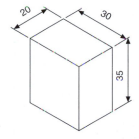

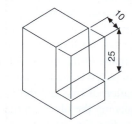

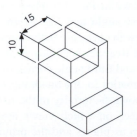

FIGURE 4.1

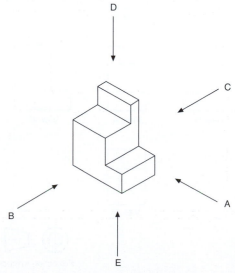

FIGURE 4.2

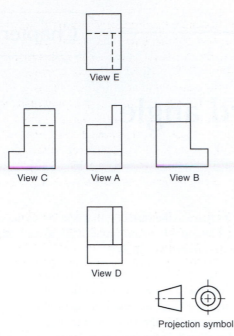

FIGURE 4.3 First angle projection arrangement. Dotted lines indicate hidden edges and corners.

Third angle projection

The difference between first and third angle projection is in the arrangement of views and, with reference to the illustration in Fig. 4.4, views are now positioned as follows:

View B from the left is placed on the left,
View C from the right is placed on the right,
View D from above is placed above, and
View E from below is placed underneath.

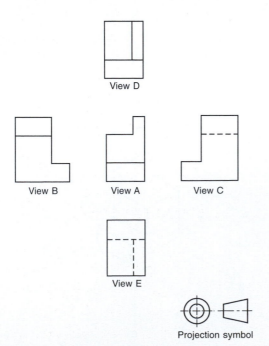

FIGURE 4.4 Third angle projection arrangement.

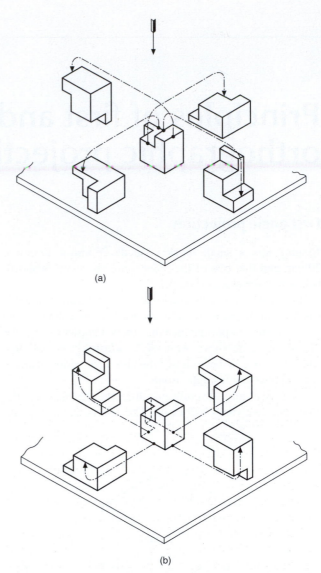

FIGURE 4.5 (a) First angle arrangement. (b) Third angle arrangement.

Study the rearrangement shown below in Fig. 4.4 and remember the above rules because it is vital that the principles of first and third angle projection are understood. The distinguishing symbol for this method is also shown.

If a model is made of the block in Fig. 4.1, and this can easily be cut from polystyrene foam used in packing, then a simple demonstration of first and third angle projection can be arranged by placing the block on the drawing board and moving it in the direction of the four chain dotted lines terminating in arrows in Fig. 4.5. Figure 4.5(a) shows the positioning for first angle and Fig. 4.5(b) for third angle projection. The view in each case in the direction of the large arrow will give the five views already explained.

The terms first and third angle correspond with the notation used in mathematics for the quadrants of a circle in Fig. 4.6 the block is shown pictorially in the first quadrant with three of the surfaces on which views are projected. The surfaces are known as planes and the principal view in direction of arrow A is projected on to the principal vertical

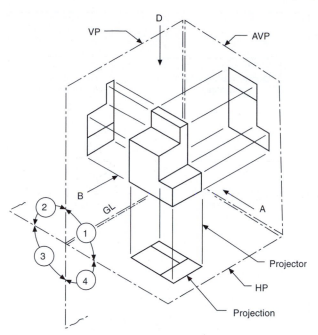

FIGURE 4.6 VP is the vertical plane. HP is the horizontal plane. AVP the auxiliary vertical plane. GL is the ground line.

plane. The view from D is projected on to a horizontal plane. View B is also projected on to a vertical plane at 90° to the principal vertical plane and the horizontal plane and this is known as an auxiliary vertical plane. Another horizontal plane can be positioned above for the projection from arrow E, also a second auxiliary vertical plane on the left for the projection of view C. Notice that the projections to each of the planes are all parallel, meeting the planes at right angles and this is a feature of orthographic projection.

The intersection of the vertical and horizontal planes give a line which is the ground line GL. This line is often referred to as the *XY* line; this is useful in projection problems since it represents the position of the horizontal plane with reference to a front view and also the position of the vertical plane with reference to a plan view. Many examples follow in the text.

If the planes containing the three views are folded back into the plane of the drawing board, then the result is shown in Fig. 4.7 where dimensions have also been added. The

draughtsman adjusts the distances between views to provide adequate spaces for the dimensions and notes.

To describe a simple object, a draughtsman does not need to draw all five views and it is customary to draw only the minimum number which completely illustrates the component. You will note in this particular case that we have omitted views which contain dotted lines in preference to those where corners and edges face the observer. Many parts do not have a definite 'front', 'top' or 'side' and the orientation is decided by the draughtsman, who selects views to give the maximum visual information.

Traditionally, front views are also known as front elevations, side views are often known as side or end elevations and the views from above or beneath are referred to as plans. All of these terms are freely used in industrial drawing offices.

Projection symbols

First angle projection is widely used throughout all parts of Europe and often called European projection. Third angle is the system used in North America and alternatively described as American projection. In the British Isles, where industry works in co-operation with the rest of the world, both systems of projection are regularly in use. The current British and ISO standards state that both systems of projection are equally acceptable but they should never be mixed on the same drawing. The projection symbol must be added to the completed drawing to indicate which system has been used.

Figure 4.8 shows the recommended proportions of the two projection symbols.

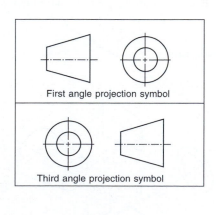

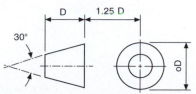

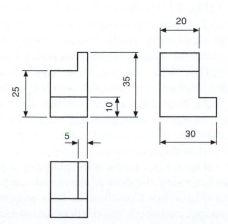

FIGURE 4.7

FIGURE 4.8

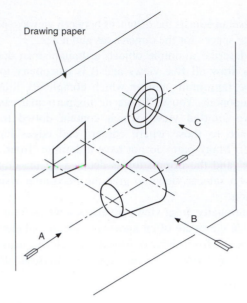

FIGURE 4.9

Figure 4.9 indicates how the first angle symbol was obtained from projections of a tapered roller. The third angle alternative is given in Fig. 4.10.

Please note the movement suggested by the arrow in Figs. 4.9, 4.10 and also in Fig. 4.8, since orientation is the main clue to understanding the fundamental differences in projection systems.

An experienced draughtsman must be fully conversant with all forms of orthographic and pictorial projection and be able to produce a drawing where no doubt or ambiguity relating to its interpretation can exist.

Drawing procedure

Generally, industrial draughtsmen do not complete one view on a drawing before starting the next, but rather work on all

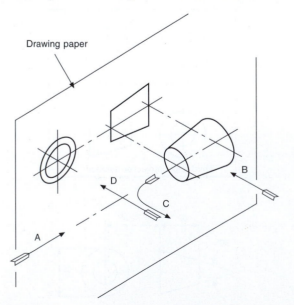

FIGURE 4.10

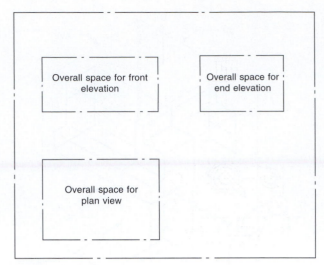

FIGURE 4.11 Stage 1.

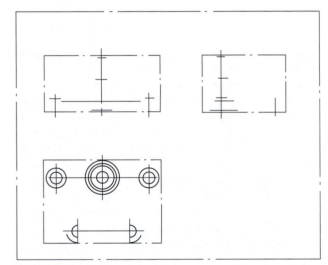

FIGURE 4.12 Stage 2.

views together. While projecting features between views, a certain amount of mental checking takes place regarding shape and form, and this assists in accuracy. The following series of drawings shows stages in producing a typical working drawing in first angle projection.

Stage 1 (Fig. 4.11): Estimate the space required for each of the views from the overall dimensions in each plane, and position the views on the available drawing sheet so that the spaces between the three drawings are roughly the same.

Stage 2 (Fig. 4.12): In each view, mark out the main centre-lines. Position any complete circles, in any view, and line them from the start, if possible. Here complete circles exist only in the plan view. The heights of the cylindrical features are now measured in the front view and are projected over to the end view.

Stage 3 (Fig. 4.13): Complete the plan view and project up into the front view the sides of the cylindrical parts.

Stage 4 (Fig. 4.14): Complete the front and end views. Add dimensions, and check that the drawing (mental check) can be redrawn from the dimensions given; otherwise the

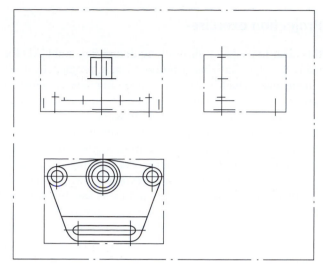

FIGURE 4.13 Stage 3.

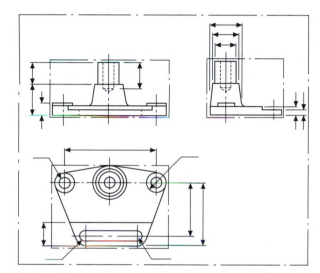

FIGURE 4.14 Stage 4.

dimensioning is incomplete. Add the title and any necessary notes.

It is generally advisable to mark out the same feature in as many views as is possible at the same time. Not only is this practice time-saving, but a continuous check on the correct projection between each view is possible, as the draughtsman then tends naturally to think in the three dimensions of length, breadth and depth. It is rarely advantageous to complete one view before starting the others.

Reading engineering drawings

The following notes and illustrations are intended to assist in reading and understanding simple drawings. In all orthographic drawings, it is necessary to project at least two views of a three-dimensional object – or one view and an adequate description in some simple cases, a typical example being the drawing of a ball for a bearing. A drawing of a circle on its own could be interpreted as the end elevation of a cylinder

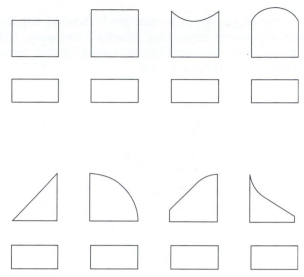

FIGURE 4.15

or a sphere. A drawing of a rectangle could be understood as part of a bar of rectangular cross-section, or it might be the front elevation of a cylinder. It is therefore generally necessary to produce at least two views, and these must be read together for a complete understanding. Figure 4.15 shows various examples where the plan views are identical and the elevations are all different.

A single line may represent an edge or the change in direction of a surface, and which will be determined only by reading both views simultaneously. Figure 4.16 shows other cases where the elevations are similar but the plan views are considerably different.

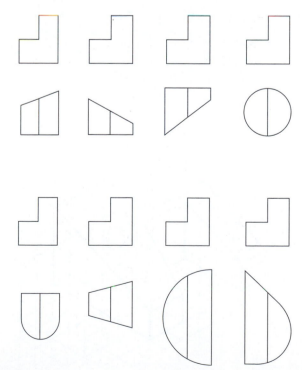

FIGURE 4.16

A certain amount of imagination is therefore required when interpreting engineering drawings. Obviously, with an object of greater complexity, the reading of three views, or more, may well be necessary.

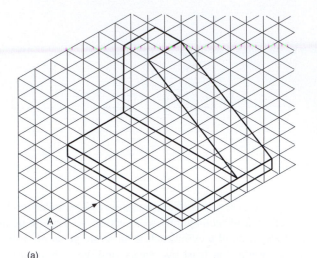

(a)

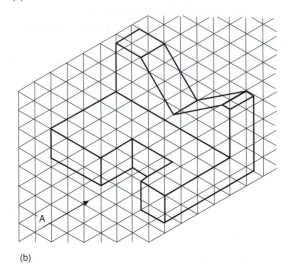

(b)

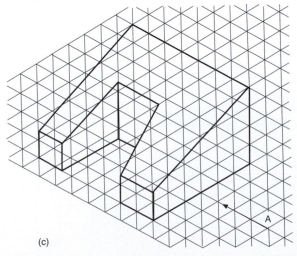

(c)

FIGURE 4.17

Projection exercises

It is clear to us that after teaching draughting and CAD for many years visualizing a proposed new product in three dimensions, which is how you naturally view a finished article, is difficult when it is necessary to read more than one complex two-dimensional drawing simultaneously. The draughtsman also ultimately needs to produce technically correct drawings, often from vague initial ideas. The very action of making proposal drawings stimulates many questions and their answers allow development to continue. Modifications to original ideas involve drawing amendments and changes to one view invariably have a 'knock on effect'. Comprehension, understanding and the ability to read technical drawings fluently comes with practice.

The following simple exercises are designed to assist in the perfection of draughting skills. They are equally suitable for CAD and the drawing board. Produce answers for each series and select standard sizes of drawing sheets, taking particular care with linework and layout.

If the CAD software program permits, move the separate views for each exercise so that they are positioned a similar distance from each other. Then experiment and position the groups to give a pleasing layout on the drawing sheet. Note how uniformity can improve presentation and give a professional appearance. Layout is a very important aspect when preparing drawings for desktop publishing applications.

Straight line examples

Figure 4.17 shows three components in which each has been machined from solid blocks. These examples have been prepared on a grid formed by equilateral triangles.

In every case, the scale is such that each side of the triangle will be 10 mm. For each component, draw five views in first angle projection, omitting hidden detail, and assume that the view in the direction of the arrow A will be the front view.

Examples involving radii and holes (Fig. 4.18)

For each example, project five views in first angle projection, taking the view in the direction of the arrow A as the front view. Hidden detail is required in the solutions to these problems, and note that in some cases the position of some of the holes will be found to coincide with centre lines. Where this occurs, the dotted line should take priority. Take each side of the grid triangle to be 10 mm in length.

If only three views of each component were required, which one would you choose? The professional draughtsman would select a front view, end view, and plan view with the

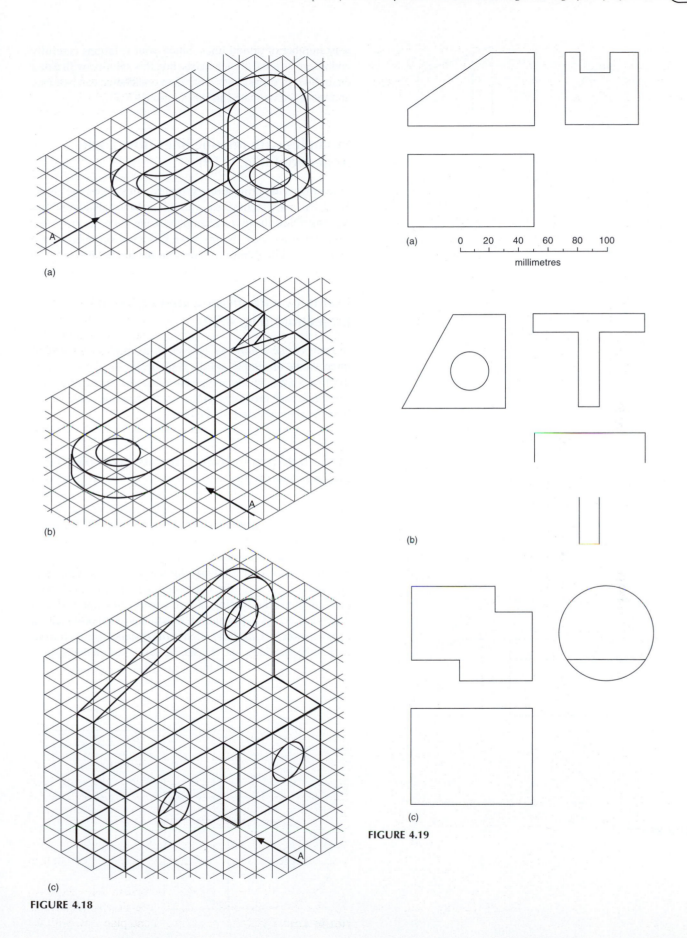

(a)

0 20 40 60 80 100

millimetres

(a)

(b)

(b)

(c)

FIGURE 4.19

(c)

FIGURE 4.18

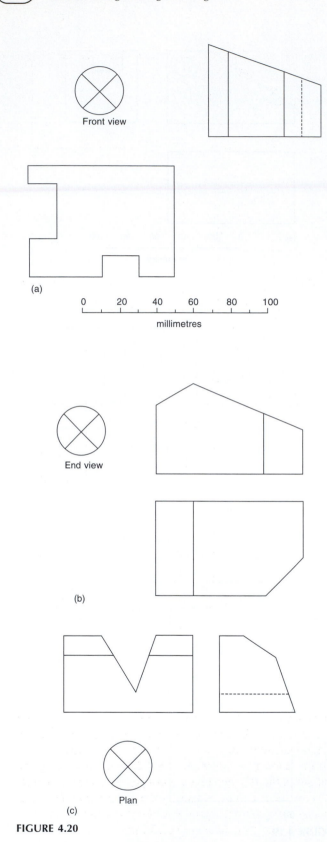

Front view

(a)

0   20   40   60   80   100

millimetres

End view

(b)

Plan

(c)

FIGURE 4.20

least number of dotted lines. Study your solutions carefully and where an ideal choice exists, box this solution with a thin chain line. In some cases more than one choice can be made and these are indicated in the solutions.

Examples with missing lines (first angle projection) (Fig. 4.19)

In the following projection examples, three views are given. Some views are incomplete with full lines and all dotted lines missing. Draw the given examples, using the scale provided. Complete each view, by inserting full lines where necessary and add all dotted lines to represent the hidden detail.

Examples with missing views (first angle projection) (Fig. 4.20)

In each of the following projection examples, two out of three views of simple solid components are shown. Draw the two views which are given using the scale provided. Complete each problem by drawing the missing view or plan in the space indicated by the cross.

First angle projection examples with plotted curves (Fig. 4.22)

In orthographic projection, all widths in the end view are equal in size to depths in the plan view, and of course the opposite is true that some dimensions required to complete end views may be obtained from given plan views. Figure 4.21 shows part of a solid circular bar which has been cut at an angle of 30° with the horizontal axis. Point A is at any position along the sloping face. If a horizontal line is drawn through A across to the end view then the width of the chord is dimension X. This dimension is the distance across the cut face in the plan view and this has been marked on the vertical

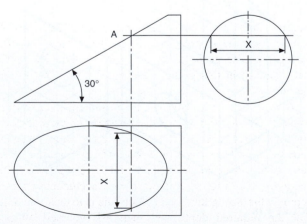

FIGURE 4.21

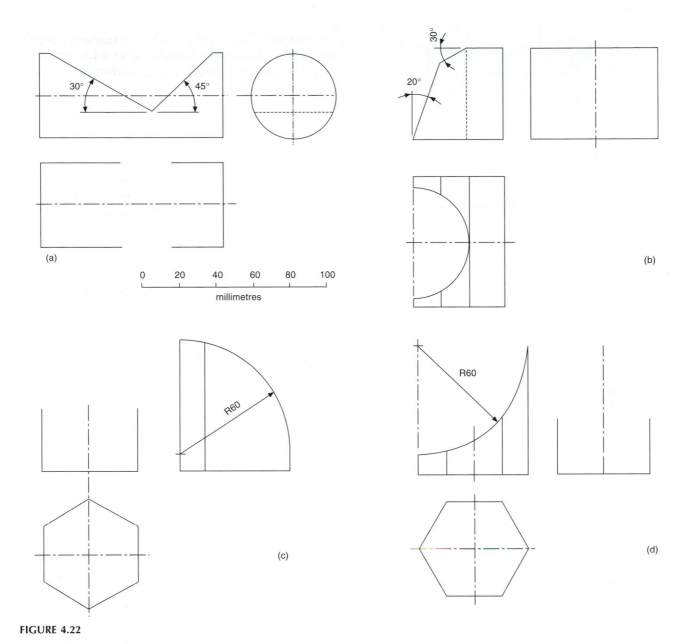

FIGURE 4.22

line from A to the plan. If this procedure is repeated for other points along the sloping face in the front view then the resulting ellipse in the plan view will be obtained. All of the examples in this group may be solved by this simple method.

A word of warning: do not draw dozens of lines from points along the sloping face across to the end view and also down to the plan view before marking any dimensions on your solution. Firstly, you may be drawing more lines than you need, and in an examination this is a waste of time. Secondly, confusion may arise if you accidentally plot a depth on the wrong line. The golden rule is to draw one line, plot the required depth and then ask yourself 'Where do I now need other points to obtain an accurate curve?'

Obviously, one needs to know in the plan view the position at the top and bottom of the slope, and the width at the horizontal centre line and at several points in between.

In the examples shown in Fig. 4.22 three views are given but one of them is incomplete and a plotted curve is required. Redraw each component using the scale provided. Commence each solution by establishing the extreme limits of the curve and then add intermediate points.

Pictorial sketching from orthographic views

Figure 4.23 shows six components in first angle projection. Make a pictorial sketch of each component and arrange that

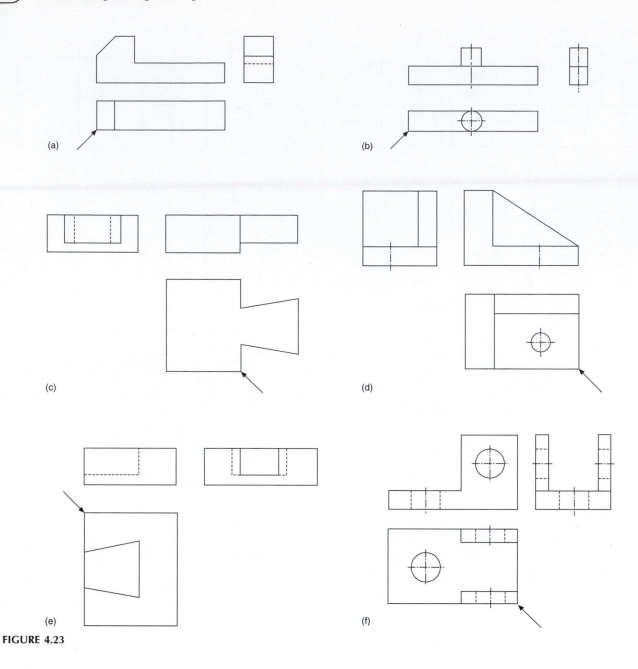

(a)

(b)

(c)

(d)

(e)

(f)

FIGURE 4.23

the corner indicated by the arrow is in the foreground. No dimensions are given but estimate the proportions of each part assuming that the largest dimension in every example is 100 mm.

Geometric solids in third angle projection

Figure 4.24 shows three views of each of three geometric solids. Sufficient dimensions are given to define the shapes but in each case two of the views are incomplete. Redraw the details provided and complete the views in third angle projection.

Sectional views in third angle projection

In Fig. 4.25 there are three components and two views are provided for each one. Copy the views given, using the scale provided, and project the missing view which will be a section. Your solution should include the section plane, cross hatching and the statement A–A.

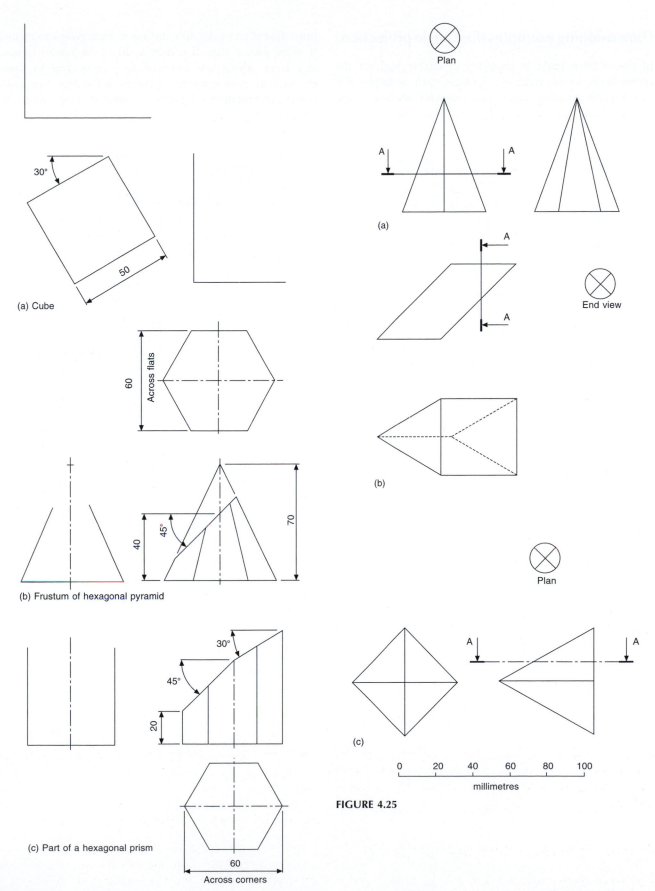

(a) Cube

(b) Frustum of hexagonal pyramid

(c) Part of a hexagonal prism

FIGURE 4.24

(a)

(b)

(c)

0   20   40   60   80   100
millimetres

FIGURE 4.25

Dimensioning examples (first angle projection)

In Fig. 4.26 a scale is provided to enable each of the components to be redrawn. Redraw each example and add any dimensions which you consider necessary and which would be required by the craftsman. Bear in mind that if an object has sufficient dimensions to enable it to be drawn, then it can most likely be made. Hence, any sizes which are required to enable you to draw the part are also required by the manufacturer. For additional information regarding dimensioning refer to Chapter 14.

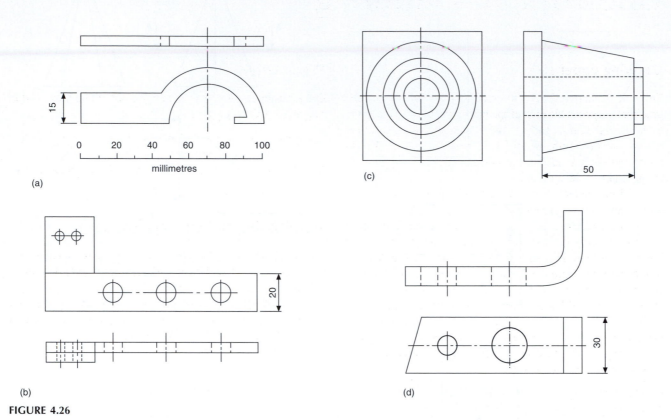

FIGURE 4.26

Chapter 5

Linework and lettering

Drawing paper sizes

The British Standard BS 8888 recommends that for normal practical purposes the area of the largest sheet is 1 m^2 and the sides are in the ratio of $1:\sqrt{2}$. The dimensions of the sheet are 841 mm × 1189 mm. For smaller sheets the longest side is progressively halved and the designations and dimensions are given in Table 5.1 and Fig. 5.1. Since the A0 size has the area of 1 m^2, paper weights are conveniently expressed in the unit 'grams per square metre'.

Drawing sheets may be obtained from a standard roll of paper or already cut to size. Cut sheets sometimes have a border of at least 15 mm width to provide a frame and this frame may be printed with microfilm registration marks, which are triangular in shape and positioned on the border at the vertical and horizontal centre lines of the sheet.

Title blocks are also generally printed in the bottom right-hand corner of cut sheets and contain items of basic information required by the drawing office or user of the drawing. Typical references are as follows:

- Name of firm,
- Drawing number,
- Component name,
- Drawing scale and units of measurement,
- Projection used (first or third angle) and or symbol,
- Draughtsman's name and checker's signature,
- Date of drawing and subsequent modifications,
- Cross references with associated drawings or assemblies.

Other information will vary according to the branch and type of industry concerned but is often standardized by particular firms for their own specific purposes and convenience.

Presentation

Drawing sheets and other documents should be presented in one of the following formats:

(a) *Landscape*—presented to be viewed with the longest side of the sheet horizontal.
(b) *Portrait*—presented to be viewed with the longest side of the sheet vertical.

Types of line and their application

Two thicknesses of line are recommended for manual and CAD (computer aided design) drawings. A wide line and a narrow line in the ratio of 2:1.

Standard lead holders, inking pens for manual use, and those for CAD plotters are all available in the following mm sizes: 0.25, 0.35, 0.5, 0.7, 1.0, 1.4 and 2.0.

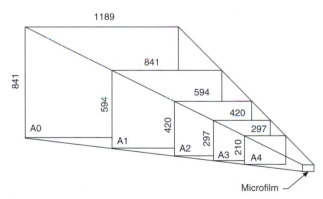

FIGURE 5.1 Standard size reductions from A0 to 35 mm microfilm.

TABLE 5.1		
Designation	Size (mm)	Area
A0	841 × 1189	1 m^2
A1	594 × 841	5000 cm^2
A2	420 × 594	2500 cm^2
A3	297 × 420	1250 cm^2
A4	210 × 297	625 cm^2

Line thicknesses of 0.7 and 0.35 are generally used and will give good quality, black, dense and contrasting lines.

Table 5.2 shows applications for different line types which are designed to obtain a good professional finish to a drawing. Line type designations are as referenced in ISO 128–24.

Various combinations of line thickness and type are shown on the mechanism in Fig. 5.2. Circled numbers relate to the line types in Table 5.2.

Figure 5.3 shows part of a cone and if the complete cone was required, for example for dimensioning purposes, then the rest would be shown by adding narrow continuous lines which intersect in a dot.

If it is necessary to show the initial outline of a part before it is bent or formed, then the initial outline can be indicated by a chain thin line which is double dashed. Figure 5.4 shows the standard applied to a metal strip.

TABLE 5.2 Types of line

Example	Type	Description and representation		Application
A	01.2	Continuous wide line	1	Visible edges and outlines
			2	Crests of screw threads and limit of length of full depth thread
			3	Main representations on diagrams, maps and flow charts
			4	Lines of cuts and sections
B	01.1	Continuous narrow line	1	Dimension, extension and projection lines
			2	Hatching lines for cross-sections
			3	Leader and reference lines
			4	Outlines of revolved sections
			5	Imaginary lines of intersection
			6	Short centre lines
			7	Diagonals indicating flat surfaces
			8	Bending lines
			9	Indication of repetitive features
			10	Root of screw threads
			11	Indication of repetitive features
C	01.1	Continuous narrow irregular line		Limits of partial views or sections provided the line is not an axis
D	02.1	Dashed narrow line		Hidden outlines and edges
E	04.1	Long-dashed dotted narrow line	1	Centre lines
			2	Lines of symmetry
			3	Pitch circle for gears
			4	Pitch circle for holes
			5	Cutting planes (see F [04.2] for ends and changes of direction)
F	04.2	Long-dashed dotted wide line	1	Surfaces which have to meet special requirements
			2	Cutting planes at ends and changes of direction (see E 04.1)
G		Long-dashed dotted narrow line with wide line at ends and at changes to indicate cutting planes		Note BS EN ISO 128–24 shows a long-dashed dotted wide line for this application
H	05.1	Long-dashed double dotted narrow line	1	Outlines of adjacent parts
			2	Extreme positions of moveable parts
			3	Initial outlines prior to forming
			4	Outline of finished parts within blanks
			5	Projected tolerance zones
			6	Parts situated at the front of a cutting plane
			7	Framing of particular fields or areas
			8	Centroidal lines
			9	Outlines of alternative executions
J	01.1	Continuous straight narrow line with zig-zags		Limits of partial or interrupted views Suitable for CAD drawings provided the line is not an axis

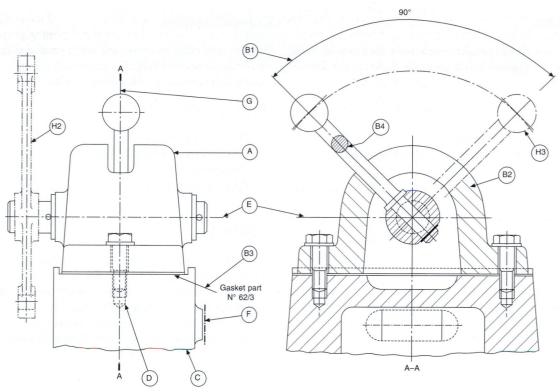

FIGURE 5.2

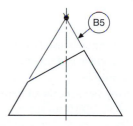

FIGURE 5.3 Example showing imaginary lines of intersection.

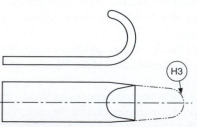

FIGURE 5.4 Initial outline applications.

Figure 5.5 shows the method of detailing a long strip of metal which has 60 holes in it at constant pitch. There would be no need to detail all of the component and this illustration gives one end only. The line to show the interruption in the drawing is narrow continuous and with the zig-zag cutting line indicated by the letter J.

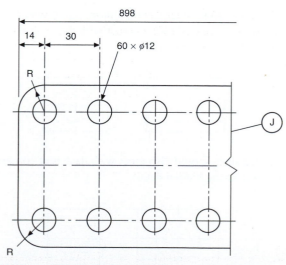

FIGURE 5.5 Interrupted view application.

Chain lines

Particular care should be taken with chain lines to ensure that they are neatly applied and attention is drawn to the following points:

(a) All chain lines should start and finish with a long dash.
(b) When centre points are defined, then the chain lines should cross one another at solid portions of the line.

(c) Centre lines should extend for a short distance beyond the feature unless they are required for dimensioning or other purpose.
(d) Centre lines should not extend through the spaces between views and should never terminate at another line on the drawing.

(e) If an angle is formed by chain lines, then the long dashed should intersect and define the angle.

(f) Arcs should meet straight lines at tangency points.

(g) When drawing hidden detail, a dashed line should start and finish with dashes in contact with the visible lines from which they originate.

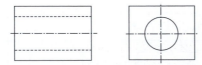

(h) Dashed lines should also meet with dashes at corners when drawing hidden detail.

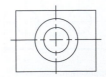

Coinciding lines

It is inevitable that at some time when producing a drawing two or more lines of differing types will coincide. The following order of priority should be applied:

(a) Visible outlines and edges.
(b) Cutting planes.
(c) Hidden outlines and edges.
(d) Centre lines and lines of symmetry.
(e) Centroidal lines.
(f) Projection lines.

Lettering

It has previously been mentioned that technical drawings are prepared using only two-line thicknesses and if reasonable care is taken a pleasing result can easily be obtained. Drawings invariably need dimensions and notes and if these are added in a careless and haphazard manner, then a very poor overall impression may be given. Remember that technical drawings are the main line of communication between the originator and the user. Between a consultant and his client, the sales manager and his customer, the designer and the manufacturer, a neat well-executed technical drawing helps to establish confidence. The professional draughtsman also takes considerable pride in his work and much effort and thought is needed with respect to lettering, and spacing, in order to produce an acceptable drawing of high standard.

The following notes draw attention to small matters of detail which we hope will assist the draughtsman's technique of lettering.

(a) Lettering may be vertical or slanted, according to the style which is customarily used by the draughtsman. The aim is to produce clear and unambiguous letters, numbers, and symbols.
(b) If slanted lettering is used, the slope should be approximately 65°–70° from the horizontal. Legibility is important. The characters should be capable of being produced at reasonable speed and in a repeatable manner. Different styles on the same drawing spoil the overall effect.
(c) Use single stroke characters devoid of serifs and embellishments.
(d) All strokes should be of consistent density.
(e) The spacing round each character is important to ensure that 'filling in' will not occur during reproduction.
(f) Lettering should not be underlined since this impairs legibility.
(g) On parts lists or where information is tabulated, the letters or numerals should not be allowed to touch the spacing lines.
(h) All drawing notes and dimensions should remain legible on reduced size copies and on the screens of microfilm viewers.

TABLE 5.3

Application	Drawing sheet size	Minimum character height (mm)
Drawing numbers etc.	A0, A1, A2 and A3	7
	A4	5
Dimensions and notes	A0	3.5
	A1, A2, A3, and A4	2.5

(i) Capital letters are preferred to lower case letters since they are easier to read on reduced size copies of drawings. Lower case letters are generally used only where they are parts of standard symbols, codes or abbreviations.

(j) When producing a manual drawing the draughtsman should take care to select the proper grade of pencil for lettering. The pencil should be sharp, but with a round point which will not injure the surface. Mechanical pencils save time and give consistent results since no resharpening is necessary.

(k) Typewritten, stencilled or letters using the 'Letraset' adhesive letter system may be used since these provide uniformity and a high degree of legibility.

Minimum character height for capital letters and numerals

Table 5.3 gives the minimum recommended character heights for different sizes of drawing sheet and it is stressed that these are *minimum* sizes. If lower case letters are used then they should be proportioned so that the body height will be approximately 0.6 times the height of a capital letter.

The stroke thickness should be approximately 0.1 times the character height and the clear space between characters should be about 0.7 mm for 2.5 mm capitals and other sizes in proportion.

The spaces between lines of lettering should be consistent and preferably not less than half the character height. In the case of titles, this spacing may have to be reduced.

All notes should be placed so that they may be read from the same direction as the format of the drawing but there are cases, for example when a long vertical object is presented, where it may be necessary to turn the drawing sheet through 90° in the clockwise direction, in effect, to position the note which is then read from the right hand side of the drawing sheet.

The shape and form of an acceptable range of letters and numbers is illustrated in Fig. 5.6.

Open styles are often used on drawings which are to be microfilmed, as increased clarity is obtainable on small reproductions.

Drawing modifications

After work has been undertaken on a drawing for a reasonable amount of time, then that drawing will possess some financial value. The draughtsman responsible for the draw-

ings must be concerned with the reproducible quality of his work as prints or photographic copies are always taken from the originals. Revisions and modifications are regularly made to update a product, due for example, to changes in materials, individual components, manufacturing techniques, operating experience and other causes outside the draughtsman's control.

When a drawing is modified, its content changes and it is vital that a note is given on the drawing describing briefly the reason for change and the date that modifications were made. Updated drawings are then reissued to interested parties. Current users must all read from a current copy. Near the title block, on a drawing will be placed a box giving the date and Issue No., i.e. XXXA, XXXB, etc. These changes would usually be of a minimal nature.

If a component drawing is substantially altered, it would be completely redrawn and given an entirely new number.

Drawings on a computer, of course, leave no trace when parts are deleted but this is not necessarily the case if the work is undertaken manually on tracing film or paper. The point to remember is that on the area covered by the erasure, part of a new drawing will be added and the quality of this drawing must be identical in standard with the original. Obviously, if the surface of the drawing sheet has been damaged in any way during erasure, then the draughtsman performing the work starts with a serious disadvantage.

(a) ABCDEFGHIJKLMNOPQRSTUVWXYZ
abcdefghijklmnopqrstuvwxyz
1234567890

(b) *ABCDEFGHIJKLMNOPQRSTUVWXYZ*
abcdefghijklmnopqrstuvwxyz
1234567890

(c) ABCDEFGHIJKLMNOPQRSTUVWXYZ
aabcdefghijklmnopqrstuvwxyz
1234567890

(d) *ABCDEFGHIJKLMNOPQRSTUVWXYZ*
aabcdefghijklmnopqrstuvwxyz
1234567890

(e) 1234567890

(f) *12334567890*

FIGURE 5.6

The following suggestions are offered to assist in the preservation of drawings when erasures have to be made.

1. Use soft erasers with much care. Line removal without damaging the drawing surface is essential.
2. An erasing shield will protect areas adjacent to modifications.
3. Thoroughly erase the lines, as a ghost effect may be observed with incomplete erasures when prints are made. If in any doubt, a little time spent performing experimental trial erasures on a sample of a similar drawing medium will pay dividends, far better than experimenting on a valuable original.

Care and storage of original drawings

Valuable drawings need satisfactory handling and storage facilities in order to preserve them in first class condition. Drawings may be used and reused many times and minimum wear and tear is essential if good reproductions and micro-films are to be obtained over a long period of time. The following simple rules will assist in keeping drawings in 'mint' condition.

1. Never fold drawings.
2. Apart from the period when the drawing is being prepared or modified, it is good policy to refer to prints at other times when the drawing is required for information purposes.
3. The drawing board should be covered outside normal office hours, to avoid the collection of dust and dirt.
4. Too many drawings should not be crowded in a filing drawer. Most drawing surfaces, paper or plastics, are reasonably heavy and damage results from careless manipulation in and out of drawers.
5. Do not roll drawings tightly since they may not lie flat during microfilming.
6. Do not use staples or drawing pins. Tape and drawing clips are freely available.
7. When using drawings, try to use a large reference table. Lift the drawings rather than slide them, to avoid smudging and wear.
8. Drawings should be stored under conditions of normal heat and humidity, about 21°C and 40–60% relative humidity.

Chapter 6

Three-dimensional illustrations using isometric and oblique projection

Isometric projection

Figure 6.1 shows three views of a cube in orthographic projection; the phantom line indicates the original position of the cube, and the full line indicates the position after rotation about the diagonal AB. The cube has been rotated so that the angle of $45°$ between side AC_1 and diagonal AB now appears to be $30°$ in the front elevation, C_1 having been rotated to position C. It can clearly be seen in the end view that to obtain this result the angle of rotation is greater than $30°$. Also, note that, although DF in the front elevation appears to be vertical, a cross check with the end elevation will confirm that the line slopes, and that point F lies to the rear of point D. However, the front elevation now shows a three-dimensional view, and when taken in isolation it is known as an *isometric projection*.

This type of view is commonly used in pictorial presentations, for example in car and motor-cycle service manuals and model kits, where an assembly has been 'exploded' to indicate the correct order and position of the component parts.

It will be noted that, in the isometric cube, line AC_1 is drawn as line AC, and the length of the line is reduced. Figure 6.2 shows an isometric scale which in principle is obtained from lines at $45°$ and $30°$ to a horizontal axis. The $45°$ line XY is calibrated in millimetres commencing from point X, and the dimensions are projected vertically on to the line XZ. By similar triangles, all dimensions are

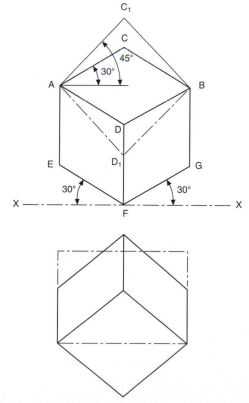

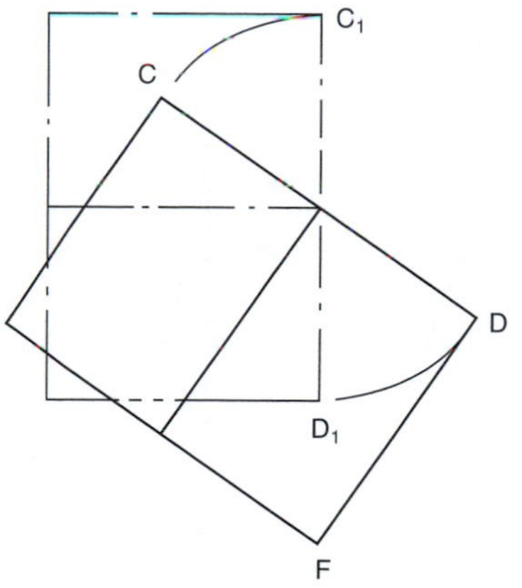

FIGURE 6.1

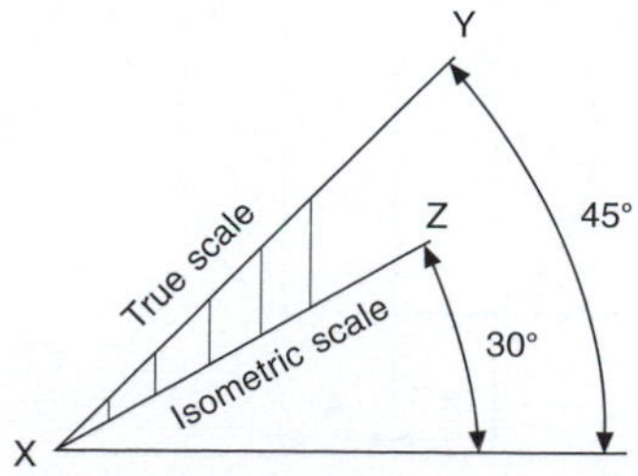

FIGURE 6.2 Isometric scale.

reduced by the same amount, and isometric lengths can be measured from point X when required. The reduction in length is in the ratio

$$\frac{\text{isometric length}}{\text{true length}} = \frac{\cos 45°}{\cos 30°} = \frac{0.7071}{0.8660} = 0.8165$$

Now, to reduce the length of each line by the use of an isometric scale is an interesting academic exercise, but commercially an isometric projection would be drawn using the true dimensions and would then be enlarged or reduced to the size required.

Note that, in the isometric projection, lines AE and DB are equal in length to line AD; hence an equal reduction in length takes place along the apparent vertical and the two axes at 30° to the horizontal. Note also that the length of the diagonal AB does not change from orthographic to isometric, but that of diagonal C_1D_1 clearly does. When setting out an isometric projection, therefore, measurements must be made only along the isometric axes EF, DF, and GF.

Figure 6.3 shows a wedge which has been produced from a solid cylinder, and dimensions A, B, and C indicate typical measurements to be taken along the principal axes when setting out the isometric projection.

Any curve can be produced by plotting a succession of points in space after taking ordinates from the X, Y, and Z axes.

Figure 6.4(a) shows a cross-section through an extruded alloy bar: the views (b), (c), and (d) give alternative isometric presentations drawn in the three principal planes of projection. In every case, the lengths of ordinates OP, OQ, P1, and Q2, etc. are the same, but are positioned either vertically or inclined at 30° to the horizontal.

Figure 6.5 shows an approximate method for the construction of isometric circles in each of the three major planes. Note the position of the points of intersection of radii RA and RB.

The construction shown in Fig. 6.5 can be used partly for producing corner radii. Figure 6.6 shows a small block with radiused corners together with isometric projection which emphasizes the construction to find the centres for the corner radii; this should be the first part of the drawing to be attempted. The thickness of the block is obtained from projecting back these radii a distance equal to the block thickness and at 30°. Line in those parts of the corners visible behind the front face, and complete the pictorial view by adding the connecting straight lines for the outside of the profile.

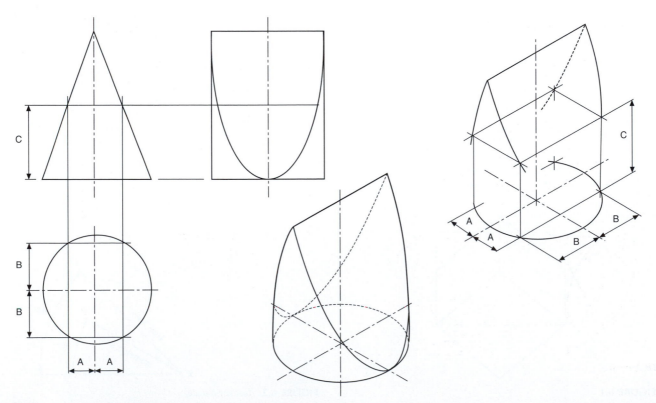

FIGURE 6.3 Construction principles for points in space, with complete solution.

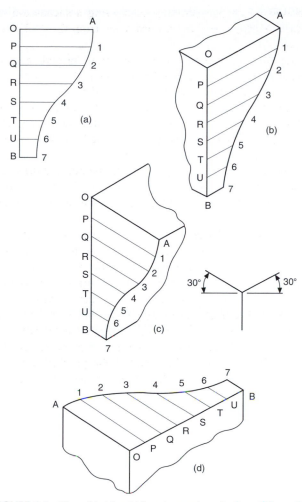

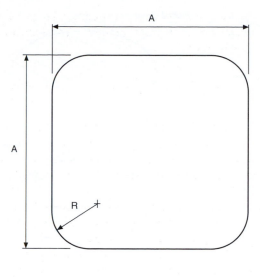

FIGURE 6.4 Views (b), (c) and (d) are isometric projections of the section in view (a).

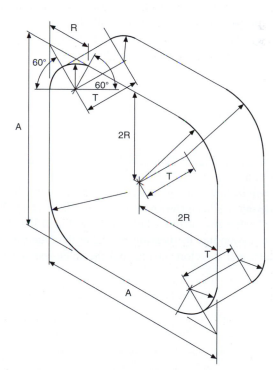

FIGURE 6.6 Isometric constructions for corner radii.

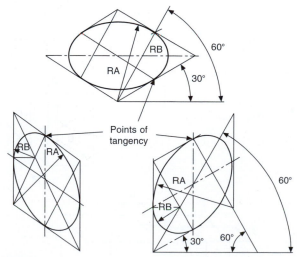

FIGURE 6.5

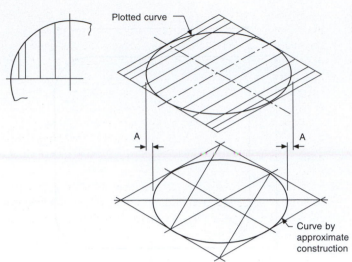

In the approximate construction shown, a small inaccuracy occurs along the major axis of the ellipse, and Fig. 6.7 shows the extent of the error in conjunction with a plotted circle. In the vast majority of applications where complete but small circles are used, for example spindles, pins, parts of nuts, bolts, and fixing holes, this error is of little importance and can be neglected.

Oblique projection

Figure 6.8 shows part of a plain bearing in orthographic projection, and Figs. 6.9 and 6.10 show alternative pictorial projections.

It will be noted in Figs. 6.9 and 6.10 that the thickness of the bearing has been shown by projecting lines at 45° back from a front elevation of the bearing. Now, Fig. 6.10 conveys the impression that the bearing is thicker than the true plan suggests, and therefore in Fig. 6.9 the thickness has been reduced to one half of the actual size. Figure 6.9 is known as an *oblique projection*, and objects which have curves in them are easiest to draw if they are turned, if possible, so that the curves are presented in the front elevation. If this proves impossible or undesirable, then Fig. 6.11 shows part of the ellipse which results from projecting half sizes back along the lines inclined at 45°.

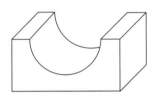

FIGURE 6.10

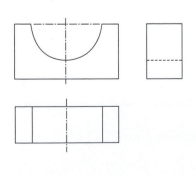

FIGURE 6.8

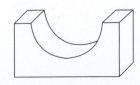

FIGURE 6.9

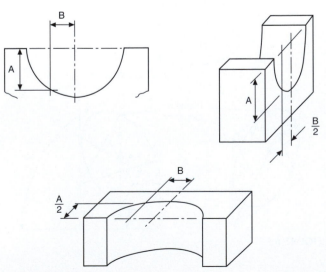

FIGURE 6.11

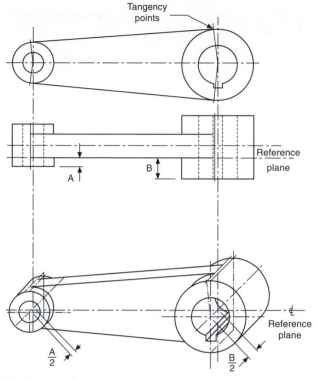

FIGURE 6.12

A small die-cast lever is shown in Fig. 6.12, to illustrate the use of a reference plane. Since the bosses are of different thicknesses, a reference plane has been taken along the side of the web; and from this reference plane, measurements are taken forward to the boss faces and backwards to the opposite sides. Note that the points of tangency are marked to position the slope of the web accurately.

With oblique and isometric projections, no allowance is made for perspective, and this tends to give a slightly unrealistic result, since parallel lines moving back from the plane of the paper do not converge.

Further information regarding pictorial representations, reference can be made to BS EN ISO 5456–3. The Standard contains details of dimetric, trimetric, cavalier, cabinet, planometric, and perspective projections.

Chapter 7

Drawing layouts and simplified methods

Single-part drawing

A single-part drawing should supply the complete detailed information to enable a component to be manufactured without reference to other sources. It should completely define shape or form and size, and should contain a specification. The number of views required depends on the degree of complexity of the component. The drawing must be fully dimensioned, including tolerances where necessary, to show all sizes and locations of the various features. The specification for the part includes information relating to the material used and possible heat-treatment required, and notes regarding finish. The finish may apply to particular surfaces only, and may be obtained by using special machining operations or, for example, by plating, painting, or enamelling. Figure 7.1 shows typical single-part drawings.

An alternative to a single-part drawing is to collect several small details relating to the same assembly and group them together on the same drawing sheet. In practice, grouping in this manner may be satisfactory provided all the parts are made in the same department, but it can be inconvenient where, for example, pressed parts are drawn with turned components or sheet–metal fabrications.

More than one drawing may also be made for the same component. Consider a sand-cast bracket. Before the bracket is machined, it needs to be cast; and before casting, a pattern needs to be produced by a patternmaker. It may therefore be desirable to produce a drawing for the patternmaker which includes the various machining allowances, and then produce a separate drawing for the benefit of the machinist which shows only dimensions relating to the surfaces to be machined and the size of the finished part. The two drawings would each have only parts of the specification which suited one particular manufacturing process (see also Figs. 14.34 and 14.35).

Collective single-part drawings

Figure 7.2 shows a typical collective single-part drawing for a rivet. The drawing covers 20 rivets similar in every respect except length; in the example given, the part number for a 30 mm rivet is S123/13. This type of drawing can also be used where, for example, one or two dimensions on a component (which are referred to on the drawing as A and B) are variable, all others being standard.

For a particular application, the draughtsman would insert the appropriate value of dimensions A and B in a table, then add a new suffix to the part number. This type of drawing can generally be used for basically similar parts.

Assembly drawings

Machines and mechanisms consist of numerous parts, and a drawing which shows the complete product with all its components in their correct physical relationship is known as an assembly drawing. A drawing which gives a small part of the whole assembly is known as a sub-assembly drawing. A sub-assembly may in fact be a complete unit in itself; for example, a drawing of a clutch could be considered as a sub-assembly of a drawing showing a complete automobile engine. The amount of information given on an assembly drawing will vary considerably with the product and its size and complexity.

If the assembly is relatively small, information which might be given includes a parts list. The parts list, as the name suggests, lists the components, which are numbered. Numbers in 'balloons' with leader lines indicate the position of the component on the drawing (see Fig. 7.3). The parts list will also contain information regarding the quantity required of each component for the assembly, its individual single-part drawing number, and possibly its material. Parts lists are not standard items, and their contents vary from one drawing office to another.

The assembly drawing may also give other information, including overall dimensions of size, details of bolt sizes and centres where fixings are necessary, weights required for shipping purposes, operating details and instructions, and also, perhaps, some data regarding the design characteristics.

Collective assembly drawing

This type of drawing is used where a range of products which are similar in appearance but differing in size is manufactured and assembled. Figure 7.4 shows a nut-and-bolt fastening used to secure plates of different combined thickness; the nut is standard, but the bolts are of different lengths. The accompanying table is used to relate the various assemblies with different part numbers.

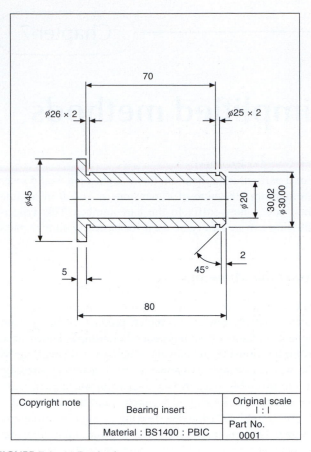

FIGURE 7.1   (a) Bearing insert.

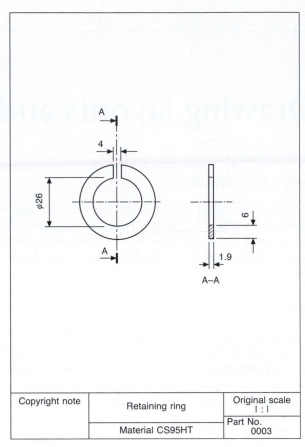

FIGURE 7.1   (c) Retaining ring.

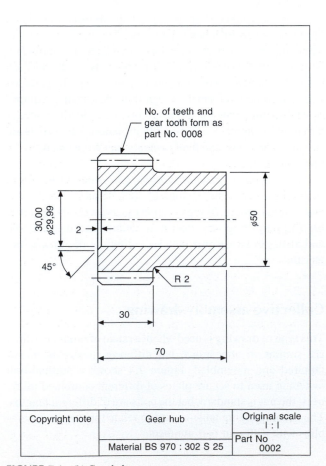

FIGURE 7.1   (b) Gear hub.

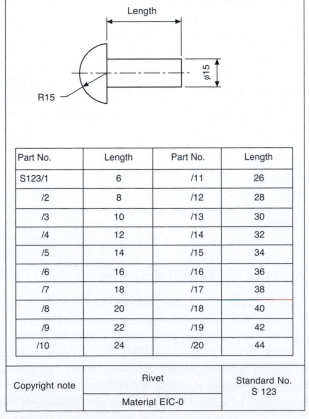

Part No.	Length	Part No.	Length
S123/1	6	/11	26
/2	8	/12	28
/3	10	/13	30
/4	12	/14	32
/5	14	/15	34
/6	16	/16	36
/7	18	/17	38
/8	20	/18	40
/9	22	/19	42
/10	24	/20	44

FIGURE 7.2   Collective single-part drawing of a rivet.

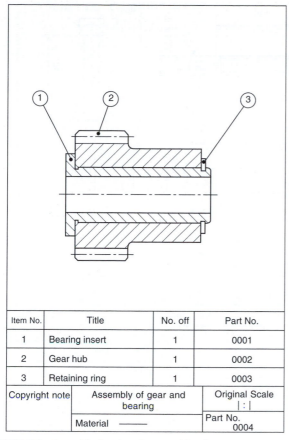

FIGURE 7.3   Assembly drawing of gear and bearing.

Item No.	Title	No. off	Part No.
1	Bearing insert	1	0001
2	Gear hub	1	0002
3	Retaining ring	1	0003

Copyright note	Assembly of gear and bearing	Original Scale 1 : 1
	Material ———	Part No. 0004

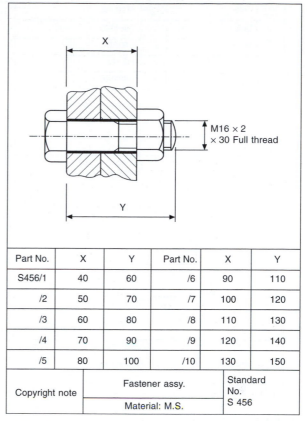

Part No.	X	Y	Part No.	X	Y
S456/1	40	60	/6	90	110
/2	50	70	/7	100	120
/3	60	80	/8	110	130
/4	70	90	/9	120	140
/5	80	100	/10	130	150

Copyright note	Fastener assy.	Standard No. S 456
	Material: M.S.	

FIGURE 7.4   Typical collective assembly drawing of a nut with bolts of various length.

Design layout drawings

Most original designs are planned in the drawing office where existing or known information is collected and used to prepare a provisional layout drawing before further detailed design work can proceed. This type of drawing is of a preliminary nature and subject to much modification so that the designer can collect his thoughts together. The drawing can be true to scale or possibly enlargements or reductions in scale depending on the size of the finished product or scheme, and is essentially a planning exercise. They are useful in order to discuss proposals with prospective customers or design teams at a time when the final product is by no means certain, and should be regarded as part of the design process.

Provisional layout drawings may also be prepared for use with tenders for proposed work where the detailed design will be performed at a later date when a contract has been negotiated, the company being confident that it can ultimately design and manufacture the end product. This confidence will be due to experience gained in similar schemes undertaken previously.

Combined detail and assembly drawings

It is sometimes convenient to illustrate details with their assembly drawing on the same sheet. This practice is particularly suited to small 'one-off' or limited-production-run assemblies. It not only reduces the actual number of drawings, but also the drawing-office time spent in scheduling and printing. Figure 7.5 shows a simple application of an assembly of this type.

Exploded assembly drawings

Figure 7.6 shows a typical exploded assembly drawing; these drawings are prepared to assist in the correct understanding of the various component positions in an assembly. Generally a pictorial type of projection is used, so that each part will be shown in three dimensions. Exploded views are invaluable when undertaking servicing and maintenance work on all forms of plant and appliances. Car manuals and do-it-yourself assembly kits use such drawings, and these are easily understood. As well as an aid to construction, an exploded assembly drawing suitably numbered can also be of assistance in the ordering of spare parts; components are more easily recognizable in a pictorial projection, especially by people without training in the reading of technical drawings.

Simplified drawings

Simplified draughting conventions have been devised to reduce the time spent drawing and detailing symmetrical components and repeated parts. Figure 7.7 shows a gasket which

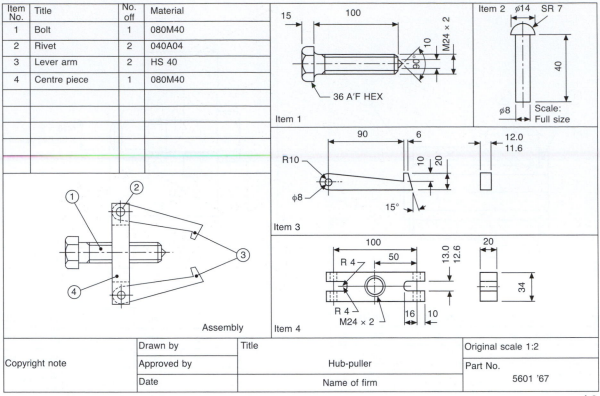

Item No.	Title	No. off	Material
1	Bolt	1	080M40
2	Rivet	2	040A04
3	Lever arm	2	HS 40
4	Centre piece	1	080M40

Copyright note	Drawn by	Title
	Approved by	Hub-puller
	Date	Name of firm

Original scale 1:2

Part No. 5601 '67

A 3

FIGURE 7.5 Combined detail and assembly drawing of hub-puller.

is symmetrical about the horizontal centre line. A detail drawing indicating the line of symmetry and half of the gasket is shown in Fig. 7.8, and this is sufficiently clear for the part to be manufactured.

If both halves are similar except for a small detail, then the half which contains the exception is shown with an explanatory note to that effect, and a typical example is illustrated in Fig. 7.9.

A joint-ring is shown in Fig. 7.10, which is symmetrical about two axes of symmetry. Both axes are shown in detail, and a quarter view of the joint-ring is sufficient for the part to be made.

The practice referred to above is not restricted to flat thin components, and Fig. 7.11 gives a typical detail of a straight lever with a central pivot in part section. Half the lever is shown, since the component is symmetrical, and a partial view is added and drawn to an enlarged scale to clarify the shape of the boss and leave an adequate space for dimensioning.

Repeated information also need not be drawn in full; for example, to detail the peg-board in Fig. 7.12 all that is required is to draw one hole, quoting its size and fixing the centres of all the others.

Similarly Fig. 7.13 shows a gauze filter. Rather than draw the gauze over the complete surface area, only a small portion is sufficient to indicate the type of pattern required.

Knurled screws are shown in Fig. 7.14 to illustrate the accepted conventions for straight and diamond knurling.

Machine drawing

The draughtsman must be able to appreciate the significance of every line on a machine drawing. He must also understand the basic terminology and vocabulary used in conjunction with machine drawings.

Machine drawings of components can involve any of the geometrical principles and constructions described in this book and in addition the accepted drawing standards covered by BS8888.

Figure 7.15 illustrates many features found on machine drawings and the notes which follow given additional explanations and revision comments.

1. *Angular dimension* – Note that the circular dimension line is taken from the intersection of the centre lines of the features.
2. *Arrowheads* – The point of an arrowhead should touch the projection line or surface, it should be neat and easily readable and normally not less than 3 mm in length.
3. *Auxiliary dimension* – A dimension given for information purposes but not used in the actual manufacturing process.
4. *Boss* – A projection, which is usually circular in cross-section, and often found on castings and forgings. A shaft boss can provide extra bearing support, for example, or a boss could be used on a thin cast surface to increase its thickness in order to accommodate a screw thread.

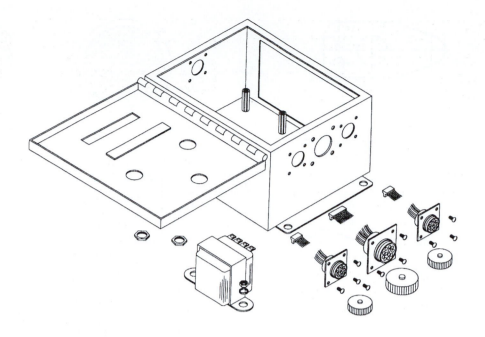

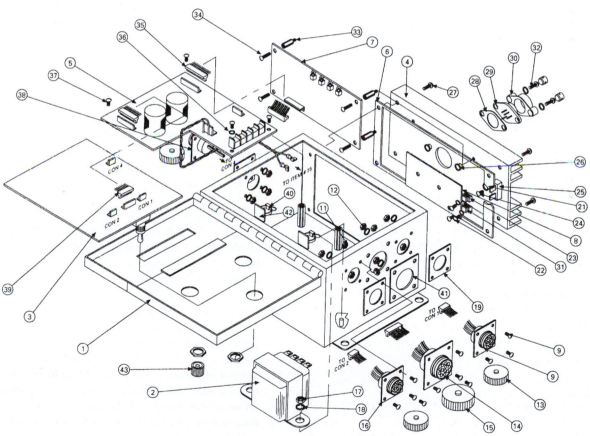

FIGURE 7.6

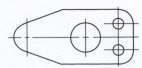

FIGURE 7.7

FIGURE 7.8

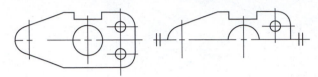

FIGURE 7.9   When dimensioning add drawing note 'slot on one side only'.

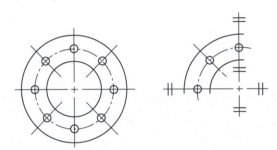

FIGURE 7.10

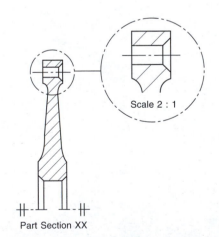

Part Section XX

FIGURE 7.11   Part of a lever detail drawing symmetrical about the horizontal axis.

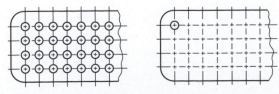

FIGURE 7.12

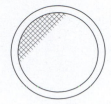

FIGURE 7.13

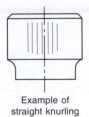

| Example of straight knurling | Example of diamond knurling |

FIGURE 7.14

5. *Centre line* – Long-dashed dotted narrow line which is used to indicate the axes of holes, components and circular parts.
6. *Long-dashed dotted wide line* – This is used to indicate surfaces which are required to meet special specifications and which differ from the remainder of the component.
7. *Chamfer* – A chamfer is machined to remove a sharp edge. The angle is generally 45°. Often referred to as a bevelled edge.
8. *Circlip groove* – A groove to accommodate a circlip. A circlip may be manufactured from spring steel wire, sheet or plate which is hardened and tempered and when applied in an assembly provides an inward or outward force to locate a component within a bore or housing.
9. *Clearance hole* – A term used in an assembly to describe a particular hole which is just a little larger and will clear the bolt or stud which passes through.
10. *Counterbore* – A counterbored hole may be used to house a nut or bolthead so that it does not project above a surface. It is machined so that the bottom surface of the larger hole is square to the hole axis.
11. *Countersink* – A hole which is recessed conically to accommodate the head of a rivet or screw so that the head will lie at the same level as the surrounding surface.
12. *Section plane* or *cutting plane* – These are alternative terms used to define the positions of planes from which sectional elevations and plans are projected.
13. *Dimension line* – This is a narrow continuous line which is placed outside the outline of the object, if possible. The arrowheads touch the projection lines. The dimension does not touch the line but is placed centrally above it.
14. *Enlarged view* – Where detail is very small or insufficient space exists for dimensions or notes then a partial view may be drawn with an increased size scale.

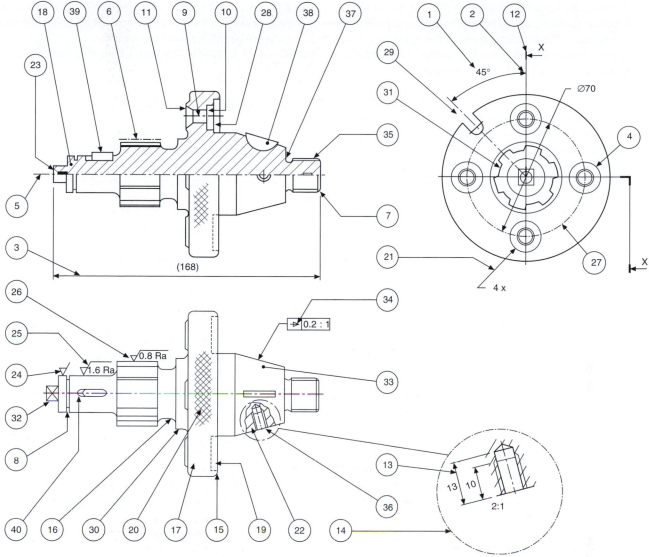

FIGURE 7.15

15. *Round* – This term is often used to describe an external radius.
16. *Fillet* – This is the term given to the radii on internal corners. Often found on castings, where its function is to prevent the formation of stress cracks, which can originate from sharp corners. Where three surfaces meet on a casting the fillet radii will be spherical.
17. *Flange* – This is a term to describe a projecting rim or an edge which is used for stiffening or for fixing. The example here is drilled for countersunk screws.
18. *Hatching* – Note that cross hatching of the component at the section plane is performed with narrow continuous lines at 45°. Spacing between the hatching lines varies with the size of the component but should not be less than 4 mm.
19. *Hidden detail* – Indicated by a narrow dashed line. Dashes of 3 mm and spaces of 2 mm are of reasonable proportion.

20. *Knurl* – A surface finish with a square or diamond pattern. Can be used in a decorative manner or to improve grip.
21. *Leader line* – Leaders are used to indicate where dimensions or notes apply and are drawn as narrow continuous lines terminating in arrowheads or dots. An arrowhead should always terminate on a line; dots should be within the outline of the object.
22. *Local section* – A local section may be drawn if a complete section or a half section is inconvenient. The local break around the section is a continuous narrow irregular line.
23. *Machining centre* – An accurately drilled hole with a good finish at each end of the component which enables the work to be located during a machining operation on a lathe.
24. *Machining symbol* – If it is desired to indicate that a particular surface is to be machined, without further

defining the actual machining process or the surface finish, a symbol is added normal to the line representing the surface. The included angle of the symbol is approximately 60°. A general note may be added to a drawing where all surfaces are to be machined as follows:

25. *Surface finish* – If a surface is to be machined and a particular quality surface texture is desired then a standard machining symbol is added to the drawing with a number which gives the maximum permissible roughness expressed numerically in micrometres.

26. *Surface finish* – If maximum and minimum degrees of roughness are required then both figures are added to the machining symbol.

27. *Pitch circle diameter* – A circle which passes through the centres of a series of holes. The circle is drawn with a long dashed dotted narrow line.

28. *Recess* – A hollow feature which is used to reduce the overall weight of the component. A recess can also be used to receive a mating part.

29. *Slot* – An alternative term to a slit, groove, channel or aperture.

30. *Spigot* – This is a circular projection which is machined to provide an accurate location between assembled components.

31. *Splined shaft* – A rotating member which can transmit a torque to a mating component. The mating component may move axially along the splines which are similar in appearance to keyways around the spindle surface.

32. *Square* – Diagonal lines are drawn to indicate the flat surface of the square and differentiate between a circular and a square section shaft. The same convention is used to show spanner flats on a shaft.

33. *Taper* – A term used in connection with a slope or incline. Rate of taper can also define a conical form.

34. *Taper symbol* – The taper symbol is shown here in a rectangular box which also includes dimensional information regarding the rate of taper on the diameter.

35. *External thread* – An alternative term used for a male thread. The illustration here shows the thread convention.

36. *Internal thread* – An alternative term for a female thread. The illustration here shows the convention for a female tapped hole.

37. *Undercut* – A circular groove at the bottom of a thread which permits assembly without interference from a rounded corner. Note in the illustration that a member can be screwed along the M20 thread right up to the tapered portion.

38. *Woodruff key* – A key shaped from a circular disc which fits into a circular keyway in a tapered shaft. The key can turn in the circular recess to accommodate any taper in the mating hub.

39. *Key* – A small block of metal, square or rectangular in cross-section, which fits between a shaft and a hub and prevents circumferential movement.

40. *Keyway* – A slot cut in a shaft or hub to accommodate a key.

Drawing scales

Small objects are sometimes drawn larger than actual size, while large components and assemblies of necessity are drawn to a reduced size. A drawing should always state the scale used. The scale on a full-size drawing will be quoted as 'ORIGINAL SCALE 1:1'.

Drawings themselves should not be scaled when in use for manufacturing purposes, and warnings against the practice are often quoted on standard drawing sheets, e.g. 'DO NOT SCALE' and 'IF IN DOUBT, ASK'. A drawing must be adequately dimensioned, or referenced sufficiently so that all sizes required are obtainable.

The recommended multipliers for scale drawings are 2, 5, and 10.

1:1 denotes a drawing drawn full-size.
2:1 denotes a drawing drawn twice full-size.
5:1 denotes a drawing drawn five times full size.

Other common scales are 10:1, 20:1, 50:1, 100:1, 200:1, 500:1, and 1000:1.

It should be pointed out that a scale drawing can be deceiving; a component drawn twice full-size will cover four times the area of drawing paper as the same component drawn full-size, and its actual size may be difficult to visualize. To assist in appreciation, it is a common practice to add a pictorial view drawn full-size, provided that the drawing itself is intended to be reproduced to the same scale and not reproduced and reduced by microfilming.

The recommended divisors for scale drawings are also 2, 5, and 10.

1:1 denotes a drawing drawn full-size.
1:2 denotes a drawing drawn half full-size.
1:5 denotes a drawing drawn a fifth full-size.

Other common scales used are 1:10, 1:20, 1:50, 1:100, 1:200, 1:500, and 1:1000.

The draughtsman will select a suitable scale to use on a standard drawing sheet and this will depend on the size of the object to be drawn. Remember that the drawing must clearly show necessary information and detail. It may be beneficial to make a local enlargement of a small area and an example is given in Fig. 7.15.

Scale used in geometric construction

Division of lines

Figure 7.16 shows the method of dividing a given line AB, 89 mm long, into a number of parts (say seven).

Draw line AC, and measure seven equal divisions. Draw line B7, and with the tee-square and set-square draw lines

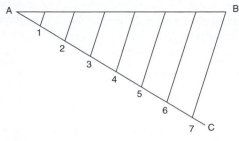

FIGURE 7.16

parallel to line B7 through points 1 to 6, to give the required divisions on AB.

Figure 7.17 shows an alternative method.

1. Draw vertical lines from A and B.
2. Place the scale rule across the vertical lines so that seven equal divisions are obtained and marked.
3. Draw vertical lines up from points 2 to 7 to intersect AB.

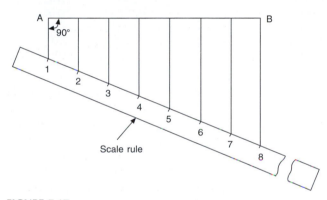

FIGURE 7.17

Diagonal scales

Figure 7.18 shows the method of drawing a diagonal scale of 40–1000 mm which can be read by 10–4000 mm. Diagonal scales are so called since diagonals are drawn in the rectangular part at the left-hand end of the scale.

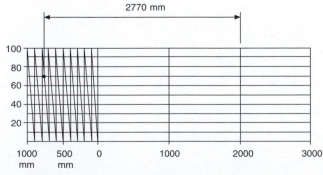

FIGURE 7.18　Diagonal scale where 40 mm represents 1000 mm.

The diagonals produce a series of similar triangles.

1. Draw a line 160 mm long.
2. Divide the line into four equal parts.
3. Draw 10 vertical divisions as shown and to any reasonable scale (say 5 mm) and add diagonals.

An example of reading the scale is given.

Plain scales

The method of drawing a plain scale is shown in Fig. 7.19. The example is for a plain scale of 30–500 mm to read by 125–2500 mm.

1. Draw a line 150 mm long and divide it into five equal parts.
2. Divide the first 30 mm length into four equal parts, and note the zero position or, the solution.

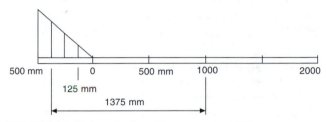

FIGURE 7.19　Plain scale where 30 mm represents 500 mm.

An example of a typical reading is given.

This method of calibration is in common use in industry, and scales can be obtained suitable for a variety of scale ratios.

Abbreviations

With the increasing globalization of design and manufacture the use of abbreviations is now discouraged. However, in some areas of industry some abbreviations are commonly used and understood and where these exist their continuing use is permitted. The abbreviations in the following list are commonly understood by English speaking nations,

Term	Abbreviation
Across flats	A/F
Assembly	ASSY
Centres	CRS
Centre line	
(a) on a view and across a centre line	℄
(b) in a note	CL
Centre of gravity	CG
Chamfer or chamfered (in a note)	CHAM
Cheese head	CH HD
Countersunk/countersink	CSK
Countersunk head	CSK HD
Counterbore	CBORE
Counterbore or Spotface	

Term	Abbreviation	Term	Abbreviation
Countersink	⋁	Pattern number	PATT NO.
Cylinder or cylindrical	CYL	Pitch circle diameter	PCD
Deep	⤓	Radius	
Diameter		(a) In a note	RAD
(a) In a note	DIA	(b) Preceding a dimension	R
(b) Preceding a dimension	Ø	Reference	REF
Dimension	DIM	Required	REQD
Drawing	DRG	Right hand	RH
Equally spaced	EQUI SP	Round head	RD HD
External	EXT	Screw or screwed	SCR
Figure	FIG	Sheet (referring to a drawing sheet)	SH
Full indicated movement	FIM	Sketch (prefix to a drawing number)	SK
Hexagon	HEX	Specification	SPEC
Hexagon head	HEX HD	Spherical diameter (only preceding	
Insulated or insulation	INSUL	a dimension)	SØ
Internal	INT	Spherical radius (only preceding	
Least material condition		a dimension)	SR
(a) In a note	LMC	Spotface	SFACE
(b) Part of a geometrical tolerance	Ⓛ	Square	
Left hand	LH	(a) In a note	SQ
Long	LG	(b) Preceding a dimension	□ or ⊠
Machine	MC	Standard	STD
Material	MATL	Taper (on diameter or width)	
Maximum	MAX	Thread	THD
Maximum material condition		Thick	THK
(a) In a note	MMC	Tolerance	TOL
(b) Part of geometrical tolerance	Ⓜ	Typical or typically	TYP
Minimum	MIN	Undercut	UCUT
Not to scale (in a note and underlined)	NTS	Volume	VOL
Number	NO.	Weight	WT

Chapter 8

Sections and sectional views

A section is used to show the detail of a component, or an assembly, on a particular plane which is known as the cutting plane. A simple bracket is shown in Fig. 8.1 and it is required to draw three sectional views. Assume that you had a bracket and cut it with a hacksaw along the line marked B–B. If you looked in the direction of the arrows then the end view B–B in the solution (Fig. 8.2), would face the viewer and the surface indicated by the cross hatching would be the actual metal which the saw had cut through. Alternatively had we cut along the line C–C then the plan in the solution would be the result. A rather special case exists along the plane A–A where in fact the thin web at this point has been sliced. Now if we were to cross hatch all the surface we had cut through on this plane we would give a false impression of solidity. To provide a more realistic drawing, the web is defined by a full line and the base and perpendicular parts only have been cross hatched. Note, that cross hatching is never undertaken between dotted lines, hence the full line between the web and the remainder of the detail. However, the boundary at this point is theoretically a dotted line since the casting is formed in one piece and no join exists here. This standard drawing convention is frequently tested on examination papers.

Cutting planes are indicated on the drawing by a long chain line 0.35 mm thick and thickened at both ends to 0.7 mm. The cutting plane is lettered and the arrows indicate the direction of viewing. The sectional view or plan must then be stated to be A–A, or other letters appropriate to the cutting plane. The cross hatching should always be at 45 to the centre lines, with continuous lines 0.35 mm thick.

If the original drawing is to be microfilmed successive lines should not be closer than 4 mm as hatching lines tend to merge with much reduced scales. When hatching very small areas, the minimum distance between lines should not be less than 1 mm.

In the case of very large areas, cross hatching may be limited to a zone which follows the contour of the hatched area. On some component detail drawings it may be necessary to add dimensions to a sectional drawing and the practice is to interrupt the cross hatching so that the letters and numbers are clearly visible.

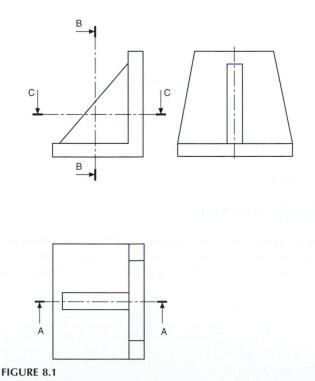

FIGURE 8.1

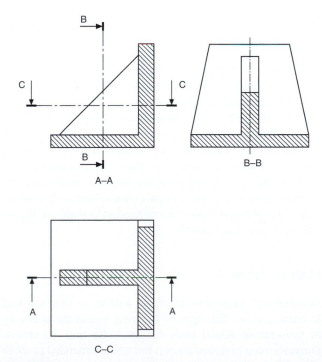

FIGURE 8.2

Figure 8.3 shows three typical cases of cross hatching. Note that the hatching lines are equally spaced and drawn at an angle of 45 to the principal centre line in each example.

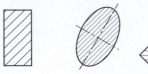

FIGURE 8.3

A bush is shown in Fig. 8.4 in a housing. There are two adjacent parts and each is cross hatched in opposite directions. It is customary to reduce the pitch between hatching lines for the smaller part.

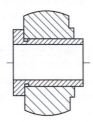

FIGURE 8.4

If the interior of a component is of an intricate nature or it contains several parts to form an assembly, then the customary orthographic drawing would contain a confusion of dotted lines, which, in addition to being difficult to draw could also be terribly difficult to understand. The reader of any engineering drawing should be able to obtain only one positive interpretation of the component, or the draughtsman has failed in his duty. Sectional drawings are prepared which cut away a portion of the component to reveal internal details and certain standard conventions have been established to cover this aspect of drawing practice.

Figure 8.5 shows some advantages of drawing a sectional view with a small cast component.

Note, that in Plan (A), the sectional plan gives clearly the exact outline along the horizontal axis where the casting has assumed to have been cut. This contrasts with the confusion in Plan (B) which obviously results from attempting to include all the detail by inserting the appropriate dotted lines.

Where the location of a single cutting plane is obvious, no indication of its position or identification is required. Figure 8.6 gives a typical example.

Half sections

Symmetrical parts may be drawn half in section and half in outside view. This type of drawing avoids the necessity of introducing dotted lines for the holes and the recess. Dimensioning to dotted lines is not a recommended practice (Fig. 8.7).

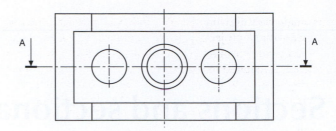

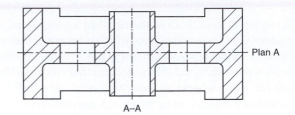

Plan A

A–A

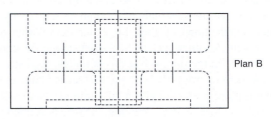

Plan B

FIGURE 8.5

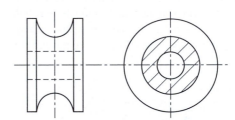

FIGURE 8.6

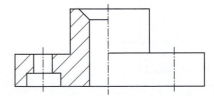

FIGURE 8.7

Revolved sections

A special spanner is illustrated in Fig. 8.8. A revolved section is shown on the handle to indicate the shape of the cross-section at that point. This is a convenient convention to use

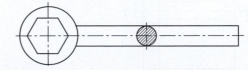

FIGURE 8.8

on single view drawings because the shape could not be confirmed without projecting a second view or an added note.

A second type of revolved section in Fig. 8.9 shows a case where it is required to indicate details on two separate intersecting planes. The elevation in section has been drawn assuming that the right-hand plane has been revolved to the horizontal position. Note that the thin web is not cross hatched.

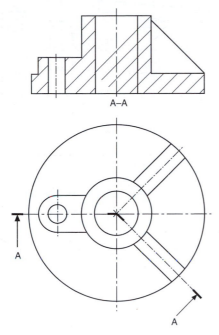

FIGURE 8.9

Figure 8.10 shows a sectioned elevation from a plan where the section line is taken along three neighbouring planes which are not at right angles to one another. The section line follows the section planes in order, and is thickened at each change of direction.

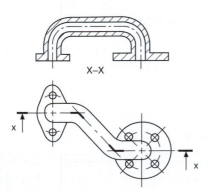

FIGURE 8.10

Removed sections

A removed section is shown in Fig. 8.11. Note that no additional background information has been included, since the

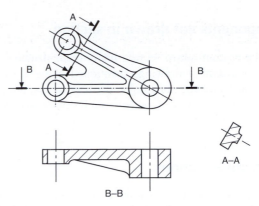

FIGURE 8.11

removed section only indicates the true shape of the casting at the point where the section has been taken. B–B gives the section along the horizontal centre line through the thin web.

Sections through thin material

Many products are manufactured from very thin materials which would be virtually impossible to cross hatch in a sectional view and in these cases it is usual to make them entirely black. However, if two or more thin sections are adjacent to each other, a gap is left so that the profile of the separate components is clearly defined. A compound stanchion used in structural steelwork and drawn to the reduced scale is shown in Fig. 8.12. The same situation applies with sections through sheet-metal fabrications, gaskets, seals and packings.

FIGURE 8.12

Local sections

It is not always necessary to draw a complete section through a component if a small amount of detail only needs to be illustrated. A typical example is shown in Fig. 8.13 where a keyway is drawn in a section. The irregular line defines the boundary of the section. It is not required to add a section plane to this type of view.

FIGURE 8.13

Components not drawn in section

It is the custom not to section many recognizable components in assembly drawings positioned along the cutting plane; these include nuts, bolts, washers, rivets, pins, keys, balls, rollers, spokes of wheels, and similar symmetrical parts.

Successive sections

Figure 8.14 shows the front and end elevations of a special purpose mounting plate where sectional plans are given at different levels to illustrate the shapes of the various cutouts and details. Now it will be noted that the presentation of this problem takes considerable vertical space since all of the plan views are in correct projection.

Note that where successive sections are drawn, each view only gives the detail at that section plane and not additional

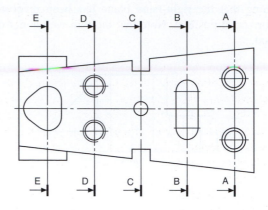

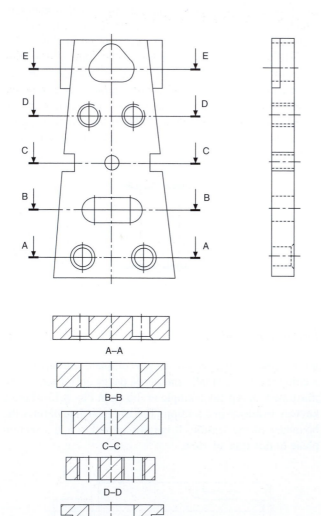

FIGURE 8.14

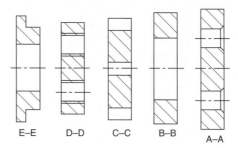

E–E   D–D   C–C   B–B
A–A

FIGURE 8.15

background information. Figure 8.15 gives the details at each of the section planes in a much closer and less remote arrangement.

Sections in two parallel planes

Figure 8.16 shows a method of presenting two sections from parallel planes along the same part.

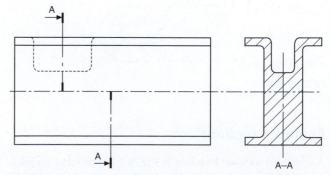

FIGURE 8.16

Chapter 9

Geometrical constructions and tangency

Students will often experience difficulty in handling problems involving two- and three-dimensional geometrical constructions. The examples in Chapters 9–13 are included in order to provide a background in solving engineering problems connected with lines, planes, and space. The separate chapters are grouped around applications having similar principles.

Copying a selection of these examples on the drawing board or on CAD equipment will certainly enable the reader to gain confidence. It will assist them to visualize and position the lines in space which form each part of a view, or the boundary, of a three-dimensional object. It is a necessary part of draughtsmanship to be able to justify every line and dimension which appears on a drawing correctly.

Many software programs will offer facilities to perform a range of constructions, for example tangents, ellipses and irregular curves. Use these features where possible in the examples which follow.

Assume all basic dimensions where applicable.

To bisect a given angle AOB (Fig. 9.1)

1. With centre O, draw an arc to cut OA at C and OB at D.
2. With centres C and D, draw equal radii to intersect at E.
3. Line OE bisects angle AOB.

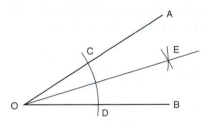

FIGURE 9.1

To bisect a given straight line AB (Fig. 9.2)

1. With centre A and radius greater than half AB, describe an arc.
2. Repeat with the same radius from B, the arcs intersecting at C and D.
3. Join C to D and this line will be perpendicular to and bisect AB.

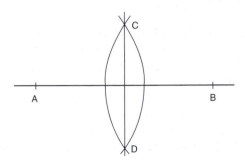

FIGURE 9.2

To bisect a given arc AB (Fig. 9.3)

1. With centre A and radius greater than half AB, describe an arc.
2. Repeat with the same radius from B, the arcs intersecting at C and D.
3. Join C to D to bisect the arc AB.

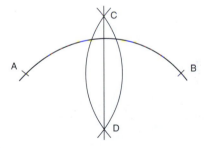

FIGURE 9.3

To find the centre of a given arc AB (Fig. 9.4)

1. Draw two chords, AC and BD.

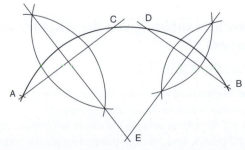

FIGURE 9.4

2. Bisect AC and BD as shown; the bisectors will intersect at E.
3. The centre of the arc is point E.

To inscribe a circle in a given triangle ABC (Fig. 9.5)

1. Bisect any two of the angles as shown so that the bisectors intersect at D.
2. The centre of the inscribed circle is point D.

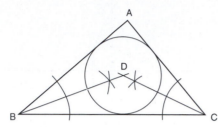

FIGURE 9.5

To circumscribe a circle around triangle ABC (Fig. 9.6)

1. Bisect any two of the sides of the triangle as shown, so that the bisectors intersect at D.
2. The centre of the circumscribing circle is point D.

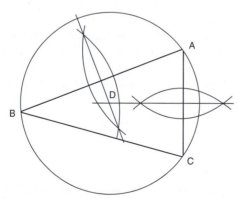

FIGURE 9.6

To draw a hexagon, given the distance across the corners

Method A (Fig. 9.7(a))

1. Draw vertical and horizontal centre lines and a circle with a diameter equal to the given distance.
2. Step off the radius around the circle to give six equally spaced points, and join the points to give the required hexagon.

(a)

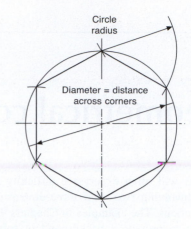

FIGURE 9.7

Method B (Fig. 9.7(b))

1. Draw vertical and horizontal centre lines and a circle with a diameter equal to the given distance.
2. With a 60° set-square, draw points on the circumference 60° apart.
3. Connect these six points by straight lines to give the required hexagon.

(b)

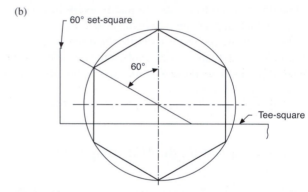

FIGURE 9.7

To draw a hexagon, given the distance across the flats (Fig. 9.8)

1. Draw vertical and horizontal centre lines and a circle with a diameter equal to the given distance.

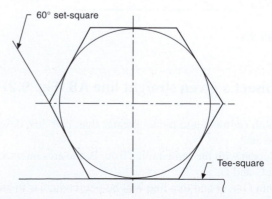

FIGURE 9.8

2. Use a 60° set-square and tee-square as shown, to give the six sides.

To draw a regular octagon, given the distance across corners (Fig. 9.9)

Repeat the instructions in Fig. 9.7(b) but use a 45° set-square, then connect the eight points to give the required octagon.

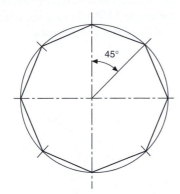

FIGURE 9.9

To draw a regular octagon, given the distance across the flats (Fig. 9.10)

Repeat the instructions in Fig. 9.8 but use a 45° set-square to give the required octagon.

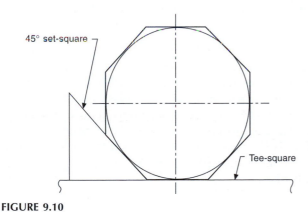

FIGURE 9.10

To draw a regular polygon, given the length of the sides (Fig. 9.11)

Note that a regular polygon is defined as a plane figure which is bounded by straight lines of equal length and which contains angles of equal size. Assume the number of sides is seven in this example.

1. Draw the given length of one side AB, and with radius AB describe a semi-circle.

2. Divide the semi-circle into seven equal angles, using a protractor, and through the second division from the left join line A2.
3. Draw radial lines from A through points 3, 4, 5, and 6.
4. With radius AB and centre on point 2, describe an arc to meet the extension of line A3, shown here as point F.
5. Repeat with radius AB and centre F to meet the extension of line A4 at E.
6. Connect the points as shown, to complete the required polygon.

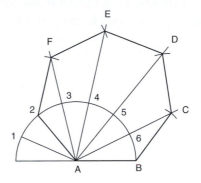

FIGURE 9.11

Tangency

If a disc stands on its edge on a flat surface it will touch the surface at one point. This point is known as the point of tangency, as shown in Fig. 9.12 and the straight line which represents the flat plane is known as a tangent. A line drawn from the point of tangency to the centre of the disc is called a normal, and the tangent makes an angle of 90° with the normal.

The following constructions show the methods of drawing tangents in various circumstances.

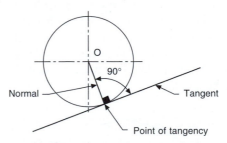

FIGURE 9.12

To draw a tangent to a point A on the circumference of a circle, centre O (Fig. 9.13)

Join OA and extend the line for a short distance. Erect a perpendicular at point A by the method shown.

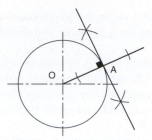

FIGURE 9.13

To draw a tangent to a circle from any given point A outside the circle (Fig. 9.14)

Join A to the centre of the circle O. Bisect line AO so that point B is the mid-point of AO. With centre B, draw a semi-circle to intersect the given circle at point C. Line AC is the required tangent.

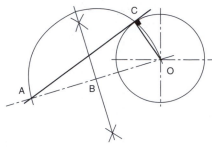

FIGURE 9.14

To draw an external tangent to two circles (Fig. 9.15)

Join the centres of the circles by line AB, bisect AB, and draw a semi-circle. Position point E so that DE is equal to the radius of the smaller circle. Draw radius AE to cut the semi-circle at point G. Draw line AGH so that H lies on the circumference of the larger circle. Note that angle AGB lies in a semi-circle and will be 90°. Draw line HJ parallel to BG. Line HJ will be tangential to the two circles and lines BJ and AGH are the normals.

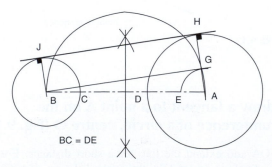

BC = DE

FIGURE 9.15

To draw an internal tangent to two circles (Fig. 9.16)

Join the centres of the circles by line AB, bisect AB and draw a semi-circle. Position point E so that DE is equal to the radius of the smaller circle BC. Draw radius AE to cut the semi-circle in H. Join AH; this line crosses the larger circle circumference at J. Draw line BH. From J draw a line parallel to BH to touch the smaller circle at K. Line JK is the required tangent. Note that angle AHB lies in a semi-circle and will therefore be 90°. AJ and BK are normals.

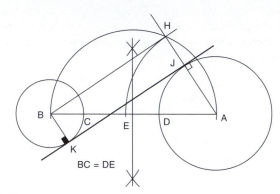

BC = DE

FIGURE 9.16

To draw internal and external tangents to two circles of equal diameter (Fig. 9.17)

Join the centres of both circles by line AB. Erect perpendiculars at points A and B to touch the circumferences of the circles at points C and D. Line CD will be the external tangent. Bisect line AB to give point E, then bisect BE to give point G. With radius BG, describe a semi-circle to cut the circumference of one of the given circles at H. Join HE and extend it to touch the circumference of the other circle at J. Line HEJ is the required tangent. Note that again the angle in the semi-circle, BHE, will be 90°, and hence BH and AJ are normals.

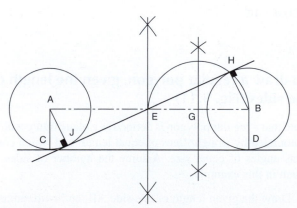

FIGURE 9.17

To draw a curve of given radius to touch two circles when the circles are outside the radius (Fig. 9.18)

Assume that the radii of the given circles are 20 and 25 mm, spaced 85 mm apart, and that the radius to touch them is 40 mm.

With centre A, describe an arc equal to 20 + 40 = 60 mm.
With centre B, describe an arc equal to 25 + 40 = 65 mm.
The above arcs intersect at point C. With a radius of 40 mm, describe an arc from point C as shown, and note that the points of tangency between the arcs lie along the lines joining the centres AC and BC. It is particularly important to note the position of the points of tangency before lining in engineering drawings, so that the exact length of an arc can be established.

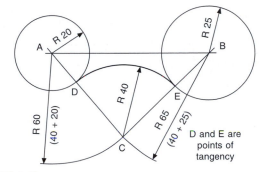

FIGURE 9.18

To draw a curve of given radius to touch two circles when the circles are inside the radius (Fig. 9.19)

Assume that the radii of the given circles are 22 and 26 mm, spaced 86 mm apart, and that the radius to touch them is 100 mm.

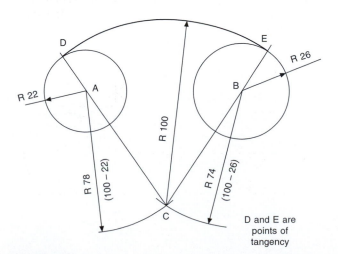

FIGURE 9.19

With centre A, describe an arc equal to 100 − 22 = 78 mm.
With centre B, describe an arc equal to 100 − 26 = 74 mm.
The above arcs intersect at point C. With a radius of 100 mm, describe an arc from point C, and note that in this case the points of tangency lie along line CA extended to D and along line CB extended to E.

To draw a radius to join a straight line and a given circle (Fig. 9.20)

Assume that the radius of the given circle is 20 mm and that the joining radius is 22 mm.

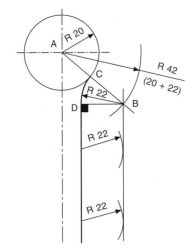

FIGURE 9.20

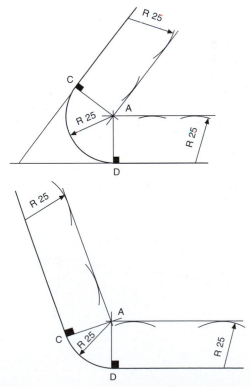

FIGURE 9.21

With centre A, describe an arc equal to 20 + 22 = 42 mm.

Draw a line parallel to the given straight line and at a perpendicular distance of 22 mm from it, to intersect the arc at point B.

With centre B, describe the required radius of 22 mm, and note that one point of tangency lies on the line AB at C; the other lies at point D such that BD is at 90° to the straight line.

To draw a radius which is tangential to given straight lines (Fig. 9.21)

Assume that a radius of 25 mm is required to touch the lines shown in the figures. Draw lines parallel to the given straight lines and at a perpendicular distance of 25 mm from them to intersect at points A. As above, note that the points of tangency are obtained by drawing perpendiculars through the point A to the straight lines in each case.

Chapter 10

Loci applications

If a point, line, or surface moves according to a mathematically defined condition, then a curve known as a *locus* is formed. The following examples of curves and their constructions are widely used and applied in all types of engineering.

Methods of drawing an ellipse

Two-circle method

Construct two concentric circles equal in diameter to the major and minor axes of the required ellipse. Let these diameters be AB and CD in Fig. 10.1.

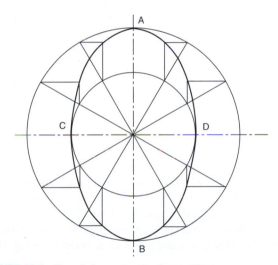

FIGURE 10.1   Two-circle construction for an ellipse.

Divide the circles into any number of parts; the parts do not necessarily have to be equal. The radial lines now cross the inner and outer circles.

Where the radial lines cross the outer circle, draw short lines parallel to the minor axis CD. Where the radial lines cross the inner circle, draw lines parallel to AB to intersect with those drawn from the outer circle. The points of intersection lie on the ellipse. Draw a smooth connecting curve.

Trammel method

Draw major and minor axes at right angles, as shown in Fig. 10.2.

Take a strip of paper for a trammel and mark on it half the major and minor axes, both measured from the same end. Let the points on the trammel be E, F, and G.

Position the trammel on the drawing so that point F always lies on the major axis AB and point G always lies on the minor axis CD. Mark the point E with each position of the trammel, and connect these points to give the required ellipse.

Note that this method relies on the difference between half the lengths of the major and minor axes, and where these axes are nearly the same in length, it is difficult to position the trammel with a high degree of accuracy. The following alternative method can be used.

Draw major and minor axes as before, but extend them in each direction as shown in Fig. 10.3.

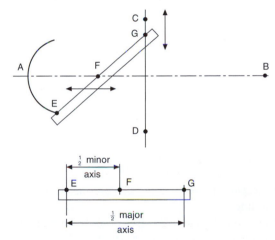

FIGURE 10.2   Trammel method for ellipse construction.

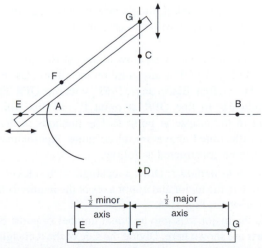

FIGURE 10.3   Alternative trammel method.

Take a strip of paper and mark half of the major and minor axes in line, and let these points on the trammel be E, F, and G.

Position the trammel on the drawing so that point G always moves along the line containing CD; also, position point E along the line containing AB. For each position of the trammel, mark point F and join these points with a smooth curve to give the required ellipse.

To draw an ellipse using the two foci

Draw major and minor axes intersecting at point O, as shown in Fig. 10.4. Let these axes be AB and CD. With a radius equal to half the major axis AB, draw an arc from centre C to intersect AB at points F_1 and F_2. These two points are the foci. For any ellipse, the sum of the distances PF_1 and PF_2 is a constant, where P is any point on the ellipse. The sum of the distances is equal to the length of the major axis.

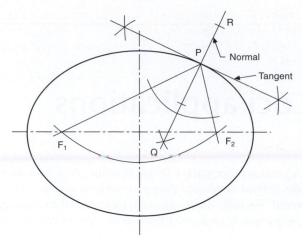

FIGURE 10.5

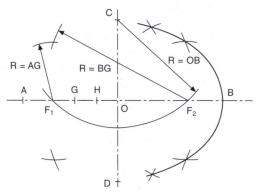

FIGURE 10.4   Ellipse by foci method.

Divide distance OF_1 into equal parts. Three are shown here, and the points are marked G and H.

With centre F_1 and radius AG, describe an arc above and beneath line AB.

With centre F_2 and radius BG, describe an arc to intersect the above arcs.

Repeat these two steps by firstly taking radius AG from point F_2 and radius BG from F_1.

The above procedure should now be repeated using radii AH and BH. Draw a smooth curve through these points to give the ellipse.

It is often necessary to draw a tangent to a point on an ellipse. In Fig. 10.5 P is any point on the ellipse, and F_1 and F_2 are the two foci. Bisect angle F_1PF_2 with line QPR. Erect a perpendicular to line QPR at point P, and this will be a tangent to the ellipse at point P. The methods of drawing ellipses illustrated above are all accurate. Approximate ellipses can be constructed as follows.

Approximate method 1: Draw a rectangle with sides equal in length to the major and minor axes of the required ellipse, as shown in Fig. 10.6.

Divide the major axis into an equal number of parts; eight parts are shown here. Divide the side of the rectangle into the same equal number of parts. Draw a line from A

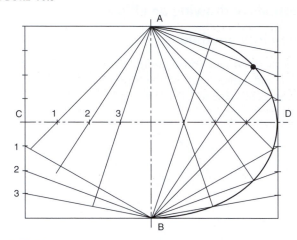

FIGURE 10.6

through point 1, and let this line intersect the line joining B to point 1 at the side of the rectangle as shown. Repeat for all other points in the same manner, and the resulting points of intersection will lie on the ellipse.

Approximate method 2: Draw a rectangle with sides equal to the lengths of the major and minor axes, as shown in Fig. 10.7.

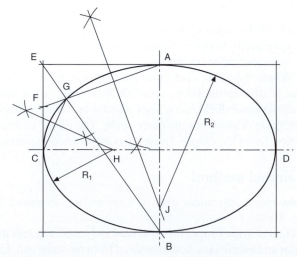

FIGURE 10.7

Bisect EC to give point F. Join AF and BE to intersect at point G. Join CG. Draw the perpendicular bisectors of lines CG and GA, and these will intersect the centre lines at points H and J.

Using radii CH and JA, the ellipse can be constructed by using four arcs of circles.

The involute

The involute is defined as the path of a point on a straight line which rolls without slip along the circumference of a cylinder. The involute curve will be required in a later Chapter for the construction of gear teeth.

Involute construction

1. Draw the given base circle and divide it into, say, 12 equal divisions as shown in Fig. 10.8. Generally only the first part of the involute is required, so the given diagram shows a method using half of the length of the circumference.

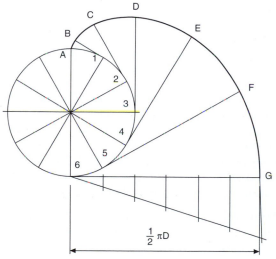

FIGURE 10.8 Involute construction.

2. Draw tangents at points 1, 2, 3, 4, 5, and 6.
3. From point 6, mark off a length equal to half the length of the circumference.
4. Divide line 6G into six equal parts.
5. From point 1, mark point B such that 1B is equal to one part of line 6G.
6. From point 2, mark point C such that 2C is equal to two parts of line 6G.
   Repeat the above procedure from points 3, 4 and 5, increasing the lengths along the tangents as before by one part of line 6G.
7. Join points A to G, to give the required involute.

Alternative method

1. As above, draw the given base circle, divide into, say, 12 equal divisions, and draw the tangents from points 1 to 6.
2. From point 1 and with radius equal to the chordal length from point 1 to point A, draw an arc terminating at the tangent from point 1 at point B.
3. Repeat the above procedure from point 2 with radius 2B terminating at point C.
4. Repeat the above instructions to obtain points D, E, F, and G, and join points A to G to give the required involute.

The alternative method given is an approximate method, but is reasonably accurate provided that the arc length is short; the difference in length between the arc and the chord introduces only a minimal error.

Archimedean spiral

The Archimedean spiral is the locus of a point which moves around a centre at uniform angular velocity and at the same time moves away from the centre at uniform linear velocity. The construction is shown in Fig. 10.9.

1. Given the diameter, divide the circle into an even number of divisions and number them.
2. Divide the radius into the same number of equal parts.
3. Draw radii as shown to intersect radial lines with corresponding numbers, and connect points of intersection to give the required spiral.

Note that the spiral need not start at the centre; it can start at any point along a radius, but the divisions must be equal.
Self-centring lathe chucks utilize Archimedean spirals.

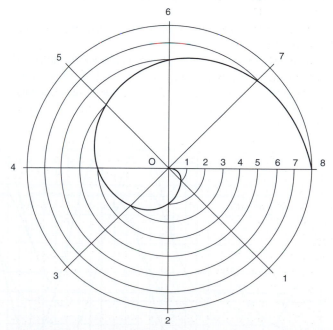

FIGURE 10.9 Archimedean spiral.

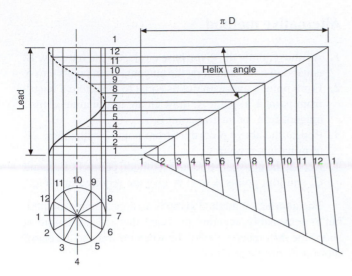

FIGURE 10.10 Right-hand cylindrical helix.

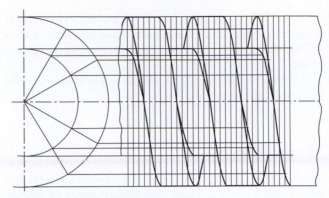

FIGURE 10.12 Single-start square thread.

Right-hand cylindrical helix

The helix is a curve generated on the surface of the cylinder by a point which revolves uniformly around the cylinder and at the same time either up or down its surface. The method of construction is shown in Fig. 10.10.

1. Draw the front elevation and plan views of the cylinder, and divide the plan view into a convenient number of parts (say 12) and number them as shown.
2. Project the points from the circumference of the base up to the front elevation.
3. Divide the lead into the same number of parts as the base, and number them as shown.
4. Draw lines of intersection from the lead to correspond with the projected lines from the base.
5. Join the points of intersection, to give the required cylindrical helix.
6. If a development of the cylinder is drawn, the helix will be projected as a straight line. The angle between the helix and a line drawn parallel with the base is known as the helix angle.

Note: If the numbering in the plan view is taken in the clockwise direction from point 1, then the projection in the front elevation will give a left-hand helix.

The construction for a helix is shown applied to a right-hand helical spring in Fig. 10.11. The spring is of square cross-section, and the four helices are drawn from the two outside corners and the two corners at the inside diameter. The pitch of the spring is divided into 12 equal parts, to correspond with the 12 equal divisions of the circle in the end elevation, although only half of the circle need be drawn. Points are plotted as previously shown.

A single-start square thread is illustrated in Fig. 10.12. The construction is similar to the previous problem, except that the centre is solid metal. Four helices are plotted, spaced as shown, since the threadwidth is half the pitch.

Right-hand conical helix

The conical helix is a curve generated on the surface of the cone by a point which revolves uniformly around the cone and at the same time either up or down its surface. The method of construction is shown in Fig. 10.13.

1. Draw the front elevation and plan of the cone, and divide the plan view into a convenient number of parts (say 12) and number them as shown.

FIGURE 10.11 Square-section right-hand helical spring.

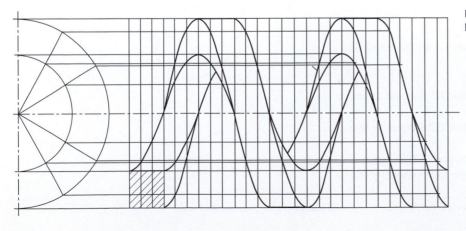

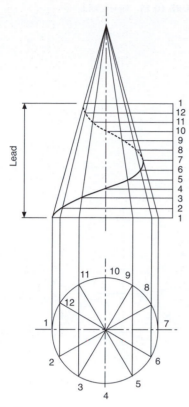

FIGURE 10.13   Right-hand conical helix.

2. Project the points on the circumference of the base up to the front elevation, and continue the projected lines to the apex of the cone.
3. The lead must now be divided into the same number of parts as the base, and numbered.
4. Draw lines of intersection from the lead to correspond with the projected lines from the base.
5. Join the points of intersection, to give the required conical helix.

The cycloid

The cycloid is defined as the locus of a point on the circumference of a cylinder which rolls without slip along a flat surface. The method of construction is shown in Fig. 10.14.

1. Draw the given circle, and divide into a convenient number of parts; eight divisions are shown in Fig. 10.14.
2. Divide line AA_1 into eight equal lengths. Line AA_1 is equal to the length of the circumference.
3. Draw vertical lines from points 2 to 8 to intersect with the horizontal line from centre O at points O_2, O_3, etc.
4. With radius OA and centre O_2, describe an arc to intersect with the horizontal line projected from B.
5. Repeat with radius OA from centre O_3 to intersect with the horizontal line projected from point C. Repeat this procedure.
6. Commencing at point A, join the above intersections to form the required cycloid.

The epicycloid

An epicycloid is defined as the locus of a point on the circumference of a circle which rolls without slip around the outside of another circle. The method of construction is shown in Fig. 10.15.

1. Draw the curved surface and the rolling circle, and divide the circle into a convenient number of parts (say six) and number them as shown.
2. Calculate the length of the circumference of the smaller and the larger circle, and from this information calculate the angle θ covered by the rolling circle.
3. Divide the angle θ into the same number of parts as the rolling circle.
4. Draw the arc which is the locus of the centre of the rolling circle.
5. The lines forming the angles in step 3 will now intersect with the arc in step 4 to give six further positions of the centres of the rolling circle as it rotates.
6. From the second centre, draw radius R to intersect with the arc from point 2 on the rolling circle. Repeat this process for points 3, 4, 5 and 6.
7. Draw a smooth curve through the points of intersection, to give the required epicycloid.

FIGURE 10.14   Cycloid.

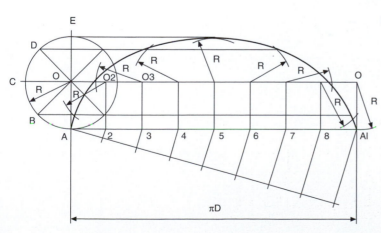

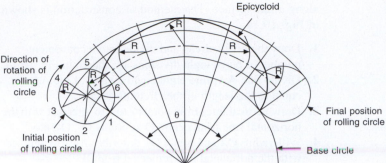

Epicycloid

FIGURE 10.15 Epicycloid.

Direction of rotation of rolling circle

Final position of rolling circle

Initial position of rolling circle

Base circle

The hypocycloid

A hypocycloid is defined as the locus of a point on the circumference of a circle which rolls without slip around the inside of another circle.

The construction for the hypocycloid (Fig. 10.16) is very similar to that for the epicycloid, but note that the rolling circle rotates in the opposite direction for this construction.

It is often necessary to study the paths taken by parts of oscillating, reciprocating, or rotating mechanisms; from a knowledge of displacement and time, information regarding velocity and acceleration can be obtained. It may also be required to study the extreme movements of linkages, so that safety guards can be designed to protect machine operators.

Figure 10.17 shows a crank OA, a connecting rod AB, and a piston B which slides along the horizontal axis BO. P is any point along the connecting rod. To plot the locus of point P, a circle of radius OA has been divided into 12 equal parts. From each position of the crank, the connecting rod is drawn,

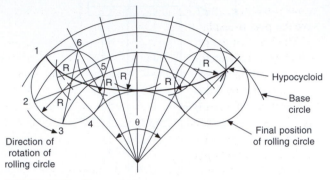

Hypocycloid

Base circle

Final position of rolling circle

Direction of rotation of rolling circle

FIGURE 10.16 Hypocycloid.

distance AP measured, and the path taken for one revolution lined in as indicated.

The drawing also shows the piston-displacement diagram. A convenient vertical scale is drawn for the crank angle and in this case clockwise rotation was assumed to start from the 9 o'clock position. From each position of the

FIGURE 10.17

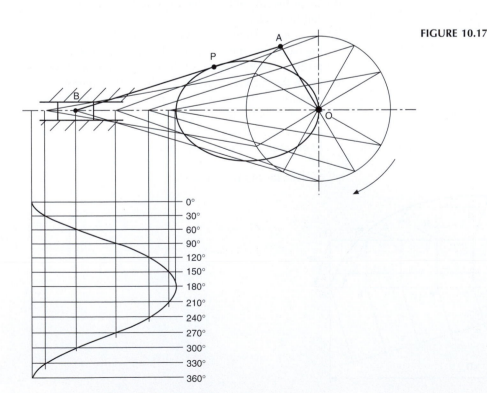

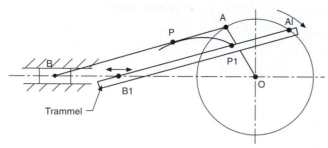

FIGURE 10.18

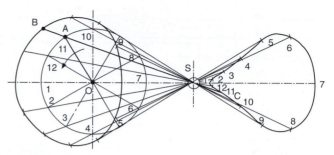

FIGURE 10.19

piston, a vertical line is drawn down to the corresponding crank-angle line, and the points of intersection are joined to give the piston-displacement diagram.

The locus of the point P can also be plotted by the trammel method indicated in Fig. 10.18. Point P_1 can be marked for any position where B_1 lies on the horizontal line, provided A_1 also lies on the circumference of the circle radius OA. This method of solving some loci problems has the advantage that an infinite number of points can easily be obtained, and these are especially useful where a change in direction in the loci curve takes place.

Figure 10.19 shows a crank OA rotating anticlockwise about centre O. A rod BC is connected to the crank at point A, and this rod slides freely through a block which is allowed to pivot at point S. The loci of points B and C are indicated after reproducing the mechanism in 12 different positions. A trammel method could also be used here if required.

Part of a shaping-machine mechanism is given in Fig. 10.20. Crank OB rotates about centre O. A is a fixed pivot point, and CA slides through the pivoting block at B. Point C moves in a circular arc of radius AC, and is connected by link CD, where point D slides horizontally. In the

position shown, angle OBA is 90°, and if OB now rotates anticlockwise at constant speed it will be seen that the forward motion of point D takes more time than the return motion. A displacement diagram for point D has been constructed as previously described.

In Fig. 10.21 the radius OB has been increased, with the effect of increasing the stroke of point D. Note also that the return stroke in this condition is quicker than before.

The outlines of two gears are shown in Fig. 10.22, where the pitch circle of the larger gear is twice the pitch circle of the smaller gear. As a result, the smaller gear rotates twice while the larger gear rotates once. The mechanism has been drawn in 12 positions to plot the path of the pivot point C, where links BC and CA are connected. A trammel method cannot be applied successfully in this type of problem.

Figure 10.23 gives an example of Watt's straight-line motion. Two levers AX and BY are connected by a link AB, and the plotted curve is the locus of the mid-point P. The levers in this instance oscillate in circular arcs. This mechanism was used in engines designed by James Watt, the famous engineer.

FIGURE 10.20

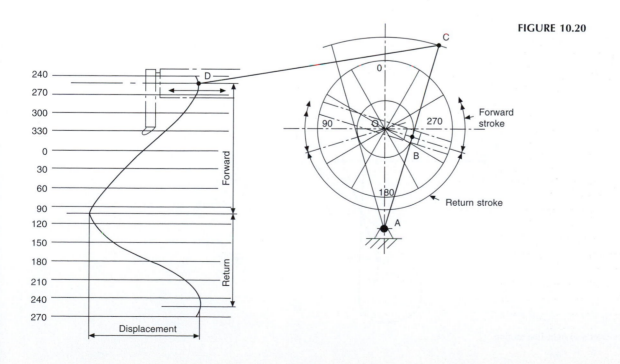

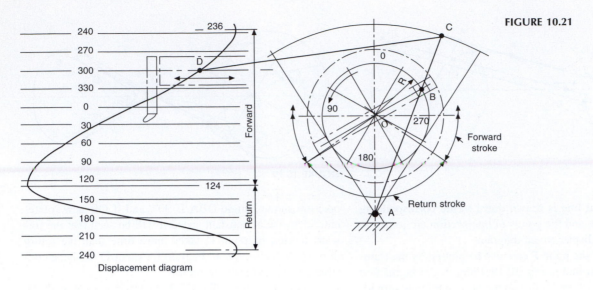

Displacement diagram

FIGURE 10.21

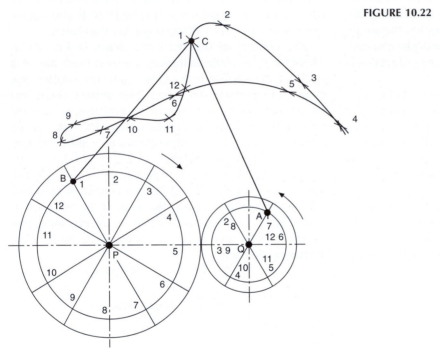

FIGURE 10.22

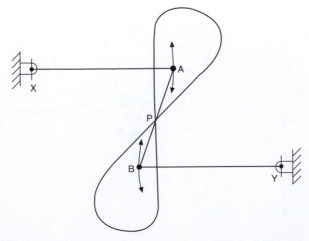

FIGURE 10.23 Watt's straight-line motion.

FIGURE 10.24

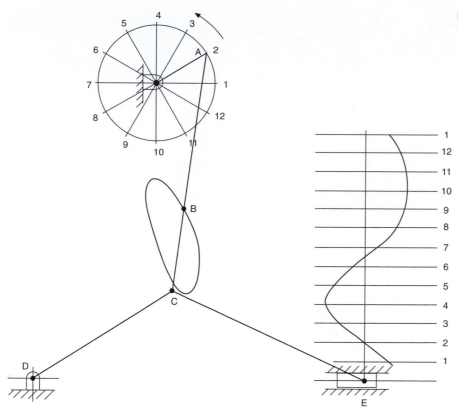

A toggle action is illustrated in Fig. 10.24, where a crank rotates anticlockwise. Links AC, CD and CE are pivoted at C. D is a fixed pivot point, and E slides along the horizontal axis. The displacement diagram has been plotted as previously described, but note that, as the mechanism at E slides to the right, it is virtually stationary between points 9, 10 and 11.

The locus of any point B is also shown.

Chapter 11

True lengths and auxiliary views

An isometric view of a rectangular block is shown in Fig. 11.1. The corners of the block are used to position a line DF in space. Three orthographic views in first-angle projection are given in Fig. 11.2, and it will be apparent that the projected length of the line DF in each of the views will be equal in length to the diagonals across each of the rectangular faces. A cross check with the isometric view will clearly show that the true length of line DF must be greater than any of the diagonals in the three orthographic views. The corners nearest to the viewing position are shown as ABCD etc.; the corners on the remote side are indicated in rings. To find the true length of DF, an auxiliary projection must be drawn, and the viewing position must be square with line DF. The first auxiliary projection in Fig. 11.2 gives the true length required, which forms part of the right-angled triangle DFG. Note that auxiliary views are drawn on planes other than the principal projection planes. A plan is projected from an elevation and an elevation from a plan. Since this is the first auxiliary view projected, and from a true plan, it is known as a *first auxiliary elevation*. Other auxiliary views could be projected from this auxiliary elevation if so required.

The true length of DF could also have been obtained by projection from the front or end elevations by viewing at 90 to the line, and Fig. 11.3 shows these two alternatives. The first auxiliary plan from the front elevation gives triangle

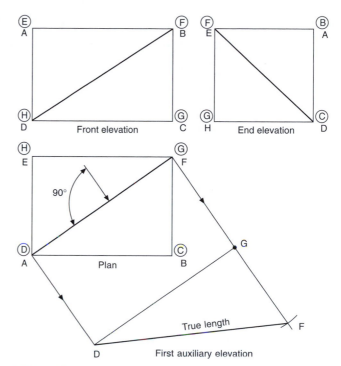

FIGURE 11.2

FDH, and the first auxiliary plan from the end elevation gives triangle FCD, both right-angled triangles.

Figure 11.4 shows the front elevation and plan view of a box. A first auxiliary plan is drawn in the direction of arrow X. Now PQ is an imaginary datum plane at right angles to the direction of viewing; the perpendicular distance from corner A to the plane is shown as dimension 1. When the first auxiliary plan view is drawn, the box is in effect turned through 90 in the direction of arrow X, and the corner A will be situated above the plane at a perpendicular distance equal to dimension 1. The auxiliary plan view is a true view on the tilted box. If a view is now taken in the direction of arrow Y, the tilted box will be turned through 90 in the direction of the arrow, and dimension 1 to the corner will lie parallel with the plane of the paper. The other seven corners of the box are projected as indicated, and are positioned by the dimensions to the plane PQ in the front elevation. A match-box can be used here as a model to appreciate the position in space for each projection.

The same box has been redrawn in Fig. 11.5, but the first auxiliary elevation has been taken from the plan view in a

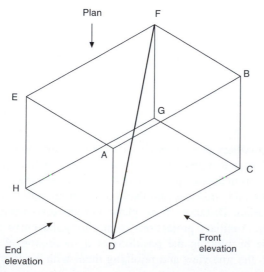

FIGURE 11.1

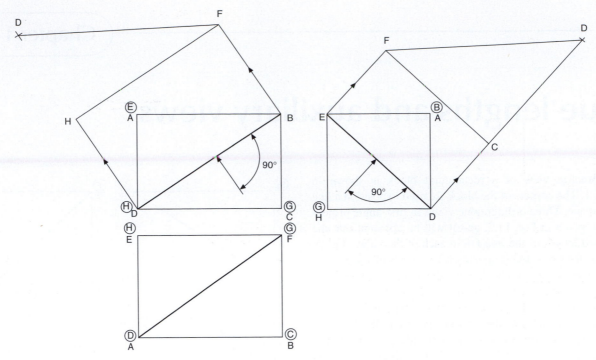

FIGURE 11.3

FIGURE 11.4

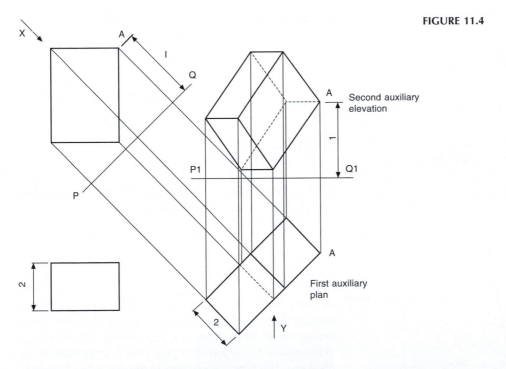

manner similar to that described in the previous example. The second auxiliary plan projected in line with arrow Y requires dimensions from plane P1Q1, which are taken as before from plane PQ. Again, check the projections shown with a match-box. All of the following examples use the principles demonstrated in these two problems.

Part of a square pyramid is shown in Fig. 11.6; the constructions for the eight corners in both auxiliary views are identical with those described for the box in Fig. 11.4.

Auxiliary projections from a cylinder are shown in Fig. 11.7; note that chordal widths in the first auxiliary plan are taken from the true plan. Each of 12 points around the circle is plotted in this way and then projected up to the auxiliary elevation. Distances from plane PQ are used from plane P_1Q_1. Auxiliary projections of any irregular curve can be made by plotting the positions of a succession of points from the true view and rejoining them with a curve in the auxiliary view.

FIGURE 11.5

FIGURE 11.6

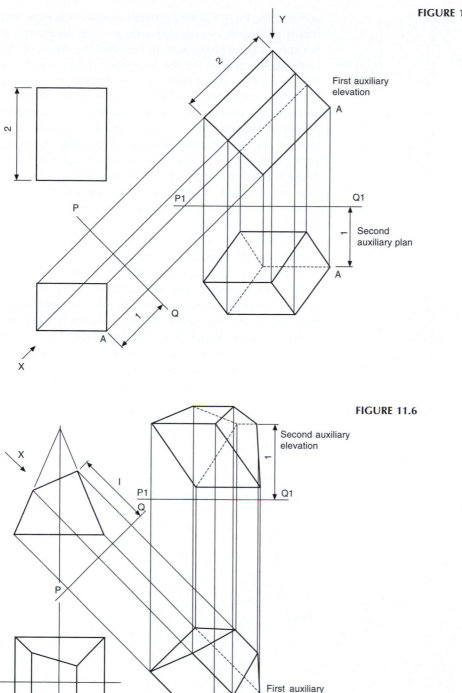

Figure 11.8 shows a front elevation and plan view of a thin lamina in the shape of the letter L. The lamina lies inclined above the datum plane PQ, and the front elevation appears as a straight line. The true shape is projected above as a first auxiliary view. From the given plan view, an auxiliary elevation has been projected in line with the arrow F, and the positions of the corners above the datum plane P_1Q_1 will be the same as those above the original plane PQ. A

typical dimension to the corner A has been added as dimension 1. To assist in comprehension, the true shape given could be cut from a piece of paper and positioned above the book to appreciate how the lamina is situated in space; it will then be seen that the height above the book of corner A will be dimension 2.

Now a view in the direction of arrow G parallel with the surface of the book will give the lamina shown projected above datum P_2Q_2. The object of this exercise is to show that if only two auxiliary projections are given in isolation, it is possible to draw projections to find the true shape of the component and also get the component back, parallel to the plane of the paper. The view in direction of arrow H has been drawn and taken at 90 to the bottom edge containing corner A; the resulting view is the straight line of true length positioned below the datum plane P_3Q_3. The lamina is situated in this view in the perpendicular position above the paper, with the lower edge parallel to the paper and at a distance equal to dimension 4 from the surface. View J is now drawn square to this projected view and positioned above the datum P_4Q_4 to give the true shape of the given lamina.

In Fig. 11.9, a lamina has been made from the polygon ACBD in the development and bent along the axis AB; again, a piece of paper cut to this shape and bent to the angle ϕ may be of some assistance. The given front elevation and plan position the bent lamina in space, and this exercise is given here since every line used to form the lamina in these two views is not a true length. It will be seen that, if a view is now drawn in the direction of arrow X, which is at right angles to the bend line AB, the resulting projection will give the true length of AB, and this line will also lie parallel with the plane of the paper. By looking along the fold in the direction of arrow Y, the two corners A and B will appear coincident; also, AD and BC will appear as the true lengths of the altitudes DE and FC. The development can now be drawn, since the positions of points E and F are known along the true length of AB. The lengths of the sides AD, DB, BC and AC are obtained from the pattern development.

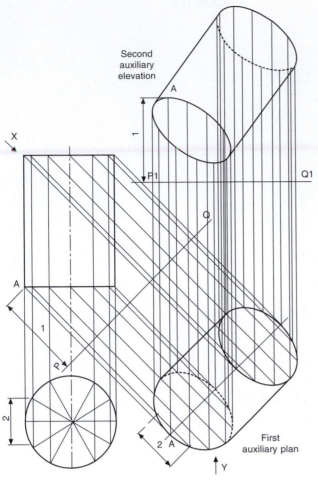

FIGURE 11.7

FIGURE 11.8

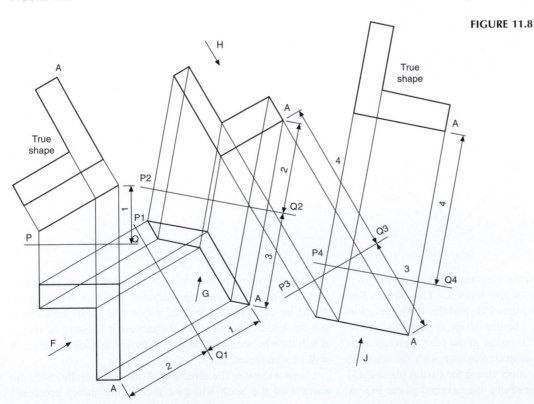

FIGURE 11.9

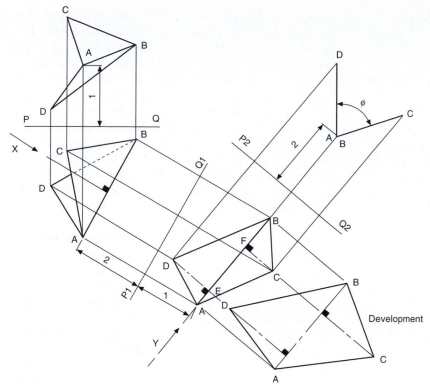

Development

Chapter 12

Conic sections and interpenetration of solids

Consider a right circular cone, i.e. a cone whose base is a circle and whose apex is above the centre of the base (Fig. 12.1).

- The true face of a section through the apex of the cone will be a triangle.
- The true face of a section drawn parallel to the base will be a circle.
- The true face of any other section which passes through two opposite generators will be an ellipse.
- The true face of a section drawn parallel to the generator will be a parabola.

If a plane cuts the cone through the generator and the base on the same side of the cone axis, then a view on the true face of the section will be a hyperbola. The special case of a section at right-angles to the base gives a rectangular hyperbola.

To draw an ellipse from part of a cone

Figure 12.2 shows the method of drawing the ellipse, which is a true view on the surface marked AB of the frustum of the given cone.

1. Draw a centre line parallel to line AB as part of an auxiliary view.
2. Project points A and B onto this line and onto the centre lines of the plan and end elevation.
3. Take any horizontal section XX between A and B and draw a circle in the plan view of diameter D.
4. Project the line of section plane XX onto the end elevation.
5. Project the point of intersection of line AB and plane XX onto the plan view.
6. Mark the chord-width W_{on} the plan, in the auxiliary view and the end elevation. These points in the auxiliary view form part of the ellipse.
7. Repeat with further horizontal sections between A and B, to complete the views as shown.

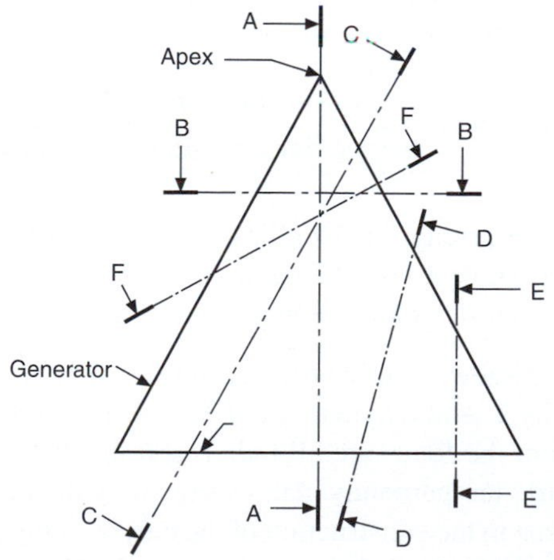

FIGURE 12.1 Conic sections: section AA – triangle; section BB – circle; section CC – parabola; section DD – hyperbola; section EE – rectangular hyperbola; section FF – ellipse.

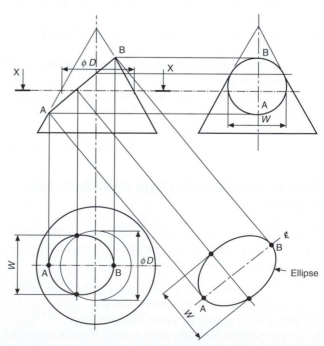

FIGURE 12.2

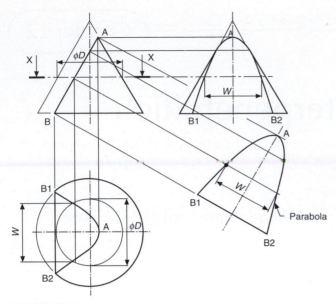

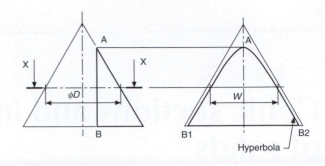

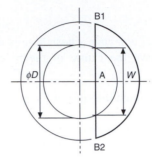

FIGURE 12.3

FIGURE 12.4

To draw a parabola from part of a cone

Figure 12.3 shows the method of drawing the parabola, which is a true view on the line AB drawn parallel to the sloping side of the cone.

1. Draw a centre line parallel to line AB as part of an auxiliary view.
2. Project point B to the circumference of the base in the plan view, to give the points B_1 and B_2. Mark chord-width B_1B_2 in the auxiliary view and in the end elevation.
3. Project point A onto the other three views.
4. Take any horizontal section XX between A and B and draw a circle in the plan view of diameter D.
5. Project the line of section plane XX onto the end elevation.
6. Project the point of intersection of line AB and plane XX to the plane view.
7. Mark the chord-width W on the plan, in the end elevation and the auxiliary view. These points in the auxiliary view form part of the parabola.
8. Repeat with further horizontal sections between A and B, to complete the three views.

To draw a rectangular hyperbola from part of a cone

Figure 12.4 shows the method of drawing the hyperbola, which is a true view on the line AB drawn parallel to the vertical centre line of the cone.

1. Project point B to the circumference of the base in the plan view, to give the points B_1 and B_2.
2. Mark points B_1 and B_2 in the end elevation.
3. Project point A onto the end elevation. Point A lies on the centre line in the plan view.

4. Take any horizontal section XX between A and B and draw a circle of diameter D in the plan view.
5. Project the line of section XX onto the end elevation.
6. Mark the chord-width W in the plan, on the end elevation. These points in the end elevation form part of the hyperbola.
7. Repeat with further horizontal sections between A and B, to complete the hyperbola.
8. The ellipse, parabola, and hyperbola are also the loci of points which move in fixed ratios from a line (the directrix) and a point (the focus). The ratio is known as the *eccentricity*.

$$\text{Eccentricity} = \frac{\text{distance from focus}}{\text{perpendicular distance from directrix}}$$

- The eccentricity for the ellipse is less than one.
- The eccentricity for the parabola is one.
- The eccentricity for the hyperbola is greater than one.

Figure 12.5 shows an ellipse of eccentricity 3/5, a parabola of eccentricity 1, and a hyperbola of eccentricity 5/3. The distances from the focus are all radial, and the distances from the directrix are perpendicular, as shown by the illustration.

To assist in the construction of the ellipse in Fig. 12.5, the following method may be used to ensure that the two dimensions from the focus and directrix are in the same ratio. Draw triangle PA1 so that side A1 and side P1 are in the ratio of 3–5 units. Extend both sides as shown. From any points B, C, D, etc., draw vertical lines to meet the horizontal at 2, 3, 4, etc.; by similar triangles, vertical lines and their corresponding horizontal lines will be in the same ratio. A similar construction for the hyperbola is shown in Fig. 12.6.

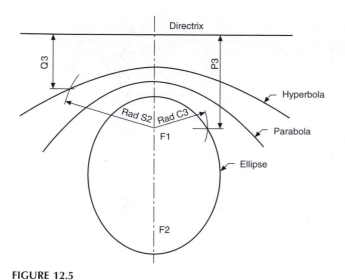

FIGURE 12.5

Commence the construction for the ellipse by drawing a line parallel to the directrix at a perpendicular distance of P3 (Fig. 12.6(a)). Draw radius C3 from point F_1 to intersect this line. The point of intersection lies on the ellipse. Similarly, for the hyperbola (Fig. 12.6(b)) draw a line parallel to the directrix at a perpendicular distance of Q2. Draw radius S2, and the hyperbola passes through the point of intersection.

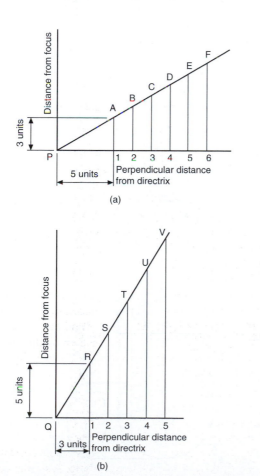

FIGURE 12.6 (a) Ellipse construction. (b) Hyperbola construction.

No scale is required for the parabola, as the perpendicular distances and the radii are the same magnitude.

Repeat the procedure in each case to obtain the required curves.

Interpenetration

Many objects are formed by a collection of geometrical shapes such as cubes, cones, spheres, cylinders, prisms, pyramids, etc., and where any two of these shapes meet, some sort of curve of intersection or interpenetration results. It is necessary to be able to draw these curves to complete drawings in orthographic projection or to draw patterns and developments.

The following drawings show some of the most commonly found examples of interpenetration. Basically, most curves are constructed by taking sections through the intersecting shapes, and, to keep construction lines to a minimum and hence avoid confusion, only one or two sections have been taken in arbitrary positions to show the principle involved; further similar parallel sections are then required to establish the line of the curve in its complete form. Where centre lines are offset, hidden curves will not be the same as curves directly facing the draughtsman, but the draughting principle of taking sections in the manner indicated on either side of the centre lines of the shapes involved will certainly be the same.

If two cylinders, or a cone and a cylinder, or two cones intersect each other at any angle, and the curved surfaces of both solids enclose the same sphere, then the outline of the intersection in each case will be an ellipse. In the illustrations given in Fig. 12.7 the centre lines of the two solids intersect at

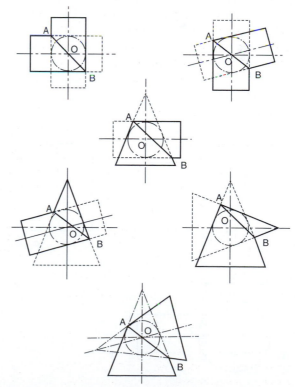

FIGURE 12.7

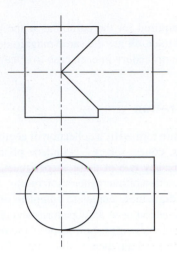

FIGURE 12.8

point O, and a true view along the line AB will produce an ellipse.

When cylinders of equal diameter intersect as shown in Fig. 12.8 the line at the intersection is straight and at 45°.

Figure 12.9 shows a branch cylinder square with the axis of the vertical cylinder but reduced in size. A section through any cylinder parallel with the axis produces a rectangle, in this case of width Y in the branch and width X in the vertical cylinder. Note that interpenetration occurs at points marked 3, and these points lie on a curve. The projection of the branch cylinder along the horizontal centre line gives the points marked 1, and along the vertical centre line gives the points marked 2.

Figure 12.10 shows a cylinder with a branch on the same vertical centre line but inclined at an angle. Instead of an end elevation, the position of section AA is shown on a part auxiliary view of the branch. The construction is otherwise the same as that for Fig. 12.9.

In Fig. 12.11 the branch is offset, but the construction is similar to that shown in Fig. 12.10.

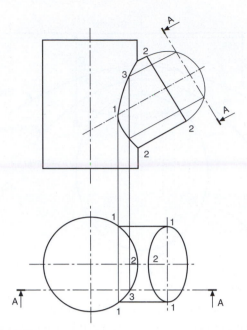

FIGURE 12.10

Figure 12.12 shows the branch offset but square with the vertical axis.

Figure 12.13 shows a cone passing through a cylinder. A horizontal section AA through the cone will give a circle of $\varnothing P$, and through the cylinder will give a rectangle of width X. The points of intersection of the circle and part of the rectangle in the plan view are projected up to the section plane in the front elevation.

The plotting of more points from more sections will give the interpenetration curves shown in the front elevation and the plan.

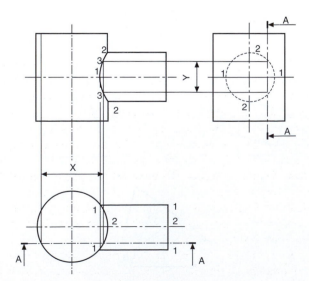

FIGURE 12.9

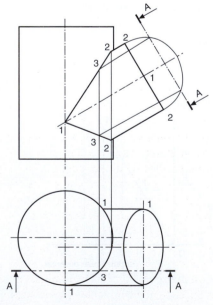

FIGURE 12.11

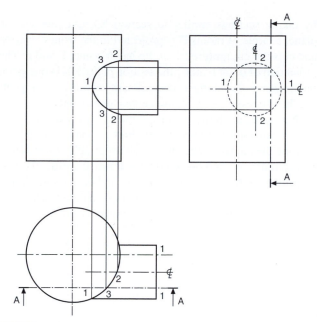

FIGURE 12.12

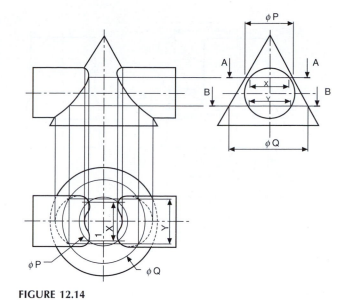

FIGURE 12.14

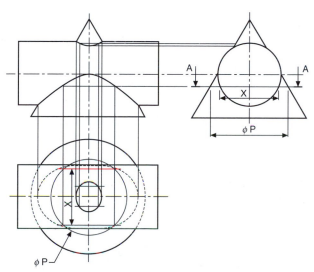

FIGURE 12.13

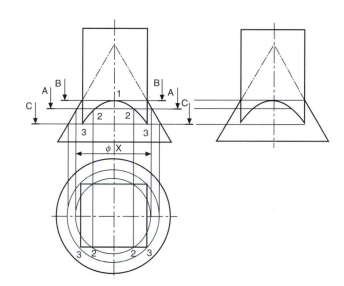

FIGURE 12.15

Figure 12.14 shows a cylinder passing through a cone. The construction shown is the same as for Fig. 12.13 in principle.

Figure 12.15 shows a cone and a square prism where interpenetration starts along the horizontal section BB at point 1 on the smallest diameter circle to touch the prism. Section AA is an arbitrary section where the projected diameter of the cone $\emptyset X$ cuts the prism in the plan view at the points marked 2. These points are then projected back to the section plane in the front elevation and lie on the curve required. The circle at section CC is the largest circle which will touch the prism across the diagonals in the plan view. Having drawn the circle in the plan view, it is projected up to the sides of the cone in the front elevation, and points 3 at the corners of the prism are the lowest points of contact.

A casting with a rectangular base and a circular-section shaft is given in Fig. 12.16. The machining of the radius R_1 in conjunction with the milling of the flat surfaces produces the curve shown in the front elevation. Point 1 is shown projected from the end elevation. Section AA produces a circle of $\emptyset X$ in the plan view and cuts the face of the casting at points marked 2, which are transferred back to the section plane. Similarly, section BB gives $\emptyset Y$ and points marked 3. Sections can be taken until the circle in the plane view increases in size to R_2; at this point, the interpenetration curve joins a horizontal line to the corner of the casting in the front elevation.

In Fig. 12.17 a circular bar of diameter D has been turned about the centre line CC and machined with a radius shown as RAD A. The resulting interpenetration curve is obtained

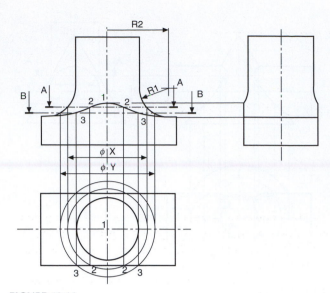

FIGURE 12.16

by taking sections similar to section XX. At this section plane, a circle of radius B is projected in the front elevation and cuts the circumference of the bar at points E and F. The projection of point F along the section plane XX is one point on the curve. By taking a succession of sections, and repeating the process described, the curve can be plotted.

Note that, in all these types of problem, it rarely helps to take dozens of sections and then draw all the circles before plotting the points, as the only result is possible confusion. It is recommended that one section be taken at a time, the first roughly near the centre of any curve, and others sufficiently far apart for clarity but near enough to maintain accuracy. More sections are generally required where curves suddenly change direction.

FIGURE 12.17

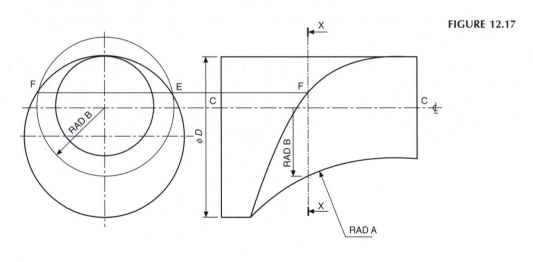

Chapter 13

Development of patterns from sheet materials

Many articles such as cans, pipes, elbows, boxes, ducting, hoppers, etc. are manufactured from thin sheet materials. Generally, a template is produced from an orthographic drawing when small quantities are required (larger quantities may justify the use of press tools), and the template will include allowances for bending and seams, bearing in mind the thickness of material used.

Exposed edges which may be dangerous can be wired or folded, and these processes also give added strength, e.g., cooking tins and pans. Some cooking tins are also formed by pressing hollows into a flat sheet. This type of deformation is not considered in this Chapter, which deals with bending or forming in one plane only. Some common methods of finishing edges, seams, and corners are shown in Fig. 13.1.

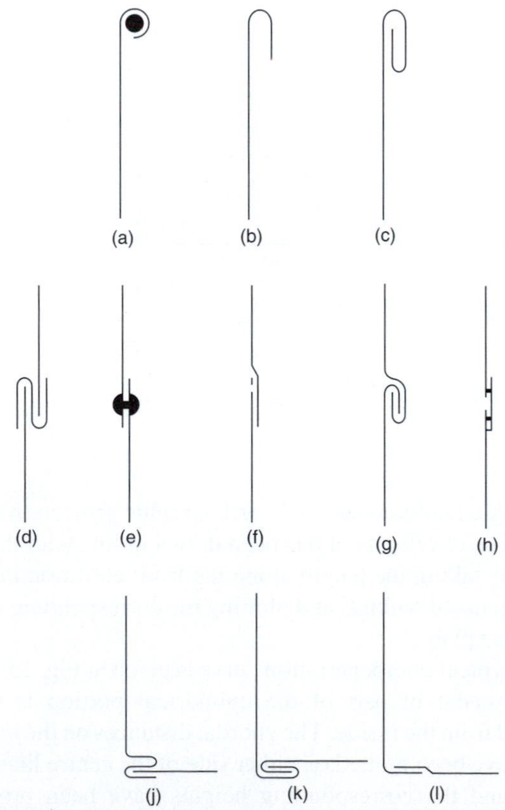

The following examples illustrate some of the more commonly used methods of development in pattern-making, but note that, apart from in the first case, no allowance has been made for joints and seams.

Where a component has its surfaces on flat planes of projection, and all the sides and corners shown are true lengths, the pattern is obtained by parallel-line or straight-line development. A simple application is given in Fig. 13.2 for an open box.

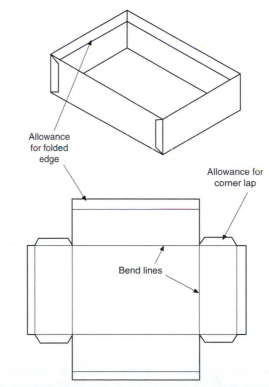

Allowance for folded edge

Allowance for corner lap

Bend lines

FIGURE 13.2

The development of a hexagonal prism is shown in Fig. 13.3. The pattern length is obtained by plotting the distances across the flat faces. The height at each corner is projected from the front elevation, and the top of the prism is drawn from a true view in the direction of arrow X.

An elbow joint is shown developed in Fig. 13.4. The length of the circumference has been calculated and divided into 12 equal parts. A part plan, divided into six parts, has the

(a)   (b)   (c)

(d)   (e)   (f)   (g)   (h)

(j)   (k)   (l)

FIGURE 13.1

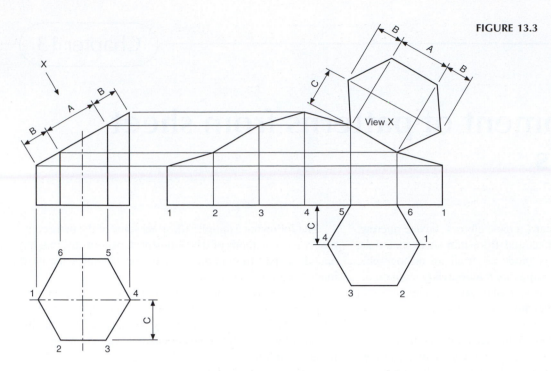

FIGURE 13.3

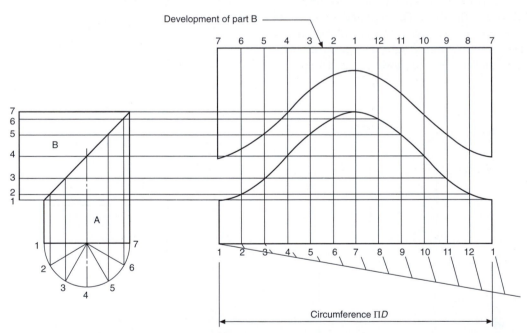

Development of part B

Circumference ΠD

FIGURE 13.4

division lines projected up to the joint, then across to the appropriate point on the pattern. It is normal practice on a development drawing to leave the joint along the shortest edge; however, on part B the pattern can be cut more economically if the joint on this half is turned through 180.

An elbow joint made from four parts has been completely developed in Fig. 13.5. Again, by alternating the position of the seams, the patterns can be cut with no waste. Note that the centre lines of the parts marked B and C are 30 apart, and that the inner and outer edges are tangential to the radii which position the elbow.

A thin lamina is shown in orthographic projection in Fig. 13.6. The development has been drawn in line with the plan view by taking the length along the front elevation in small increments of width C and plotting the corresponding depths from the plan.

A typical interpenetration curve is given in Fig. 13.7. The development of part of the cylindrical portion is shown viewed from the inside. The chordal distances on the inverted plan have been plotted on either side of the centre line of the hole, and the corresponding heights have been projected from the front elevation. The method of drawing a pattern

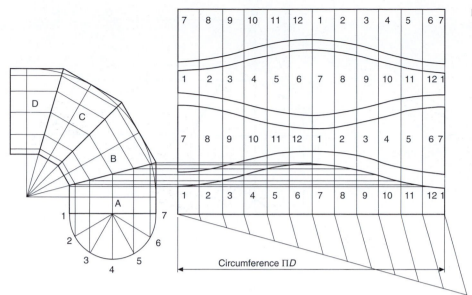

FIGURE 13.5

FIGURE 13.6

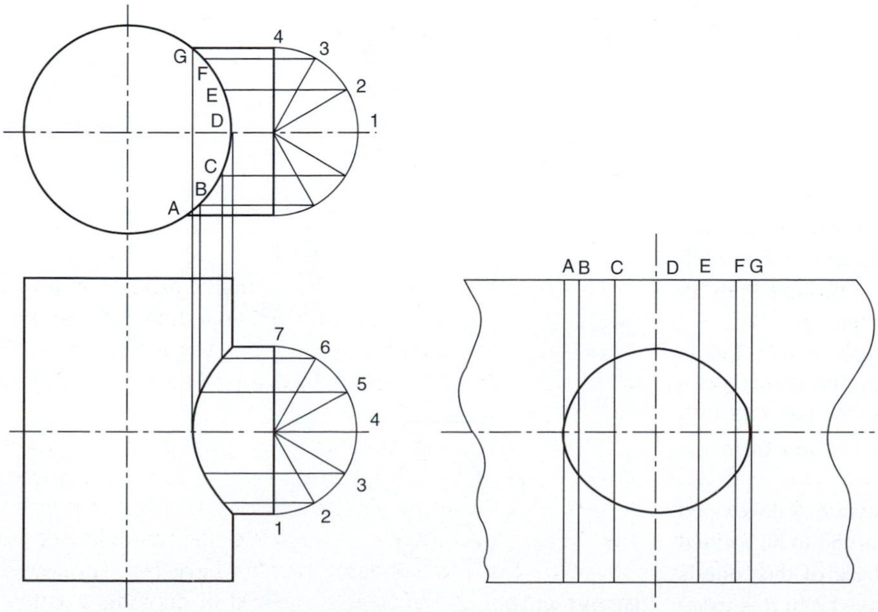

FIGURE 13.7

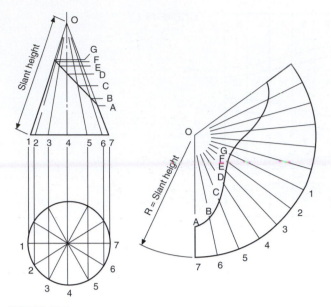

FIGURE 13.8

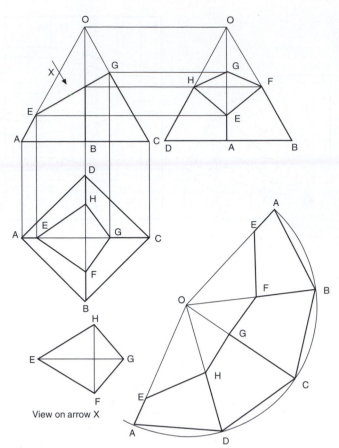

View on arrow X

FIGURE 13.9

for the branch is identical to that shown for the two-piece elbow in Fig. 13.4.

An example of radial-line development is given in Fig. 13.8. The dimensions required to make the development are the circumference of the base and the slant height of the cone. The chordal distances from the plan view have been used to mark the length of arc required for the pattern; alternatively, for a higher degree of accuracy, the angle can be calculated and then sub-divided. In the front elevation, lines O1 and O7 are true lengths, and distances OG and OA have been plotted directly onto the pattern. The lines O2–O6 inclusive are not true lengths, and, where these lines cross the sloping face on the top of the conical frustum, horizontal lines have been projected to the side of the cone and been marked B, C, D, E, and F. True lengths OF, OE, OD, OC, and OB are then marked on the pattern. This procedure is repeated for the other half of the cone. The view on the sloping face will be an ellipse, and the method of projection has been described in Chapter 12.

Part of a square pyramid is illustrated in Fig. 13.9. The pattern is formed by drawing an arc of radius OA and stepping off around the curve the lengths of the base, joining the points obtained to the apex O. Distances OE and OG are true lengths from the front elevation, and distances OH and OF are true lengths from the end elevation. The true view in direction of arrow X completes the development.

The development of part of a hexagonal pyramid is shown in Fig. 13.10. The method is very similar to that given in the previous example, but note that lines OB, OC, OD, OE, and OF are true lengths obtained by projection from the elevation.

Figure 13.11 shows an oblique cone which is developed by triangulation, where the surface is assumed to be formed from a series of triangular shapes. The base of the cone is divided into a convenient number of parts (12 in this case)

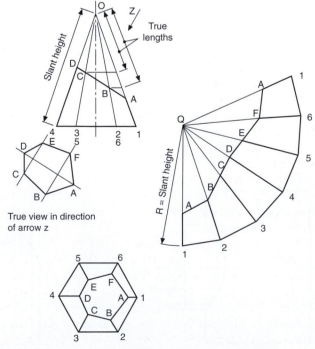

True view in direction of arrow z

FIGURE 13.10

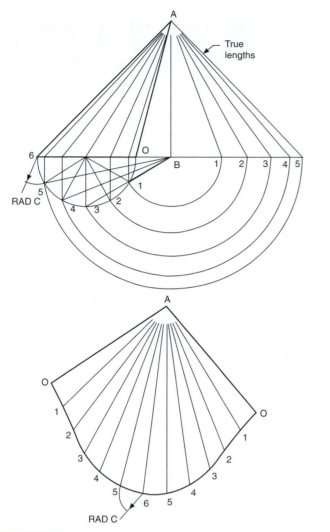

FIGURE 13.11

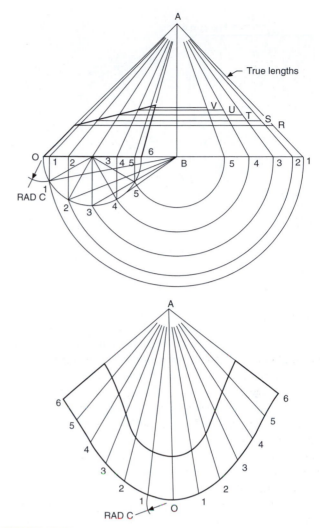

FIGURE 13.12

numbered 0–6 and projected to the front elevation with lines drawn up to the apex A. Lines 0A and 6A are true-length lines, but the other five shown all slope at an angle to the plane of the paper. The true lengths of lines 1A, 2A, 3A, 4A, and 5A are all equal to the hypotenuse of right-angled triangles where the height is the projection of the cone height and the base is obtained from the part plan view by projecting distances B1, B2, B3, B4, and B5 as indicated.

Assuming that the join will be made along the shortest edge, the pattern is formed as follows. Start by drawing line 6A, then from A draw an arc on either side of the line equal in length to the true length 5A. From point 6 on the pattern, draw an arc equal to the chordal distance between successive points on the plan view.

This curve will intersect the first arc twice at the points marked 5. Repeat by taking the true length of line 4A and swinging another arc from point A to intersect with chordal arcs from points 5. This process is continued as shown on the solution.

Figure 13.12 shows the development of part of an oblique cone where the procedure described above is followed. The points of intersection of the top of the cone with lines 1A, 2A,

3A, 4A, and 5A are transferred to the appropriate true-length constructions, and true-length distances from the apex A are marked on the pattern drawing.

A plan and front elevation is given in Fig. 13.13 of a transition piece which is formed from two halves of oblique cylinders and two connecting triangles. The plan view of the base is divided into 12 equal divisions, the sides at the top into six parts each. Each division at the bottom of the front elevation is linked with a line to the similar division at the top. These lines, P1, Q2, etc., are all the same length. Commence the pattern construction by drawing line S4 parallel to the component. Project lines from points 3 and R, and let these lines intersect with arcs equal to the chordal distances C, from the plan view, taken from points 4 and S. Repeat the process and note the effect that curvature has on the distances between the lines projected from points P, Q, R, and S. After completing the pattern to line P1, the triangle is added by swinging an arc equal to the length B from point P, which intersects with the arc shown, radius A. This construction for part of the pattern is continued as indicated.

Part of a triangular prism is shown in Fig. 13.14, in orthographic projection. The sides of the prism are

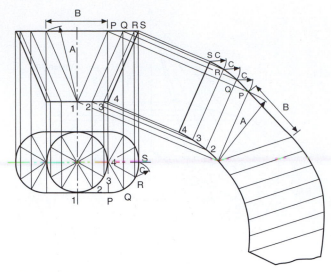

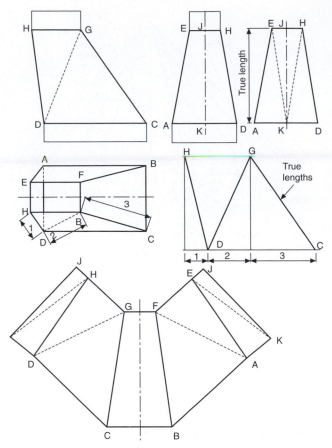

FIGURE 13.13

FIGURE 13.15

constructed from a circular arc of true radius OC in the end elevation. Note that radius OC is the only true length of a sloping side in any of the three views. The base length CA is marked around the circumference of the arc three times, to obtain points A, B, and C.

True length OE can be taken from the end elevation, but a construction is required to find the true length of OD. Draw an auxiliary view in direction with arrow Y, which is square to line OA as shown. The height of the triangle, OX, can be

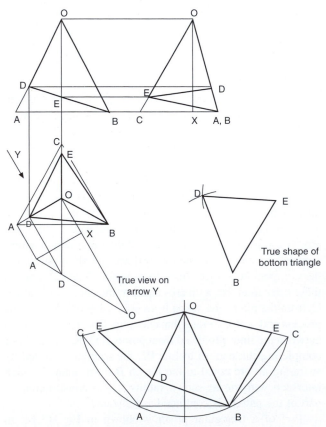

FIGURE 13.14

taken from the end elevation. The projection of point D on the side of the triangle gives the true length OD. The true shape at the bottom can be drawn by taking lengths ED, DB, and BE from the pattern and constructing the triangle shown.

A transition piece connecting two rectangular ducts is given in Fig. 13.15. The development is commenced by drawing the figure CBFG, and the centre line of this part can be obtained from the front elevation which appears as line CG, the widths being taken from the plan. The next problem is to obtain the true lengths of lines CG and DH and position them on the pattern; this can be done easily by the construction of two triangles, after the insertion of line DG. The true lengths can be found by drawing right-angled triangles where the base measurements are indicated as dimensions 1, 2, and 3, and the height is equal to the height of the front elevation. The length of the hypotenuse in each case is used as the radius of an arc to form triangles CDG and GDH. The connecting seam is taken along the centre line of figure ADHE and is marked JK. The true length of line JK appears as line HD in the front elevation, and the true shape of this end panel has been drawn beside the end elevation to establish the true lengths of the dotted lines EK and HK, since these are used on the pattern to draw triangles fixing the exact position of points K and J.

A transition piece connecting square and circular ducts is shown in Fig. 13.16. The circle is divided into 12 equal

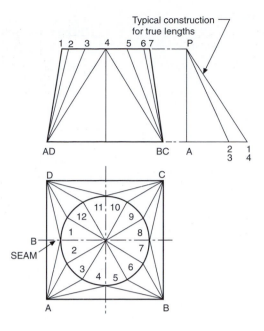

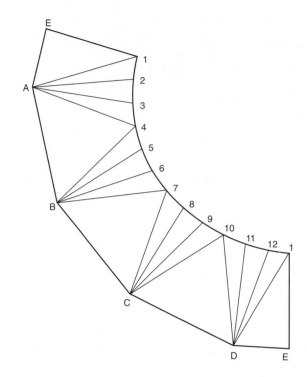

FIGURE 13.16

divisions, and triangles are formed on the surface of the component as shown. A construction is required to establish the true lengths of lines A1, A2, A3, and A4. These lengths are taken from the hypotenuse of right-angled triangles whose height is equal to the height of the front elevation, and the base measurement is taken from the projected lengths in the plan view. Note that the lengths A2 and A3 are the same, as are A1 and A4, since the circle lies at the centre of the square in the plan. The constructions from the other three corners are identical to those from corner A. To form the pattern, draw a line AB, and from A describe an arc of radius A4. Repeat from end B, and join the triangle. From point 4, swing an arc equal to the chordal length between points 4 and 3 in the plan view, and let this arc intersect with the true length A3, used as a radius from point A. Mark the intersection as point 3. This process is repeated to form the pattern shown. The true length of the seam at point E can be measured from the front elevation. Note that, although chordal distances are struck between successive points around the pattern, the points are themselves joined by a curve; hence no ultimate error of any significance occurs when using this method.

Figure 13.17 shows a similar transition piece where the top and bottom surfaces are not parallel. The construction is generally very much the same as described above, but two separate true-length constructions are required for the corners marked AD and BC. Note that, in the formation of the pattern, the true length of lines AB and CD is taken from the front elevation when triangles AB4 and DC10 are formed. The true length of the seam is also the same as line A1 in the front elevation.

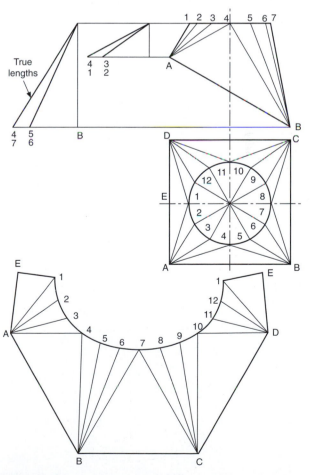

FIGURE 13.17

Chapter 14

Dimensioning principles

A drawing should provide a complete specification of the component to ensure that the design intent can be met at all stages of manufacture. Dimensions specifying features of size, position, location, geometric control and surface texture must be defined and appear on the drawing once only. It should not be necessary for the craftsman either to scale the drawing or to deduce dimensions by the subtraction or addition of other dimensions. Double dimensioning is also not acceptable.

Theoretically any component can be analysed and divided into a number of standard common geometrical shapes such as cubes, prisms, cylinders, parts of cones, etc. The circular hole in Fig. 14.1 can be considered as a cylinder through the plate. Dimensioning a component is the means of specifying the design intent in the manufacture and verification of the finished part.

A solid block with a circular hole in it is shown in Fig. 14.1 and to establish the exact shape of the item we require to know the dimensions which govern its length, height, and thickness, also the diameter and depth of the hole and its position in relation to the surface of the block. The axis of the hole is shown at the intersection of two centre

lines positioned from the left-hand side and the bottom of the block and these two surfaces have been taken as datums. The length and height have also been measured from these surfaces separately and this is a very important point as errors may become cumulative and this is discussed later in the chapter.

Dimensioning therefore should be undertaken with a view to defining the shape or form and overall size of the component carefully, also the sizes and positions of the various features, such as holes, counterbores, tappings, etc., from the necessary datum planes or axes.

The completed engineering drawing should also include sufficient information for the manufacture of the part and this involves the addition of notes regarding the materials used, tolerances of size, limits and fits, surface finishes, the number of parts required and any further comments which result from a consideration of the use to which the completed component will be put. For example, the part could be used in sub-assembly and notes would then make reference to associated drawings or general assemblies.

British Standard 8888 covers all the ISO rules applicable to dimensioning and, if these are adhered to, it is reasonably easy to produce a drawing to a good professional standard.

1. Dimension and extension lines are narrow continuous lines 0.35 mm thick, if possible, clearly placed outside the outline of the drawing. As previously mentioned, the drawing outline is depicted with wide lines of 0.7 mm thick. The drawing outline will then be clearly defined and in contrast with the dimensioning system.
2. The extension lines should not touch the outline of the drawing feature and a small gap should be left, about 2–3 mm, depending on the size of the drawing. The extension lines should then continue for the same distance past the dimension line.
3. Arrowheads should be approximately triangular, must be of uniform size and shape and in every case touch the dimension line to which they refer. Arrowheads drawn manually should be filled in. Arrowheads drawn by machine need not be filled in.
4. Bearing in mind the size of the actual dimensions and the fact that there may be two numbers together where limits of size are quoted, then adequate space must be left between rows of dimensions.

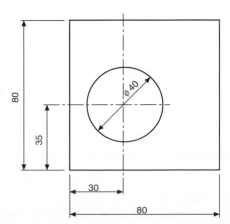

FIGURE 14.1

5. Centre lines must never be used as dimension lines but must be left clear and distinct. They can be extended, however, with the use of extension lines.
6. Dimensions are quoted in millimetres to the minimum number of significant figures. For example, 19 and not 19.0. In the case of a decimal dimension, always use a nought before the decimal marker, which might not be noticed on a drawing print that has poor line definition. We write 0.4 and not .4. It should be stated here that on metric drawings the decimal marker is a comma positioned on the base line between the figures, for example, 5,2 but never 5·2 with a decimal point midway.
7. To enable dimensions to be read clearly, figures are placed so that they can be read from the bottom of the drawing, or by turning the drawing in a clockwise direction, so that they can be read from the right-hand side.
8. Leader lines are used to indicate where specific indications apply. The leader line to the hole is directed towards the centre point, terminating at the circumference in an arrow. A leader line for a part number terminates in a dot within the outline of the component. The gauge plate here is assumed to be part number six of a set of inspection gauges.

Figure 14.2 shows a partly completed drawing of a gauge to illustrate the above aspects of dimensioning.

When components are drawn in orthographic projection, a choice often exists where to place the dimensions and the following general rules will give assistance.

1. Start by dimensioning the view which gives the clearest understanding of the profile or shape of the component.

2. If space permits, and obviously this varies with the size and degree of complexity of the subject, place the dimensions outside the profile of the component as first choice.
3. Where several dimensions are placed on the same side of the drawing, position the shortest dimension nearest to the component and this will avoid dimension lines crossing.
4. Try to ensure that similar spacings are made between dimension lines as this gives a neat appearance on the completed drawing.
5. Overall dimensions which are given for surfaces that can be seen in two projected views are generally best positioned between these two views.

Remember, that drawings are the media to communicate the design intent used for the manufacturing and verification units. Therefore always check over your drawing, view it and question yourself. Is the information complete? Ask yourself whether or not the machinist or fitter can use or work to the dimension you have quoted to make the item. Also, can the inspector verify the figure, in other words, is it a measurable distance?

Figure 14.3 shows a component which has been partly dimensioned to illustrate some of the principles involved.

Careless and untidy dimensioning can spoil an otherwise sound drawing and it should be stated that many marks are lost in examinations due to poor quality work.

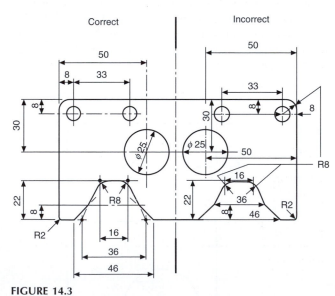

FIGURE 14.3

Dimensioning of features not drawn to scale

This method of indication is by underlining a particular dimension with a wide line as indicated in Fig. 14.4. This practice is very useful where the dimensional change does not impair the understanding of the drawing.

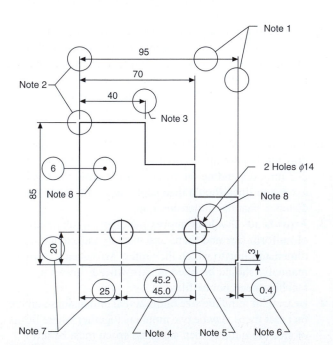

FIGURE 14.2

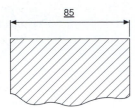

FIGURE 14.4

Chain dimensioning and auxiliary dimensioning

Chains of dimensions should only be used where the possible accumulation of tolerances does not endanger the function of the part.

A plan view of a twist drill stand is given in Fig. 14.5 to illustrate chain dimensioning. Now each of the dimensions in the chain would be subject to a manufacturing tolerance since it is not possible to mark out and drill each of the centre distances exactly. As a test of drawing accuracy, start at the left-hand side and mark out the dimensions shown in turn. Measure the overall figure on your drawing and check with the auxiliary dimension given. Note the considerable variation in length, which results from small errors in each of the six separate dimensions in the chain, which clearly accumulate. Imagine the effect of marking out say 20 holes for rivets in each of two plates, how many holes would eventually line up? The overall length is shown in parentheses (157) and is known as an auxiliary dimension. This dimension is not the one which is worked to in practice but is given purely for reference purposes. You will now appreciate that it will depend on the accuracy with which each of the pitches in the chain is marked out.

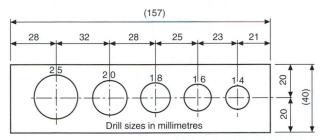

FIGURE 14.5

Parallel dimensioning

Improved positional accuracy is obtainable by dimensioning more than one feature from a common datum, and this method is shown in Fig. 14.6. The selected datum is the left-hand side of the stand. Note that the overall length is not an auxiliary dimension, but as a dimensional length in its own right.

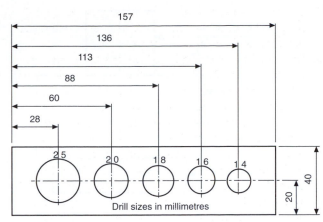

FIGURE 14.6

Running dimensioning

Is a simplified method of parallel dimensioning having the advantage that the indication requires less space. The common origin is indicated as shown (Fig. 14.7) with a narrow continuous circle and the dimensions placed near the respective arrowheads.

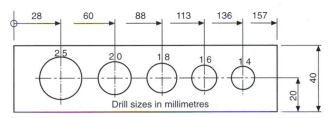

FIGURE 14.7

Staggered dimensions

For greater clarity, a number of parallel dimensions may be indicated as shown in Figs. 14.8 and 14.9.

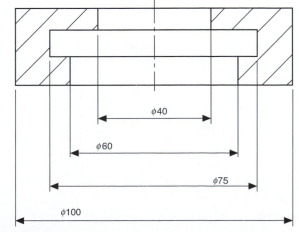

FIGURE 14.8

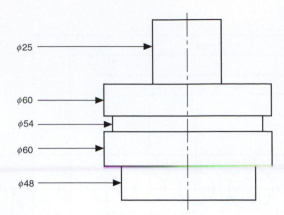

FIGURE 14.9

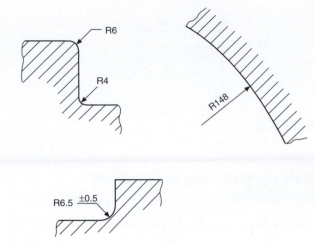

FIGURE 14.11

Dimensioning circles

The symbol Ø preceding the figure is used for specifying diameters and it should be written as large as the figures which establish the size, e.g. Ø65. Alternative methods of dimensioning diameters are given in Fig. 14.10. The size of hole and space available on the drawing generally dictates which method the draughtsman chooses.

Dimensioning radii

Alternative methods are shown in Fig. 14.11 where the position of the centre of the arc need not be located. Note that the dimension line is drawn through the arc centre or lies in a line with it in the case of short distances and the arrowhead touches the arc.

Dimensioning spherical radii and diameters

Spherical radii and diameters are dimensioned as shown in Fig. 14.12. The letter S preceding the Ø symbol (diameter) or letter R (radii).

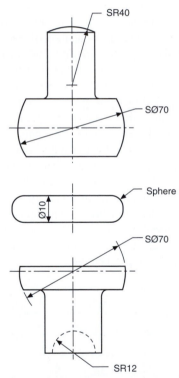

FIGURE 14.12

Dimensioning curves

A curve formed by the blending of several radii must have the radii with their centres of curvature clearly marked as indicated in Fig. 14.13.

Dimensioning irregular curves

Irregular curves may be dimensioned by the use of ordinates. To illustrate the use of ordinates, a section through the hull of a boat is shown (Fig. 14.14). Since the hull is symmetrical about the vertical centre line, it is not necessary to draw both

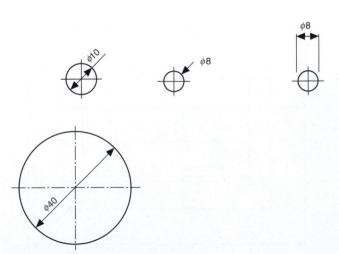

FIGURE 14.10

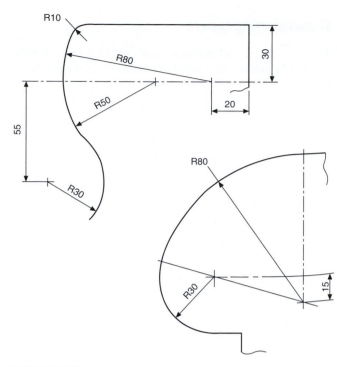

FIGURE 14.13

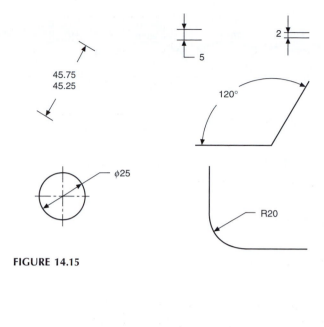

FIGURE 14.15

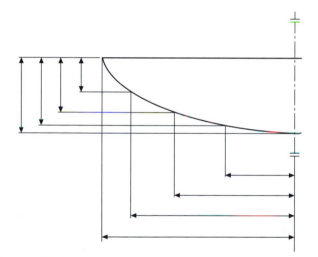

FIGURE 14.14

halves in full and if the curve is presented in this manner then two short thick parallel lines are drawn at each end of the profile at right angles to the centre line. The outline is also extended slightly beyond the centre line to indicate that the shape is to be continued. Ordinates are then positioned on the drawing and the outline passes through each of the chosen fixed points (Fig. 14.15).

Unidirectional and aligned dimensions

Both methods are in common use.

1. Unidirectional dimensions are drawn parallel with the bottom of the drawing sheet, also any notes which refer to the drawing use this method.

2. Aligned dimensions are shown in parallel with the related dimension line and positioned so that they can be read from the bottom of the drawing or from the right-hand side (Fig. 14.16).

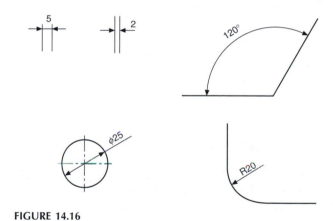

FIGURE 14.16

Angular dimensions

Angular dimensions on engineering drawings are expressed as follows:

(a) Degrees, e.g. 30°.
(b) Degrees and minutes, e.g. 30° 40′.
(c) Degrees, minutes and seconds, e.g. 30° 40′ 20″.

For clarity a full space is left between the degree symbol and the minute figure and also between the minute symbol and the second figure.

In the case of an angle less than 1° it should be preceded by 0°, e.g. 0° 25′.

Figure 14.17 shows various methods of dimensioning angles.

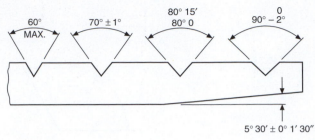

FIGURE 14.17

Tapers

In Fig. 14.18 the difference in magnitude between dimensions X and Y (whether diameters or widths) divided by the length between them defines a ratio known as a *taper*.

$$\text{Taper} = \frac{X - Y}{\text{length}} = 2 \tan \frac{\theta}{2}.$$

For example, the conical taper in Fig. 14.19

$$= \frac{20 - 10}{40} = \frac{10}{40} = 0.25$$

and may be expressed as rate of taper 0.25:1 on diameter.

The ISO recommended symbol for taper is, and this symbol can be shown on drawings accompanying the rate of taper,

i.e. $\rightarrow$ 0.25:1

The arrow indicates the direction of taper.

When a taper is required as a datum, it is enclosed in a box as follows:

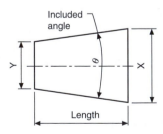

FIGURE 14.18

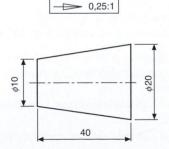

FIGURE 14.19

Dimensioning tapers

The size, form, and position of a tapered feature can be defined by calling for a suitable combination of the following:

1. the rate of taper, or the included angle;
2. the diameter or width at the larger end;
3. the diameter or width at the smaller end;
4. the length of the tapered feature;
5. the diameter or width at a particular cross-section, which may lie within or outside the feature concerned;
6. the locating dimension from the datum to the cross-section referred to above.

Care must be taken to ensure that no more dimensions are quoted on the drawing than are necessary. If reference dimensions are given to improve communications, then they must be shown in brackets, e.g. (1:5 taper).

Figure 14.20 gives four examples of the methods used to specify the size, form, and position of tapered features.

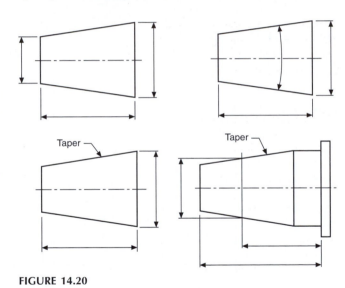

FIGURE 14.20

Dimensioning two mating tapers

When the fit to a mating part or gauge is necessary, a tried and successful method used in manufacturing units is to add the following information to the feature(s).

1. 'To FIT PART NO. YYY'.
2. 'TO FIT GAUGE (PART NO. GG)'.

When note 2 is added to the drawing, this implies that a 'standard rubbing gauge' will give an acceptable even marking when 'blued'. The functional requirement whether the end-wise location is important or not, will determine the method and choice of dimensioning.

An example of dimensioning two mating tapers when end-wise location is important is shown in Fig. 14.21.

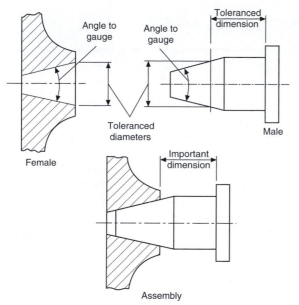

FIGURE 14.21

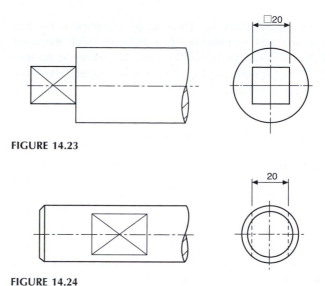

FIGURE 14.23

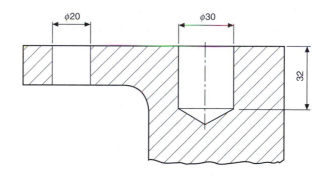

FIGURE 14.24

For more accurate repeatability of location, the use of Geometric Tolerancing and a specific datum is recommended. Additional information on this subject may be found in BS ISO 3040.

Dimensioning chamfers

Alternative methods of dimensioning internal and external chamfers are shown in Fig. 14.22.

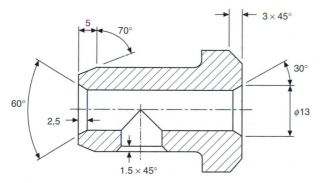

FIGURE 14.22

Dimensioning holes

The depth of drilled holes, when stated in note form, refers to the depth of the cylindrical portion and not to the point left by the drill. If no other indication is given they are assumed to go through the material. Holes in flanges or bosses are generally positioned around a pitch circle and may be spaced on the main centre lines of the component, or as shown below (Fig. 14.25) equally. It is a default condition that hole patterns on a pitch circle diameter are equally spaced unless

Dimensioning squares or flats

Figure 14.23 shows a square machined on the end of a shaft so that it can be turned by means of a spanner.

The narrow diagonal lines are added to indicate the flat surface.

Part of a spindle which carries the chain wheel of a cycle, secured by a cotter pin, illustrates a flat surface which is not at the end of the shaft (Fig. 14.24).

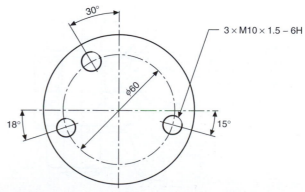

FIGURE 14.25

dimensioned otherwise. There is no requirement (as was past practice) to indicate this condition using the word 'EQUISPACED' in the hole call out. The angles of the spacings may be omitted when the intent is evident as shown in Figs. 20.28 and 20.29.

Dimensioning counterbores

A drilling machine is used for this operation, and a typical counterboring tool is shown in Fig. 14.26. The operation involves enlarging existing holes, and the depth of the enlarged hole is controlled by a stop on the drilling machine. The location of the counterbored hole is assisted by a pilot at the tip of the tool which is a clearance fit in the previously drilled hole. A typical use for a counterbored hole is to provide a recess for the head of a screw, as shown in Fig. 14.27 or a flat surface for an exposed nut or bolt, as in Fig. 14.28. The flat surface in Fig. 14.28 could also be obtained by spotfacing.

Figure 14.29 shows methods of dimensioning counterbores. Note that, in every case, it is necessary to specify the size of counterbore required. It is not sufficient to state 'COUNTERBORE FOR M10 RD HD SCREW', since

FIGURE 14.26

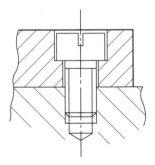

FIGURE 14.27

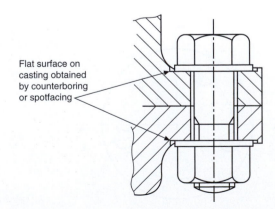

Flat surface on casting obtained by counterboring or spotfacing

FIGURE 14.28

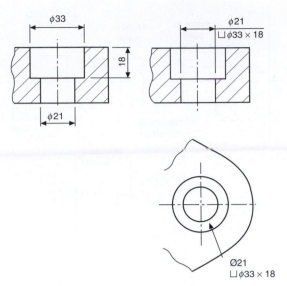

FIGURE 14.29

obviously the head of the screw will fit into any counterbore which is larger than the head.

Dimensioning countersunk holes

Countersinking is also carried out on a drilling machine, and Fig. 14.30 shows typical tools. Included angles of 60° and 90° are commonly machined, to accommodate the heads of screws and rivets to provide a flush finish (Fig. 14.31).

Note: Refer to manufacturers' catalogues for dimensions of suitable rivets and screws.

Dimensioning spotfaces

Spotfacing is a similar operation to counterboring, but in this case the metal removed by the tool is much less. The process is regularly used on the surface of castings, to provide a flat seating for fixing bolts. A spotfacing tool is shown in Fig. 14.32, where a loose cutter is used. The length of cutter controls the diameter of the spotface. As in the counterboring operation, the hole must be previously drilled, and the pilot at the tip of the spotfacing tool assists in location.

Figure 14.33 shows the method of dimensioning. Note that, in both cases, the depth of spotface is just sufficient to

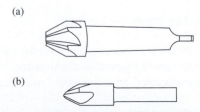

FIGURE 14.30 (a) Taper-shank countersink (with 60° or 90° included angle of countersink). (b) Straight-shank machine countersink (with 60° or 90° included angle of countersink).

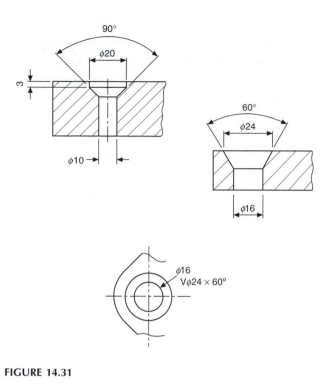

FIGURE 14.31

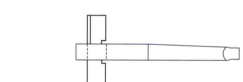

FIGURE 14.32

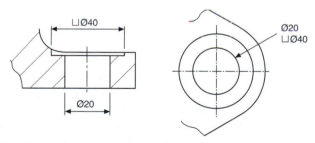

FIGURE 14.33

remove the rough surface of the casting over the 40 mm diameter area.

Dimensioning for manufacture

It should be emphasized that dimensioning must be performed with the user of the drawing very much in mind. In the case of the finished bearing housing shown in Fig. 14.35 two different production processes are involved in its manufacture namely: casting and machining of the component. It is sometimes preferable to produce two separate drawings, one to show the dimensions of the finished casting and the other to show the dimensions which are applicable to the actual machining operation. Figure 14.34 shows a suitable

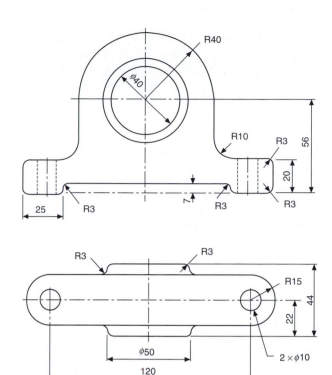

FIGURE 14.34

drawing for the casting patternmaker. Allowances are made for machining and also for the fact that the casting will shrink when it cools. The machinist will take the rough casting and remove metal to produce the finished component, all other surfaces having a rough finish. Figure 14.35 shows the required dimensions for machining. Note that the bore of the casting is required to be finished between the two sizes quoted for functional purposes.

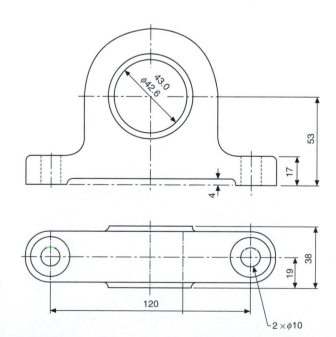

FIGURE 14.35

Chapter 15

Screw threads and conventional representations

Screw threads

The most common application of the helix is in a screw thread which follows the path of the helix. Screw threads may be either left or right hand and these are shown pictorially in Fig. 15.1. Notice the slope of the thread and the position of the index finger on each hand. The left-hand thread is used for special applications and the right-hand thread is the one normally used on nuts and bolts. The thread illustrated has a vee-section.

The following terms are associated with screw threads:

- The *thread pitch* is the distance between corresponding points on adjacent threads. Measurements must be taken parallel to the thread axis.
- The *major diameter* or *outside diameter* is the diameter over the crests of the thread, measured at right angles to the thread axis.
- The *crest* is the most prominent part of the thread, internal or external.
- The *root* lies at the bottom of the groove between two adjacent threads.
- The *flank* of the thread is the straight side of the thread between the crest and root.
- The *minor diameter, root diameter* or *core diameter* is the smallest diameter of the thread measured at right angles to the thread axis.
- The *effective diameter* is measured at right angles to the thread axis and is the diameter on which the width of the spaces is equal to the width of the threads.
- The *lead* of a thread is the axial movement of the screw in one revolution.

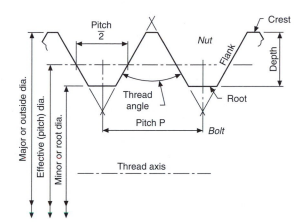

FIGURE 15.2 External form to illustrate thread terms.

The terms are illustrated in Fig. 15.2.

ISO metric threads

Figure 15.3 shows the ISO metric thread form for a nut (internal) and for a bolt (external). In the case of the nut, the root is rounded in practice. For the mating bolt, the crest of the thread may be rounded within the maximum outline, as shown, and the root radiused to the given dimension. Both male and female threads are subject to manufacturing tolerances and for complete information, reference should be made to BS 3643–1.

BS 3643–2 defines two series of diameters with graded pitches for general use in nuts, bolts and screwed fittings, one series with coarse and the other with fine

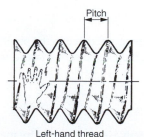

Left-hand thread   Right-hand thread

FIGURE 15.1

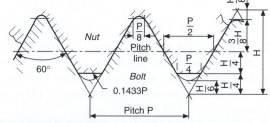

FIGURE 15.3 ISO metric thread
$H = 0.86603P$, $H/4 = 0.21651P$, $(3/8)H = 0.32476P$, $^{(5/8)H} = 0.54127P$, where P is the pitch of the thread.

TABLE 15.1

1 Basic major diameters Choice 1st	2 2nd	3 3rd	4 Coarse series with graded pitches	5 Fine series with constant pitches 6	6 4	7 3	8 2	9 1.5	10 1.25	11 1	12 0.75	13 0.5	14 0.35	15 0.25	16 0.2
1.6	–	–	0.35	–	–	–	–	–	–	–	–	–	–	–	0.2
–	1.8	–	0.35	–	–	–	–	–	–	–	–	–	–	–	0.2
2	–	–	0.4	–	–	–	–	–	–	–	–	–	–	0.25	–
–	2.2	–	0.45	–	–	–	–	–	–	–	–	–	–	0.25	–
2.5	–	–	0.45	–	–	–	–	–	–	–	–	–	0.35	–	–
3	–	–	0.5	–	–	–	–	–	–	–	–	–	0.35	–	–
–	3.5	–	0.6	–	–	–	–	–	–	–	–	–	0.35	–	–
4	–	–	0.7	–	–	–	–	–	–	–	–	0.5	–	–	–
–	4.5	–	0.75	–	–	–	–	–	–	–	–	0.5	–	–	–
5	–	–	8.8	–	–	–	–	–	–	–	–	0.5	–	–	–
–	–	5.5	–	–	–	–	–	–	–	–	–	0.5	–	–	–
6	–	–	1	–	–	–	–	–	–	–	0.75	–	–	–	–
–	–	7	1	–	–	–	–	–	–	–	0.75	–	–	–	–
8	–	–	1.25	–	–	–	–	–	–	1	0.75	–	–	–	–
–	–	9	1.25	–	–	–	–	–	–	1	0.75	–	–	–	–
10	–	–	1.5	–	–	–	–	–	1.25	1	0.75	–	–	–	–
–	–	11	1.5	–	–	–	–	–	–	1	0.75	–	–	–	–
12	–	–	1.75	–	–	–	–	1.5	1.25	1	–	–	–	–	–
–	14	–	2	–	–	–	–	1.5	1.25*	1	–	–	–	–	–
–	–	15	–	–	–	–	–	1.5	–	1	–	–	–	–	–
16	–	–	2	–	–	–	–	1.5	–	1	–	–	–	–	–
–	–	17	–	–	–	–	–	1.5	–	1	–	–	–	–	–
–	18	–	2.5	–	–	–	2	1.5	–	1	–	–	–	–	–
20	–	–	2.5	–	–	–	2	1.5	–	1	–	–	–	–	–
–	22	–	2.5	–	–	–	2	1.5	–	1	–	–	–	–	–
24	–	–	3	–	–	–	2	1.5	–	1	–	–	–	–	–

Note: For preference, choose the diameters given in Column 1. If these are not suitable, choose from Column 2, or finally from Column 3.
* The pitch of 1.25 mm for 14 mm diameter is to be used only for sparking plugs.

pitches. The extract given in Table 15.1 from the standard gives thread sizes from 1.6 to 24 mm diameter. Note that first, second and third choices of basic diameters are quoted, to limit the number of sizes within each range.

On a drawing, a thread will be designated by the letter M followed by the size of the nominal diameter and the pitch required, e.g. M10 × 1.

If a thread is dimensioned without reference to the pitch, e.g. M16, then it is assumed that the coarse series thread is required.

Unified threads

The Unified system of screw threads was introduced by the United Kingdom, Canada and the United States to provide a

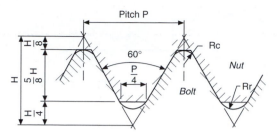

FIGURE 15.4   Unified screw thread
$H = 0.86603P$, $R_c = 0.108P$ and $R_r = 0.144P$ where P is the pitch of the thread.

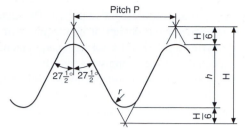

FIGURE 15.6   Basic Whitworth form
$H = 960491P$, $h = 0.640327P$, $r = 0.137329P$.

common standard thread for use by the three countries. The International Standards Organisation (ISO) recommends the system as an international system of screw threads in inch units, in parallel with a similar system in metric units. Both of these systems use a similar form of thread profile which is illustrated basically in Fig. 15.4.

Unified threads are covered by BS 1580. Types in common use include the following:

(a) UNC is a Unified coarse pitch thread, with progressive pitch sizes (i.e. the pitch varies with the diameter).
(b) UNF is a Unified fine pitch thread, also with progressive pitch sizes.
(c) UN is a Unified thread with a constant pitch (e.g. an 8 UN thread has eight threads to the inch regardless of the diameter).

Different classes of fit are obtainable by manufacture within alternative tolerance ranges and these are specified in BS 1580. Normally the same class of internal and external thread are used together.

Sellers or American thread (Fig. 15.5). This type was the American National thread in common use before the introduction of the Unified National thread, as it is described in USA and Canada, or the Unified screw-thread in Great Britain.

The *British Standard Pipe* threads are used internally and externally on the walls of pipes and tubes. The thread pitch is relatively fine, so that the tube thickness is not unduly weakened.

Pipe threads are covered by BS 21, which was adopted as the basis for ISO 7/1 where the metric values were conversions of the inch values, to obtain interchangeability.

The basic forms of the British Standard taper and parallel pipe threads are based on that of the British Standard Whitworth thread.

The Whitworth thread form is shown in Fig. 15.6. The thread angle of 55° is measured in an axial plane section, also the vee-section is truncated, at top and bottom, by one-sixth, with crest and root rounded by equal circular arcs. The theoretical thread depth is shown as $h = 0.640327P$ where P is the thread pitch. In the taper pipe thread (Fig. 15.7), a taper of 1 in 16 is used, measured on diameter.

British Association thread (Fig. 15.8). Generally used in sizes of less than $1/4$ in. on small mechanisms. This range of threads extends down to a thread size of 0.25 mm and is covered by BS 93.

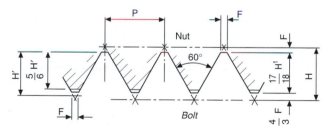

FIGURE 15.5   Sellers or American thread
$H = 0.866P$, $H' = 0.6495P$, $F = 0.1083P = H/8 = H/6$.

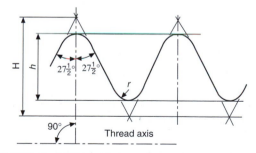

FIGURE 15.7   Basic Whitworth form of taper pipe thread
$H = 0.960273P$, $h = 0.640327P$, $r = 0.137278P$.

Whitworth thread (Fig. 15.6). The general shape of the thread shown has been used in a standard BSW (British Standard Whitworth) thread, in fine form as the BSF (British Standard Fine) thread, and as a pipe thread in the BSP (British Standard Pipe) thread.

The *British Standard Whitworth* thread was the first standardized British screw-thread.

The *British Standard Fine* thread is of Whitworth section but of finer pitch. The reduction in pitch increases the core diameter; also, small adjustments of the nut can easily be made.

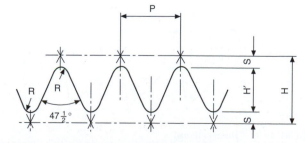

FIGURE 15.8   British Association (BA) thread
$H = 1.1363P$, $H = 0.6P$ (approx.), $R = 0.18P$, $S = 0.268P$.

Note: BS 93 is an 'obsolescent' standard. The standard is not recommended for the design of new equipment, but is retained to provide a standard for servicing equipment in use which is expected to have a long working life.

BS 4827 specifies the requirements for ISO miniature screw threads from 0.30 to 1.4 mm diameter with 60° form and are used in delicate instruments and watch making.

Threads for power transmission

Square thread (Fig. 15.9)

Used to transmit force and motion since it offers less resistance to motion than 'V' thread forms.

Widely used on lathes, this thread form is sometimes slightly modified by adding a small taper of about 5° to the sides as an aid to production.

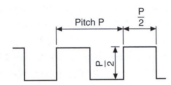

FIGURE 15.9

Acme thread (Fig. 15.10)

More easily produced than the square thread and often used in conjunction with split nuts for engagement purposes. It is applied in valve operating spindles.

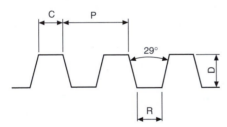

FIGURE 15.10   Acme thread
$C = 0.3707P, R = C, D = (P/2) + 0.01.$

Buttress thread (Fig. 15.11)

Used for transmitting power in one direction only. In its original form, the pressure face now sloping at 7° was perpendicular with the thread axis. A common application of the thread can be found in workshop vices.

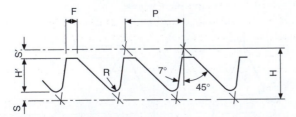

FIGURE 15.11   Buttress thread
$H = 0.8906P, H' = 0.5058P, S = 0.1395P,$
$S' = 0.245P, F = 0.2754P, R = 0.1205P.$

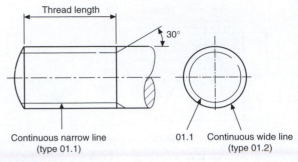

FIGURE 15.12

Draughting conventions associated with threads

Threads are so regularly used on engineering drawings that recognizable conventions are required to save draughting time. Figure 15.12 shows the convention for a male thread. The thread runout along the shank of the stud is indicated by a line drawn at 30° to the thread axis. The minor diameter of the thread is shown by parallel lines and in the end elevation the projected circle is not continuous. The break in the inside circle distinguishes the end elevation of a male thread from a female thread. Line thicknesses are given for each part of the thread. The actual dimensions of the minor diameter for any particular thread size can be approximated at 80% of the major diameter for the purposes of conventional representation.

Figure 15.13 shows the convention for a female thread applied to a blind tapped hole in a sectional view. Note that the minor diameter is drawn as a complete circle in the end elevation, the major diameter is broken and the different line thicknesses also help to distinguish the female from the male thread. The effective length of the thread is again shown by parallel lines and the runout by the taper at 30° to the thread axis. In the sectional elevation, the section lines drawn at 45°, continue through the major to the minor diameter. The included angle left by the tapping drill is 120°. Line thicknesses are indicated in the circles.

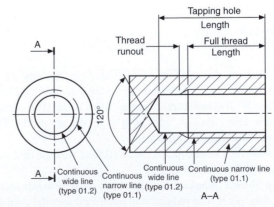

FIGURE 15.13

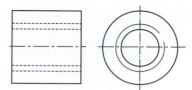

FIGURE 15.14

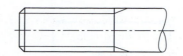

FIGURE 15.17

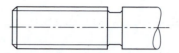

FIGURE 15.18

Note: The line at 30° indicating the runout was common practice. However, it is permitted to omit the line if there is no functional need for it. The tapered lines for incomplete threads are shown here for completeness of text.

A female thread through a collar is shown in Fig. 15.14. Note that the projection of the major and minor diameters drawn as hidden detail will be indicated by dashed narrow lines.

A section through the collar is given in Fig. 15.15. The projection of the major diameter is drawn by a continuous narrow line and the cross hatching extends to the minor diameter.

A section through a sealing cap in Fig. 15.16 illustrates a typical application where an internal thread terminates in an undercut.

Screw threads are produced by cutting or rolling. A cut thread can be made by the use of a tap for female threads, or a die in the case of a male thread. Figure 15.17 shows a male thread cut by a die and terminating in a runout. In this application, the bar diameter has been drawn equal in size with the major diameter of the thread.

Screwcutting may be undertaken on the lathe and the cutting tool is shaped to match the thread angle. Generally the thread terminates in an undercut and this feature is illustrated in Fig. 15.18. It is a normal draughting practice to draw an undercut in line with the minor diameter of the thread. Too narrow an undercut or the demand for perfect or full threads up to shoulders, or to the bottom of blind holes, increases manufacturing costs.

A rolled thread application is indicated in Fig. 15.19. The thread is formed by deformation and runs out to a bar diameter which is approximately equal to the effective diameter of the thread.

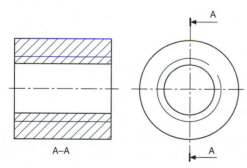

A–A

FIGURE 15.15

FIGURE 15.19

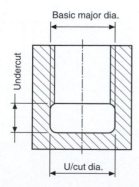

Basic major dia.

Undercut

U/cut dia.

FIGURE 15.16

Multiple threads

Generally, screws have single threads and unless it is designated otherwise, it is fair to assume that a thread will be single. This means that the thread is formed from one continuous helix. The lead of a thread is the distance moved by a mating nut in one complete revolution of the nut. In a single thread the lead is equal to the pitch.

When a two start thread is manufactured, there are two continuous helices and to accommodate the grooves, the lead is twice the thread pitch. Multiple threads are used where a quick axial movement is required with a minimum number of turns.

Now the standard drawing conventions do not differentiate between single and multiple threads. Details of the thread must be quoted with the drawing dimensions.

Triple start right-hand thread (Fig. 15.20).

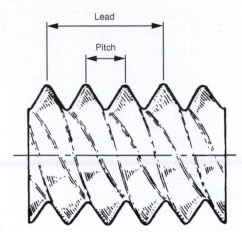

FIGURE 15.20

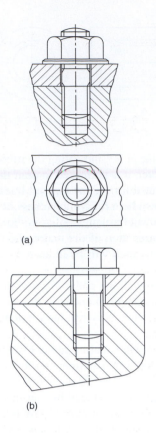

(a)

(b)

FIGURE 15.21

The application of thread conventions

Figure 15.21(a) shows an elevation of a stud in a tapped hole. When a mating thread is screwed into a tapped hole, the section lines do not cover the assembled threads, also, the threads on the stud terminate at the start of the hole to signify that the stud is screwed to maximum depth. Adjacent parts in the sectional view are cross hatched in opposite directions. It is not a normal practice to section nuts, bolts and washers.

Exercise – Draw a similar assembly using a M24 stud. The stud is screwed for 30 mm into the tapped hole and the thickness of the component with the clearance hole is 25 mm. Use Table 16.1 for other dimensions and assume proportions of unimportant detail.

Figure 15.21(b) shows part of a component which is held by a set-bolt in a tapped hole. Only part of the shank of the bolt is threaded and this must terminate above the joined line between the components for correct assembly.

Exercise – Draw a similar arrangement using a M20 set-bolt which engages with the tapped thread for a length of 25 mm. Component thickness is 22 mm. Assume other dimensions.

Note: The thickness of a nut is slightly greater than the thickness of a bolthead. For approximate constructions of a bolthead, use a thickness of 0.7D, where D is the shank diameter.

Tapping drill

The diameter of a tapping drill is equal to the minor diameter of a thread to be tapped for drawing purposes. The function of a tapping drill is to remove the bulk of the material for a female thread, leaving the tap to cut only the thread form between the major and minor diameters.

Clearance drill

A clearance drill has a diameter slightly greater than the major diameter of a male thread and its function is to provide a clearance hole to permit the free passage of a bolt through a component.

Chapter 16

Nuts, bolts, screws and washers

ISO metric precision hexagon bolts, screws and nuts are covered by BS 3643 and ISO 272. The standard includes washer faced hexagon head bolts and full bearing head bolts. In both cases there is a small radius under the bolt-head which would not normally be shown on drawings, due to its size, but is included here for completeness of the text. With an M36 bolt, the radius is only 1.7 mm. Bolts may be chamfered at 45° at the end of the shank, or radiused. The rounded end has a radius of approximately one and one quarter times the shank diameter and can also be used if required to draw the rolled thread end. The washer face under the head is also very thin and for a M36 bolt is only 0.5 mm.

Figure 16.1(a) shows the bolt proportions and Table 16.1 the dimensions for bolts in common use. Dimensions of suitable nuts are also given and illustrated in Fig. 16.1(b).

Included in Table 16.1 and shown in Fig. 16.1(c) are typical washers to suit the above bolts and nuts and these are covered by BS 4320. Standard washers are available in two different thicknesses, in steel or brass, and are normally plain, but may be chamfered.

Table 16.1 gives dimensions of commonly used bolts, nuts and washers so that these can be used easily on assembly

drawings. For some dimensions maximum and minimum values appear in the standards and we have taken an average figure rounded up to the nearest 0.5 mm and this will be found satisfactory for normal drawing purposes. Reference should be made to the relevant standards quoted for exact dimensions if required in design and manufacture.

Drawing nuts and bolts

It is often necessary to draw nuts and bolts and a quick easy method is required to produce a satisfactory result.

Nuts and bolts are not normally drawn on detail drawings unless they are of a special type. They are shown on assembly drawings and, provided they are standard stock sizes, are called up in parts lists and schedules. A description of the head, the thread and the length being generally sufficient. Templates are available for drawing nuts and bolts and can be recommended for their time saving advantages.

It is conventional drawing practice to show, as first choice, nuts and bolts in the across corners position if a single view only is illustrated since this is instantly recognizable.

TABLE 16.1

Nominal size thread diameter D	Thread pitch	Minor diameter of thread	Width across corners (A/C)	Width across flats (A/F)	Diameter of washer face D_f	Height bolt head H	Thickness of normal nut T	Thickness of thin nut t	Washer inside diameter	Washer outside diameter	Washer thickness form A	Washer thickness form B
M1.6	0.35	1.1	3.5	3.0		1.0	1.25		1.7	4.0	0.3	
M2	0.4	1.4	4.5	4.0		1.5	1.5		2.2	5.0	0.3	
M2.5	0.45	1.9	5.5	5.0		1.75	2.0		2.7	6.5	0.5	
M3	0.5	2.3	6.0	5.5	5.0	2.0	2.25		3.2	7.0	0.5	
M4	0.7	3.0	8.0	7.0	6.5	2.75	3.0		4.3	9.0	0.8	
M5	0.8	3.9	9.0	8.0	7.5	3.5	4.0		5.3	10.0	1.0	
M6	1.0	4.7	11.5	10.0	9.0	4.0	5.0		6.4	12.5	1.6	0.8
M8	1.25	6.4	15.0	13.0	12.0	5.5	6.5	5.0	8.4	17	1.6	1.0
M10	1.5	8.1	19.5	17.0	16.0	7.0	8.0	6.0	10.5	21	2.0	1.25
M12	1.75	9.7	21.5	19.0	18.0	8.0	10.0	7.0	13.0	24	2.5	1.6
M16	2.0	13.5	27.0	24.0	23.0	10.0	13.0	8.0	17.0	30	3.0	2.0
M20	2.5	16.7	34.0	30.0	29.0	13.0	16.0	9.0	21.0	37	3.0	2.0
M24	3.0	20.0	41.5	36.0	34.5	15.0	19.0	10.0	25.0	44	4.0	2.5
M30	3.5	25.5	52.0	46.0	44.5	19.0	24.0	12.0	31.0	56	4.0	2.5
M36	4.0	31.0	62.5	55.0	53.5	23.0	29.0	14.0	37.0	66	5.0	3.0

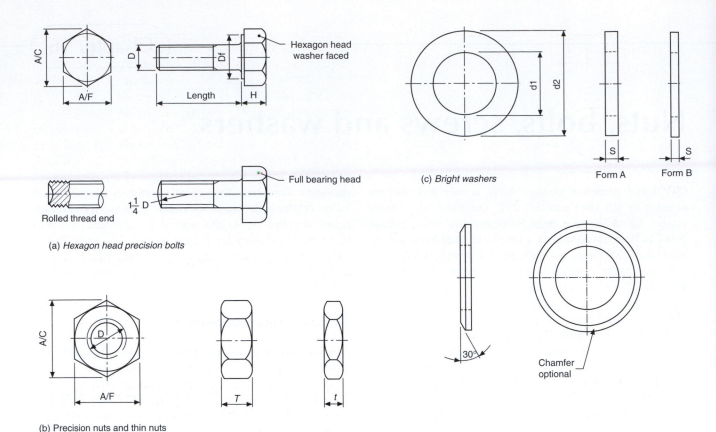

(c) *Bright washers*

(a) *Hexagon head precision bolts*

(b) Precision nuts and thin nuts

FIGURE 16.1 Proportions of bolts, nuts and washers. A/C means across corners. A/F means across flats.

Approximate construction for nuts and bolts (Figs. 16.2 and 16.3)

Stage 1

1. Draw a circle in the plan position, 2D in diameter, where D is equal to the thread size. In this example let us assume that the thread size is M20.
2. Draw a hexagon inside the 40 mm diameter circle and inside the hexagon draw another circle tangential to the hexagon on the six sides. This circle is the projection of the chamfer which can be seen on the front elevation.
3. The nut thickness is 0.8D. Project the four corners of the hexagon to the front elevation.
4. Project three corners of the hexagon in the end elevation and note, that the width of the end elevation is given by dimension W.
5. Line in the projected diameter of the chamfer circle and the base in the front elevation.
6. As an approximation, draw a radius to show the chamfer on the front elevation. The radius should equal the thread size D.
7. Add the female convention to the plan view.

Stage 2

1. The projection of the curve on the chamfered faces of the hexagon that lie at an angle would produce ellipses in the front elevation. In their place we usually show

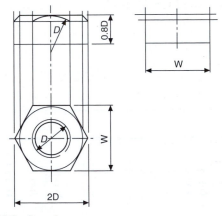

FIGURE 16.2 Stage 1.

small circular arcs, their radii can be found by trial, but are approximately 0.25D.

2. The end elevation of the nut has square corners and the projection of the corner which coincides with the centre line terminates at the bottom of the chamfer curve.
3. Complete the view by drawing circular arcs on the two chamfered faces. Find by trial, the radius of an arc which will touch the top of the nut and the projection lines from the corner in the front elevation.

Reference to Fig. 16.1a and b will show that the constructions in Figs. 16.2 and 16.3 can be used for the bolthead and

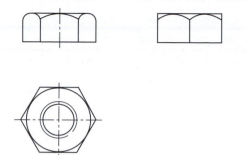

FIGURE 16.3 Stage 2.

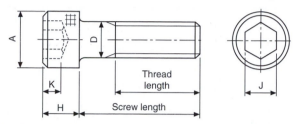

FIGURE 16.4 ISO metric hexagon socket shoulder screws. Dimensions in Table 16.3.

locknut where proportions for thickness can be approximated to 0.7*D* and 0.5*D*.

For exact dimensions however, please refer to Table 16.1.

Socket head screws manufactured to BS EN ISO 4762 and BS 3643-2

It is often required to draw these screws and although the head type and the length are generally quoted in parts lists it is necessary to know the proportions of the head. Dimensions follow for each of the most commonly used screws.

Before specifying screws, it is advisable to consult a manufacturers list for availability. In the interest of standard-

ization and economy, designers are urged to use stock lengths wherever possible and standard lengths of screws include the following: 3, 4, 5, 6, 8, 10, 12, 16, 20, 25, 30, 35, 40, 45, 50, 55, 60, 65, 70, 75, 80, 90, 100, 110, 120, 130, 140, 150, 160, 170, 180, 190, and 200 mm. If lengths over 200 mm are required, then increments of 20 mm are the preferred ISO lengths. It should be understood that not all diameters of screw are available in the above lengths. For example, the range of lengths for an M3 screw lies between 5 and 35 mm, for an M10 screw between 12 and 100 mm for one particular type of head. The same range will also not cover different types of head, hence the necessity to check stock lists (Fig. 16.4).

ISO metric socket cap screws (dimensions in Table 16.2)

These screws are distinguished by square knurling on the heads. Generally, the lengths of standard screws increase in increments of 5 mm and 10 mm, but the exact range should be checked from the manufacturer's catalogue.

ISO metric hexagon socket shoulder screws (dimensions in Table 16.3)

See Fig. 16.5.

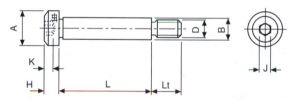

FIGURE 16.5 ISO metric hexagon socket button head screws. Dimensions in Table 16.4.

TABLE 16.2

Nominal size *D*	M3	M4	M5	M6	M8	M10	M12	M16	M20
Head diameter *A*	5.5	7	8.5	10	13	16	18	24	30
Head depth *H*	3	4	5	6	8	10	12	16	20
Key engagement *K*	1.3	2	2.7	3.3	4.3	5.5	6.6	8.8	10.7
Socket size *J*	2.5	3	4	5	6	8	10	14	17

TABLE 16.3

Screw thread diameter *D*	M5	M6	M8	M10	M12
Nominal shoulder diameter *B*	6	8	10	12	16
Head diameter *A*	10	13	16	18	24
Head height *H*	4.5	5.5	7	8	10
Socket size *J*	3	4	5	6	8
Nominal thread length L_t	9.75	11.25	13.25	16.4	18.4
Key engagement *K*	2.45	3.3	4.15	4.92	6.62

TABLE 16.4

Nominal size D	M3	M4	M5	M6	M8	M10	M12
Head diameter A	5.5	7.5	9.5	10.5	14	18	21
Head depth H	1.6	2.1	2.7	3.2	4.3	5.3	6.4
Key engagement K	2	2.5	3	4	5	6	8
Socket size J	1.04	1.3	1.56	2.08	2.6	3.12	4.16
Fillet radius							
F-minimum	0.1	0.2	0.2	0.25	0.4	0.4	0.6
d_a-maximum	3.6	4.7	5.7	6.8	9.2	11.2	14.2
S	0.38	0.38	0.5	0.8	0.8	0.8	0.8

ISO metric hexagon socket button head screws (dimensions in Table 16.4)

See Fig. 16.6.

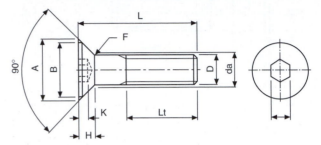

FIGURE 16.7

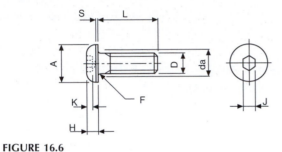

FIGURE 16.6

ISO metric socket countersunk head screws (dimensions in Table 16.5)

The basic requirement for countersunk head screws is that the head should fit into a countersunk hole with as great a degree of flushness as possible. Figure 16.7 and Table 16.5 give dimensions. To achieve this it is necessary for both the head of the screw and the countersunk hole to be controlled within prescribed limits. The maximum or design size of the head is controlled by a theoretical diameter to a sharp corner and the minimum head angle of 90°. The minimum head size is controlled by a minimum

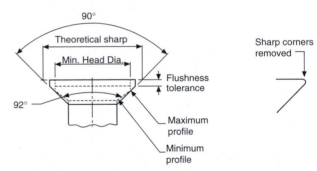

FIGURE 16.8 ISO metric hexagon socket set screws. Dimensions in Table 16.6.

head diameter, the maximum head angle of 92° and a flushness tolerance. The edge of the head may be flat, as shown in Fig. 16.8, or rounded but not sharp edged (see below).

TABLE 16.5

Nominal size D	M3	M4	M5	M6	M8	M10	M12	M16	M20
Head diameter									
A – maximum	6.72	8.96	11.2	13.44	17.92	22.4	26.88	33.6	40.32
B – minimum	5.82	7.78	9.78	11.73	15.73	19.67	23.67	29.67	35.61
Head depth H	1.86	2.48	3.1	3.72	4.96	6.2	7.44	8.8	10.16
Socket size J	2	2.5	3	4	5	6	8	10	12
Key engagement K	1.05	1.49	1.86	2.16	2.85	3.60	4.35	4.89	5.49
Fillet radius									
F – minimum	0.1	0.2	0.2	0.25	0.4	0.4	0.6	0.6	0.8
d_a – maximum	3.4	4.5	5.5	6.6	9	11	14	18	22

TABLE 16.6

Nominal size D	M3	M4	M5	M6	M8	M10	M12	M16	M20
Socket size J	1.5	2	2.5	3	4	5	6	8	10
Key engagement K	1.2	1.6	2	2.4	3.2	4	4.8	6.4	8
Dog point diameter P	2	2.5	3.5	4.5	6	7	9	12	15
Dog point length Q	0.88	1.12	1.38	1.62	2.12	2.62	3.12	4.15	5.15
Cup point and 'W' point diameter C	1.4	2	2.5	3	5	6	8	10	14

Note: The cone point angle shown as Y° is generally 118° for short screws and 90° for longer lengths.

ISO metric hexagon socket set screws (dimensions in Table 16.6)

These screws are available with a variety of pointed ends. In all cases the overall length includes the chamfer at the socket end and the point.

Machine screws

Head shapes for machine screws have been rationalized in BS EN ISO 1580 and BS EN ISO 7045. For the purpose of this British Standard, the generic term 'screws' applies to products which are threaded up to the head or having an unthreaded portion of the shank.

The length of the thread is defined as the distance from the end of the screw, and this includes any chamfer, radius or cone point, to the leading face of the nut which has been screwed as far as possible onto the screw by hand. Note on the illustrations which follow that in the case of the countersunk head types of screw, the length of the screw includes the countersunk part of the head. For pan and cheese head screws, the screw length does not include the head.

The Standard should be consulted for manufacturing dimensional tolerances, also concentricity tolerances for the heads of the screws.

The illustrations which follow show each of the screws and tables are also given showing the dimensions of regularly used sizes.

The sizes quoted in the tables are for screws manufactured in steel. Standard screws are also available in brass but generally the range is not quite so extensive.

FIGURE 16.10 Typical socket screws and wrench.

For all of the machine screws illustrated here, the countersunk head types have an included angle of 90° (Figs. 16.9–16.17).

'Posidriv' machine screws, countersunk and raised countersunk head (dimensions in Table 16.7)

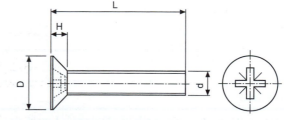

FIGURE 16.11 Countersunk head.

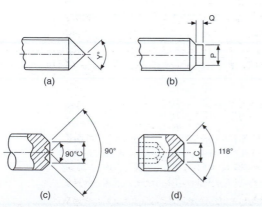

(a)   (b)

(c)   (d)

FIGURE 16.9 (a) Cone point (b) Dog point (c) W point (d) Cup point.

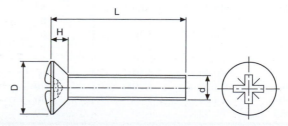

FIGURE 16.12 Raised countersunk head.

'Posidriv' machine screws, pan head (dimensions in Table 16.8)

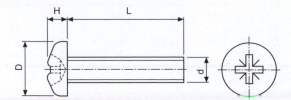

FIGURE 16.13   Pan head.

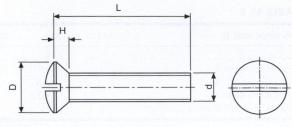

FIGURE 16.15   Raised countersunk head.

Slotted machine screws, countersunk and raised countersunk head (dimensions in Table 16.9)

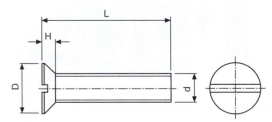

FIGURE 16.14   Countersunk head.

Slotted machine screws, pan head (dimensions in Table 16.10)

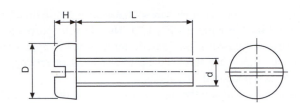

FIGURE 16.16   Pan head.

TABLE 16.7

Diameter d	Diameter of head D	Depth of head H	Driver number	Length L											
M2	4.40	1.20	1	4	5	6	8	10	12						
M2.5	5.50	1.50	1	5	6	8	10	12	16	20	25				
M3	6.30	1.65	1	5	6	8	10	12	16	20	25				
M3.5	7.35	1.93	2	6	8	10	12	16	20	25	30				
M4	8.40	2.20	2	6	8	10	12	16	20	25	30				
M5	10.00	2.50	2	6	8	10	12	16	20	25	30	35	40	45	50
M6	12.00	3.00	3	10	12	16	20	25	30	35	40	45	50	55	60
M8	16.00	4.00	4	12	16	20	25	30	40	50	60				
M10	20.00	5.00	4	16	20	25	30	40	50	60					

TABLE 16.8

Diameter d	Diameter of head D	Depth of head H	Driver number	Length L											
M2	4.00	1.60	1	4	5	6	8	10	12						
M2.5	5.00	1.75	1	5	6	8	10	12	16	20	25				
M3	6.00	2.10	1	5	6	8	10	12	16	20	25				
M3.5	7.00	2.45	2	6	8	10	12	16	20	25	30				
M4	8.00	2.80	2	5	6	8	10	12	16	20	25	30			
M5	10.00	3.50	2	6	8	10	12	16	20	25	30	35	40	45	50
M6	12.00	4.20	3	10	12	16	20	25	30	35	40	45	50	55	60
M8	16.00	5.60	4	16	20	25	30	40	50						
M10	20.00	7.00	4	20	25	30	40								

TABLE 16.9

Diameter d	Diameter of head D	Depth of head H	Length L
M2	4.40	1.20	5 6 8 10 12
M2.5	5.50	1.50	5 6 8 10 12 16 20 25
M3	6.30	1.65	5 6 8 10 12 16 20 25 30 35 40 45 50
M3.5	7.35	1.93	5 6 8 10 12 16 20 25 30 35 40 45 50 60
M4	8.40	2.20	5 6 8 10 12 16 20 25 30 35 40 45 50 60 70
M5	10.00	2.50	6 8 10 12 16 20 25 30 35 40 45 50 60 70 80
M6	12.00	3.00	8 10 12 16 20 25 30 35 40 45 50 60 70 80 90
M8	16.00	4.00	10 12 16 20 25 30 35 40 45 50 60 70 80 90
M10	20.00	5.00	16 20 25 30 35 40 45 50 55 60 70 80 90

TABLE 16.10

Diameter d	Diameter of head D	Depth of head H	Length L
M2	4.00	1.20	4 5 6 8 10 12
M2.5	5.00	1.50	5 6 8 10 12 16 20 25
M3	6.00	1.80	5 6 8 10 12 16 20 25 30 35 40 45 50
M3.5	7.00	2.10	6 8 10 12 16 20 25 30 35 40 45 50
M4	8.00	2.40	5 6 8 10 12 16 20 25 30 35 40 45 50
M5	10.00	3.00	6 8 10 12 16 20 25 30 35 40 45 50 55 60 70 80
M6	12.00	3.60	8 10 12 16 20 25 30 35 40 45 50 55 60 70 80
M8	16.00	4.80	10 12 16 20 25 30 40 50 60 70 80 90
M10	20.00	6.00	16 20 25 30 40 50 60 70

TABLE 16.11

Diameter d	Diameter of head D	Depth of head H	Length L
M2	3.80	1.30	3 4 5 6 8 10 12 16 20 25
M2.5	4.50	1.60	5 6 8 10 12 16 20 25 30
M3	5.50	2.00	4 5 6 8 10 12 16 20 25 30 35 40 45 50
M3.5	6.00	2.40	5 6 8 10 12 16 20 25 30 35 40 45 50 60 70
M4	7.00	2.60	5 6 8 10 12 16 20 25 30 35 40 45 50 60 70
M5	8.50	3.30	6 8 10 12 16 20 25 30 35 40 45 50 60 70 80 90
M6	10.00	3.90	8 10 12 16 20 25 30 35 40 45 50 60 70 80 90
M8	13.00	5.00	10 12 16 20 25 30 35 40 45 50 60 70 80 90
M10	16.00	6.00	16 20 25 30 35 40 45 50 60 70 80 90

Slotted machine screws, cheese head (dimensions in Table 16.11)

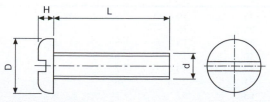

FIGURE 16.17   Cheese head.

Machine screw nuts

A range of machine screw nuts is covered by BS EN ISO 1580 and BS EN ISO 7045 and these nuts are manufactured in two different patterns, square and hexagon. The table shows typical nuts for use with the screws previously described (Table 16.12 and Figs. 16.18 and 16.19).

TABLE 16.12 Machine screw nuts, pressed type, square and hexagonal

Nominal size of nut *d*	Width across flats *s*	Width across corners *e*		Thickness *M*
		Square	Hexagon	
M2	4.0	5.7	4.6	1.2
M2.5	5.0	7.1	5.8	1.6
M3	5.5	7.8	6.4	1.6
M3.5	6.0	8.5	6.9	2.0
M4	7.0	9.9	8.1	2.0
M5	8.0	11.3	9.2	2.5
M6	10.0	14.1	11.5	3.0
M8	13.0	18.4	15.0	4.0
M10	17.0	24.0	19.6	5.0

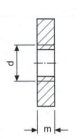
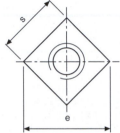

FIGURE 16.18   Square nut.

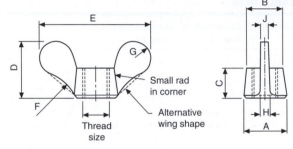

FIGURE 16.20   Sizes are in Table 16.13.

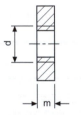
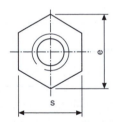

FIGURE 16.19   Hexagon nut.

Wing nuts

Figure 16.20 shows the dimensions of preferred sizes of wing nuts manufactured in brass or malleable iron by a hot stamping or casting process (Table 16.13).

An alternative wing nut is available in brass or malleable iron and manufactured by cold forging.

Locking and retaining devices

The function of a locking device is to prevent loosening or disengagement of mating components which may be operating under varying conditions of stress, temperature and vibration. The effectiveness of the locking device may be vital to safety.

One of the simplest locking devices is a locknut and these are generally thin plain nuts which are tightened against ordinary plain nuts or against components into which male threaded items are assembled. To ensure efficient locking, the bearing surfaces of the nut and component must bed together evenly and the correct degree of tightness obtained by applying the designed torque loading. The locknut should not be overtightened as this may result in the stripping of the nut threads or overstressing of the male component. In cases

TABLE 16.13

Thread size	A	B	C	D	E	F	G	H	J
M3	9	6.5	7	13.5	22	19	3.5	2.5	1.5
M4 and M5	10	8	9	15	25.5	19	4	2.5	1.5
M6	13	9.5	11	18	30	19	5	2.5	1.5
M8	16	12	13	23	38	19	6.5	3	2.5
M10	17.5	14	14	25.5	44.5	19	7	5	3
M12	19	16	15	28.5	51	25.5	8	5	3
M16	25.5	20.5	19	36.5	63.5	32	10	6.5	5

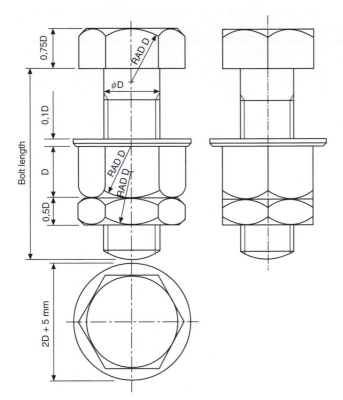

FIGURE 16.21   Bolts, nut and locknut.

where rotation can occur, the plain nut must be held stationary whilst the locknut is tightened (Fig. 16.21).

Slotted nuts and castle nuts

One method of preventing nuts from coming loose is to drill the bolt and use a pin through the assembly. Suitable nuts are shown in Fig. 16.22. Slotted nuts are available for sizes M4 to M39 and have six slots. Castle nuts are also available with six slots between sizes M12 to M39 and eight slots between sizes M42 and M68. For convenience in drawing both types of nuts, the total thickness can be approximated to the thread diameter plus 2 mm. The dimensions for the hexagons can be taken from Table 16.1.

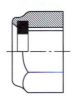

(a) Slotted nut   (b) Castle nut

FIGURE 16.22

Slotted nuts are reusable but difficult to apply where access is limited.

Simmond's locknut

This type of locknut incorporates a collar manufactured from nylon or fibre and the collar is slightly smaller in diameter than the internal thread diameter. The section in Fig. 16.23 shows the collar in black. On assembly, the stud or bolt forces its way through the resilient collar which provides a frictional lock. The locknut is a little thicker than a conventional nut.

FIGURE 16.23

Spring washers

This type of washer is produced as a single or a double coil spring. The cross-section is rectangular. Generally this type of washer dispenses with the simple plain washer although a plain washer can be used at the same time with assemblies where the component is manufactured from relatively soft-light alloys. The free height of double coil washers before compression is normally about five times the thickness of the steel section (Table 16.14 and Fig. 16.24).

TABLE 16.14 Double coil rectangular section spring washers to BS 4464

Nominal thread diameter	Maximum inside diameter I/D	Maximum outside diameter O/D	Thickness S	Free height H
M2	2.4	4.4	0.50	2.50
M3	3.6	6.2	0.80	4.00
M4	4.6	8.0	0.80	4.00
M5	5.6	9.8	0.90	4.50
M6	6.6	12.9	1.00	5.00
M8	8.6	15.1	1.20	6.00
M10	10.8	18.2	1.20	6.00
M12	12.8	20.2	1.60	8.00
M16	17.0	27.4	2.00	10.00
M20	21.5	31.9	2.00	10.00
M24	26.0	39.4	3.25	16.25
M30	33.0	49.5	3.25	16.25
M36	40.0	60.5	3.25	16.25

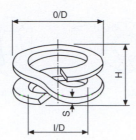

FIGURE 16.24

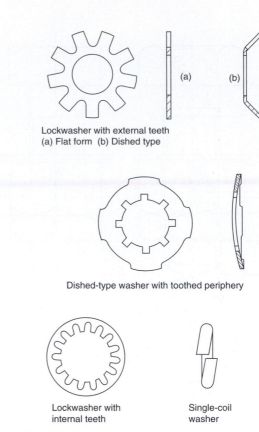

Lockwasher with external teeth
(a) Flat form (b) Dished type

Dished-type washer with toothed periphery

Lockwasher with internal teeth   Single-coil washer

FIGURE 16.25   Types of locking washer.

Shakeproof washers

This type of washer is generally made from spring steel and serrations are formed on either the internal or external diameters. These serrations then bite into the pressure faces between the nut and the component when the nut is assembled. Some slight disfiguration of the component may result on assembly but this is of little significance except where anti-corrosion treatment of the component surface has previously been carried out. Some screws are pre-assembled with conical lockwashers which are free to rotate but do not come off (Fig. 16.25).

Toothed lockwashers combat vibration and are especially suited to rough parts or surfaces.

Wire locking

Non-corrodible steel and brass wire, of the appropriate gauge, are normally used for wire locking. Generally, a hole is provided for this purpose in the component to be locked and the wire is passed through and twisted. The lay of the wire between the anchorage and the component must always be such as to resist any tendency of the locked part or parts to become loose.

Figure 16.26 shows the plan view of a pressurized cylinder and the cover is held down by four bolts which are wire locked. The operation is performed with a single strand of wire. The wire is passed in sequence through the holes in the bolts and the ends are twisted together to tension the wire loop. Note, that in order to become loose, the bolts must turn in an anticlockwise direction but this will have the effect of increasing the tension in the wire loop. The locking wire should only be used once.

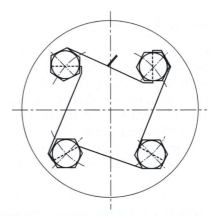

FIGURE 16.26

Tab washers

Tab washers are thin metal washers designed with two or more tabs which project from the external diameter. On assembly, a tab is bent against the component or sometimes into a hole in the component. Other tabs are then bent up against the correctly tightened nut. Another pattern has a tab projecting from the inside diameter and this is intended to fit into a slot machined in the bolt, whilst the external tabs are again bent against the flat sides of the nut. The deformation

of the tab washer is such that it is intended to be used only once.

Three different types of tab washer are shown in Fig. 16.27 together with, a typical assembly.

Locking plates

Locking plates are manufactured usually from mild steel and fit over hexagonal nuts after these have been tightened on assembly. The locking plate is then secured on the component by a separate screw which may itself be fitted with a shakeproof or spring type of washer.

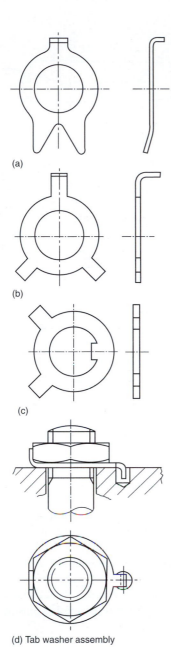

(a)

(b)

(c)

(d) Tab washer assembly

FIGURE 16.27

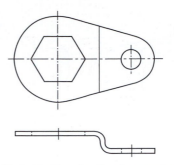

FIGURE 16.28 Typical locking plate for a hexagonal nut.

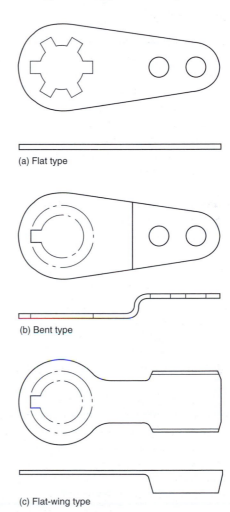

(a) Flat type

(b) Bent type

(c) Flat-wing type

FIGURE 16.29 Locking terminals.

Locking plates may be used repeatedly, provided they remain a good fit, around the hexagon of the nut or the bolthead. Locking plates may be cranked, as in Fig. 16.28 or flat.

Figure 16.29 shows a selection of locking terminals where a 'shakeproof' washer and a soldering lug are combined into one unit, thus saving assembly time. The locking teeth anchor the terminal to the base to prevent shifting of the terminal in handling, while the twisted teeth produce a multiple bite which penetrates an oxidized or painted surface to ensure good conductivity. All three types of locking terminal are generally made from phosphor bronze with a hot-tinned finish.

Taper pins and parallel pins

Taper pins, with a taper of 1 in 50, and parallel pins are used on both solid and tubular sections to secure, for example, levers to torque shafts and control rods to fork ends. Some taper pins are bifurcated, or split, and the legs can be opened out for locking. Plain taper pins and parallel pins may also be locked by peening. To prevent slackness, these pins are assembled in accurate reamed holes. Undue force should not be used during the peening process or the security of the fittings may be impaired if the pin is bent.

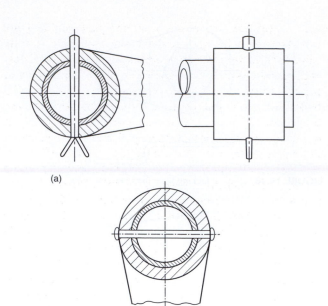

(a)

(b)

FIGURE 16.30

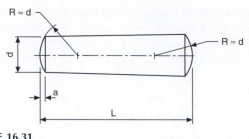

FIGURE 16.31

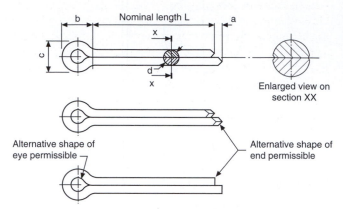

FIGURE 16.32 Proportions of split cotter pins to BS 1574.

Figure 16.30(a) shows part of a lever which is fixed to a hollow operating shaft by a bifurcated taper pin. On assembly, a hole is drilled which is slightly smaller than the diameter at the small end of the taper pin and this is enlarged by a taper pin reamer so that the small end of the taper pin, when pushed through the assembly, is flush with the surface. The pin is then driven into position. If the pin is of the bifurcated type, then the legs are spread to form an included angle of about 60°. Figure 16.30(b) shows the same operating lever assembled, but using a parallel pin, which has been peened over after ensuring that the component is adequately supported.

Figure 16.31 shows the general shape of a taper pin. Parallel sides are substituted for tapered sides in parallel pins.

Split cotter pins

Ferrous and non-ferrous split cotter pins are covered by BS 1574. The designating size of a split cotter pin is the size of the hole for which it is intended to fit. When reference is made to a split cotter pin in a parts list, this nominal dimension is followed by the length required. The closed legs of the shank of the pin form a circular cross-section. The legs should be straight and parallel throughout their nominal length. Figure 16.32 shows alternative pins in detail (Table 16.15).

TABLE 16.15 Split cotter pin dimensions

Nominal diameter of pin (hole) diameter	Shank diameter d	Outside diameter of eye C	Length of eye B	Length extended prong a
1.0	0.9	1.8	3.0	1.6
1.2	1.0	2.0	3.0	2.5
1.6	1.4	2.8	3.2	2.5
2.0	1.8	3.6	4.0	2.5
2.5	2.3	4.6	5.0	2.5
3.2	2.9	5.8	6.4	3.2
4.0	3.7	7.4	8.0	4.0
5.0	4.6	9.2	10.0	4.0
6.3	5.9	11.8	12.6	4.0
8.0	7.5	15.0	16.0	4.0
10.0	9.5	19.0	20.0	6.3
13.0	12.4	24.8	26.0	6.3

Exact dimensions are given in BS 1574 but for drawing purposes, the following extracts will be useful where the maximum allowable sizes for dimensions *a*, *b*, *c*, and *d* are quoted. Many standard lengths are obtainable for each pin size and as a rough guide between 5 and 25 times the shank diameter.

Locking by adhesives

Small components found in, for example, instruments and switches may be locked by the application of Shellac, Araldite, Loctite, or similar materials. Shellac and Loctite are usually applied to the threads of nuts, bolts, screws and studs and the components are assembled while still wet. The parts should be free from grease to achieve maximum strength. Araldite is applied to the outside of the nut face and the protruding screw thread, after tightening. Araldite is an adhesive which hardens, after mixing, within a specified time period.

Peening

This operation prevents re-use of the screw or bolt but locking can be carried out by peening over about $1\frac{1}{2}$ threads. This practice can be used in the case of screwed pivots and a simple example is often found in a pair of scissors. In the case of nuts and bolts, peening is carried down to the nut to prevent it from slackening.

Countersunk screws may be locked by peening metal from the surroundings into the screw slot. This practice is sometimes adopted when the thread is inaccessible.

Thread-cutting screws

Barber and Colman Ltd. are the manufacturers of 'shakeproof' thread-cutting screws and washers.

'Shakeproof' thread-cutting screws made from carbon steel are subjected to a special heat-treatment which provides a highly carburized surface with a toughened resilient core. The additional strength provided enables higher tightening torques to be used, and will often permit the use of a smaller size thread-cutting screw than would normally be specified for a machine screw. Thread-cutting screws actually cut their own mating thread; in any thickness of material a perfect thread-fit results in greatly increased holding power, extra vibration–resistance, and a faster assembly. The hard, keen cutting edge produces a clean-cut thread, from which the screw can be removed, if desired, without damage to screw or the cut thread. The most suitable drill sizes for use with these screws are generally larger than standard tapping-drill sizes, but this apparent loss of thread engagement is more than offset by the perfect thread-fit obtained.

Both the screws shown in Fig. 16.33 are interchangeable with standard machine screws. Type 1 is recommended for use in steel and non-ferrous sheet and plate, and they are manufactured with a wide shank slot and are eminently suitable for paint-clearing applications, as they completely eliminate the need for expensive pre-production tapping of painted assemblies. Type 23 screws incorporate a special wide cutting slot with an acute cutting angle for fast, easy thread-cutting action and ample swarf clearance. These screws are specially designed for application into soft metals or plastics where a standard thread form is required.

The Type 25 thread-cutting screw has a specially spaced thread form which is designed for fast efficient fastening into plastics and sheet–metal applications (Fig. 16.34).

Figure 16.35 illustrates a 'Teks' self-drilling screw which, with a true drilling action, embodies three basic operations in one device. It (1) prepares its own hole, (2) either cuts or forms a mating thread, and (3) makes a complete fastening in a single operation. These screws consist of an actual drill point to which a threaded screw-fastener has been added. Several different head styles are available. During the drilling stage, Teks must be supported rigidly from the head. Some bench-mounted, automatically fed screwdrivers provide a holding means which retracts as the screw is finally driven home. Other drivers connect with the fastener

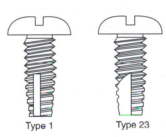

Type 1   Type 23

FIGURE 16.33

FIGURE 16.34 Type 25 thread-cutting screw.

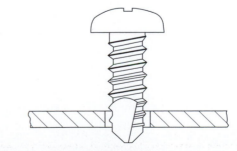

FIGURE 16.35 'Teks' self-drilling screw.

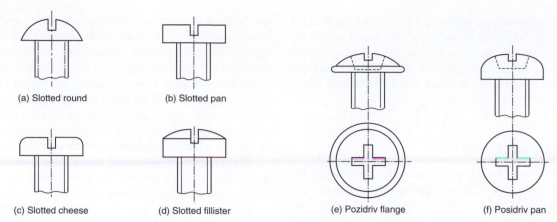

(a) Slotted round   (b) Slotted pan

(c) Slotted cheese   (d) Slotted fillister   (e) Pozidriv flange   (f) Posidriv pan

FIGURE 16.36

only through the bit or socket. A good-fitting Phillips or Posidriv bit will normally drive several thousands of these screws, and a hex socket, for hex-head designs, will drive even more.

For long screws or applications requiring absolutely guaranteed driving stability, a special chuck is available which holds the screw with three fingers and retracts upon contacting the work surface. These screws are suitable for fastening sheet steel of 16 gauge, or thicker, within 5 s maximum while using a power tool.

Figure 16.36 shows alternative head styles available for thread-cutting screws.

The examples and dimensions of nuts, bolts, screws and washers given here are intended especially to be of use to students engaged on design projects. There are however literally hundreds of industrial fastening systems available, associated with automobile, construction, electronics and aerospace developments. Manufacturers' catalogues are freely available to provide technical specifications and necessary details for designers. One further advantage of CAD systems is that such information can be used to build a library of useful data and drawings, which are invaluable, where contract drawings use a repetition of similar parts.

Chapter 17

Keys and key ways

A key, Fig. 17.1, is usually made from steel and is inserted between the joint of two parts to prevent relative movement; it is also inserted between a shaft and a hub in an axial direction, to prevent relative rotation. A keyway, Figs. 17.2, 17.3 and 17.4, is a recess in a shaft or hub to receive a key, and these recesses are commonly cut on key-seating machines or by broaching, milling, planning, shaping and slotting. The proportions of cross-sections of keys vary with

FIGURE 17.4   Keyway in hub.

the shaft size, and reference should be made to BS 4235 for the exact dimensions. The length of the key controls the area of the cross-section subject to shear, and will need to be calculated from the knowledge of the forces being transmitted or, in the case of feather keys, the additional information of the length of axial movement required.

Sunk keys

Examples of sunk keys are shown in Fig. 17.5, where the key is sunk into the shaft for half its thickness. This measurement

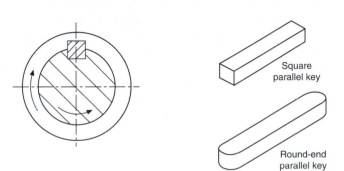

Square parallel key

Round-end parallel key

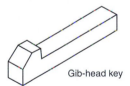

Gib-head key

FIGURE 17.1

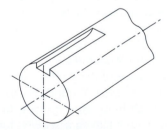

FIGURE 17.2   Edge-milled keyway.

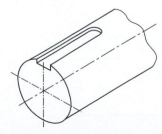

FIGURE 17.3   End-milled keyway.

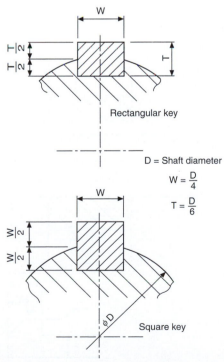

Rectangular key

D = Shaft diameter

$$W = \frac{D}{4}$$

$$T = \frac{D}{6}$$

Square key

FIGURE 17.5

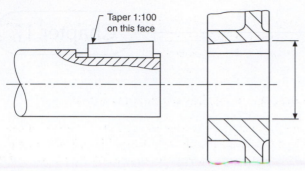

FIGURE 17.6

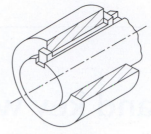

FIGURE 17.8   Double-headed feather key.

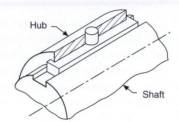

FIGURE 17.9   Peg feather key.

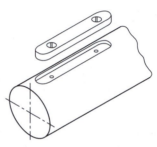

FIGURE 17.10   Feather key.

is taken at the side of the key, and not along the centre line through the shaft axis. Figure 17.5 shows useful proportions used for assembly drawings.

Square and rectangular keys may be made with a taper of 1 in 100 along the length of the key; Fig. 17.6 shows such an application. Note that, when dimensioning the mating hub, the dimension into the keyway is taken across the maximum bore diameter.

A *gib head* may be added to a key to facilitate removal, and its proportions and position when assembled are given in Fig. 17.7.

A *feather key* is attached to either the shaft or the hub, and permits relative axial movement while at the same time enabling a twisting moment to be transmitted between shaft and hub or vice versa. Both pairs of opposite faces of the key are parallel.

A *double-headed feather key* is shown in Fig. 17.8 and allows a relatively large degree of sliding motion between shaft and hub. The key is inserted into the bore of the hub, and the assembly is then fed on to the shaft, thus locking the key in position.

A *peg feather key* is shown in Fig. 17.9, where a peg attached to the key is located in a hole through the hub.

Figure 17.10 illustrates a feather key which is screwed in position in the shaft keyway by two countersunk screws.

Woodruff keys

A Woodruff key, Fig. 17.11, is a segment of a circular disc and fits into a circular recess in the shaft which is machined by a Woodruff key way cutter. The shaft may be parallel or tapered, Figs. 17.12 and 17.13 showing the method of dimensioning shafts for Woodruff keys where the depth of the recess from the outside of the shaft is given, in addition to the diameter of the recess. A Woodruff key has the advantage that it will turn itself in its circular recess to accommodate any taper in the mating hub on assembly; for this reason it cannot be used as a feather key, since it would jam. Woodruff

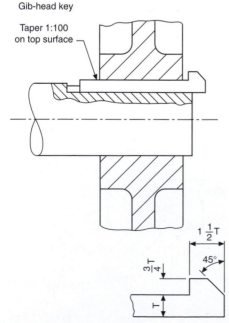

FIGURE 17.7

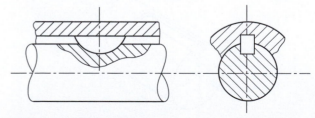

FIGURE 17.11   Woodruff key.

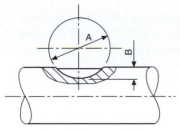

FIGURE 17.12 Dimensions required for a Woodruff key in a parallel shaft.

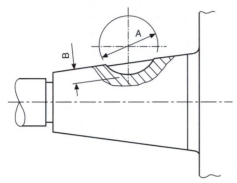

FIGURE 17.13 Dimensions required for a Woodruff key in a tapered shaft.

keys are commonly used in machine tools and, for example, between the flywheel and the crankshaft of a small internal-combustion engine where the drive depends largely on the fit between the shaft and the conically bored flywheel hub. The deep recess for a Woodruff key weakens the shaft, but there is little tendency for the key to turn over when in use.

Where lighter loads are transmitted and the cost of cutting a keyway is not justified, round keys and flat or hollow saddle keys as shown in Fig. 17.14 can be used.

Saddle keys are essentially for light duty only, overloading tending to make them rock and work loose on the shaft. Both flat and hollow saddle keys may have a taper of 1 in 100 on the face in contact with the hub. The round key may either

be tapered or, on assembly, the end of the shaft and hub may be tapped after drilling and a special threaded key be screwed in to secure the components.

Dimensioning keyways (parallel keys)

The method of dimensioning a parallel shaft is shown in Fig. 17.15, and a parallel hub in Fig. 17.16. Note that in each case it is, essential to show the dimension to the bottom of the keyway measured across the diameter of the shaft and the bore of the hub. This practice cannot be used where, either the shaft or hub is tapered, and Fig. 17.17 shows the method

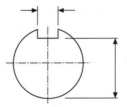

FIGURE 17.15 Keyway in parallel shaft.

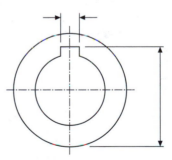

FIGURE 17.16 Keyway in parallel hub.

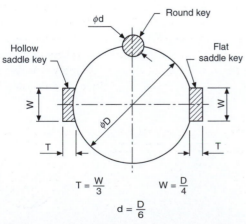

$$T = \frac{W}{3} \qquad W = \frac{D}{4}$$
$$d = \frac{D}{6}$$

FIGURE 17.14

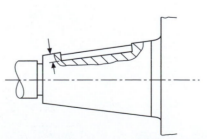

FIGURE 17.17 Keyway for square or rectangular parallel key in tapered shaft.

of dimensioning a keyway for a square or rectangular parallel key in a tapered shaft, where the keyway depth is shown from the outside edge of the shaft and is measured vertically into the bottom of the slot. Figure 17.18 shows a tapered hub with a parallel keyway where the dimension to the bottom of the slot is taken across the major diameter. A parallel hub utilizing a tapered key is also dimensioned across the major diameter, as indicated in Fig. 17.19.

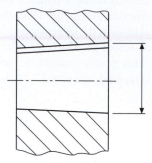

FIGURE 17.18 Tapered hub with parallel keyway.

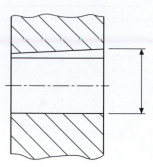

FIGURE 17.19 Parallel hub with tapered keyway.

Chapter 18

Worked examples in machine drawing

Examination questions generally relate to single parts or assemblies of detailed components, and test the students' ability to draw sectional and outside views. British Standards refer to 'views' but other terms have been traditionally used in technical drawing. The front or side view of a house is quite likely to be known as an elevation and a 'bird's eye view' as a plan. These expressions are freely used.

The examples which follow are of examination standard and a student draughtsman would be expected to produce a reasonably complete solution for each problem in about 2–3 h.

Before commencing, try and estimate the areas covered by the views so that they can be presented with reasonably equal spaces horizontally and vertically on the drawing sheet. Include a border about 15 mm width, add a title block and parts list if necessary. Note that attention to small details

will gradually enable you to improve the quality of your draughtsmanship.

In an industrial situation, before commencing a drawing, the draughtsman will make a mental picture of how to orient the component, or arrangement, so that the maximum amount of information can be indicated with the minimum number of views necessary to produce a clear unambiguous solution. However, this is easier said than done in the case of the student, and especially where the drawing is being made on CAD (computer aided design) equipment since the size of the screen often means that part of the drawing is temporarily out of sight. This is part of the learning experience.

Copy the following solutions and try to appreciate the reasons for the position of every line which contributes to the finished drawing.

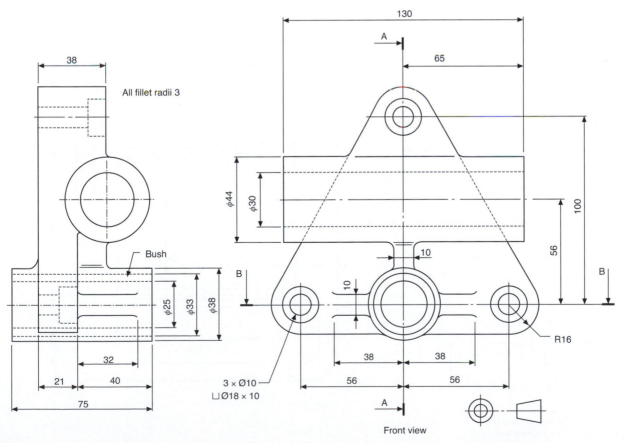

Front view

FIGURE 18.1

Bushed bearing bracket

Front- and end-views of a bushed bearing bracket are shown in Fig. 18.1. Copy the given front view and project from it a sectional-end view and a sectional-plan view taken from cutting planes A–A and B–B. Draw your solution in first angle projection. Note that the question is presented in third angle projection.

Drill table

Figure 18.2 shows details of a table for a drilling machine. Draw half-full size the following views:

(a) A front view taken as a section along the cutting plane A–A.

(b) The given plan view with hidden detail.

(c) An end view projected to the left of the front view with hidden detail included.

Draw your solution in first angle projection and add the title and projection symbol in a suitable title block.

Cam operated clamp

Details of a cam operated clamp are shown in Fig. 18.3 together with a key showing the position of the various components in the assembly. Draw the following views in first angle projection:

(a) A front view taken as a section through the vertical centre line of the clamp and generally as outlined in the key.

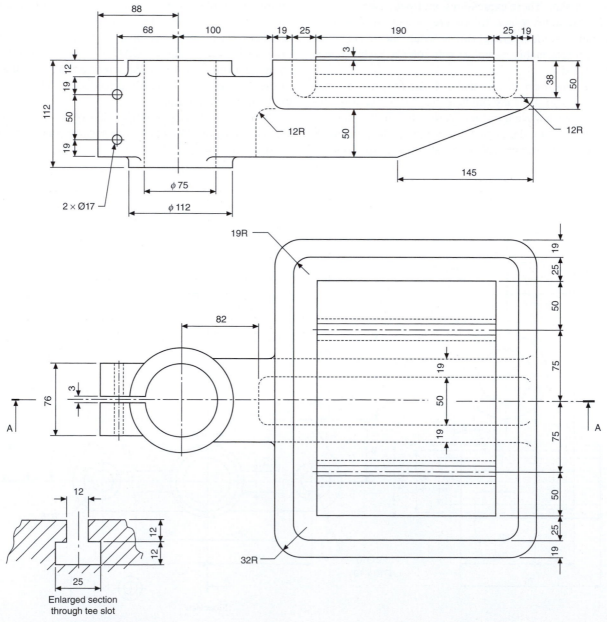

FIGURE 18.2 *Note*: All unspecified radii 6.

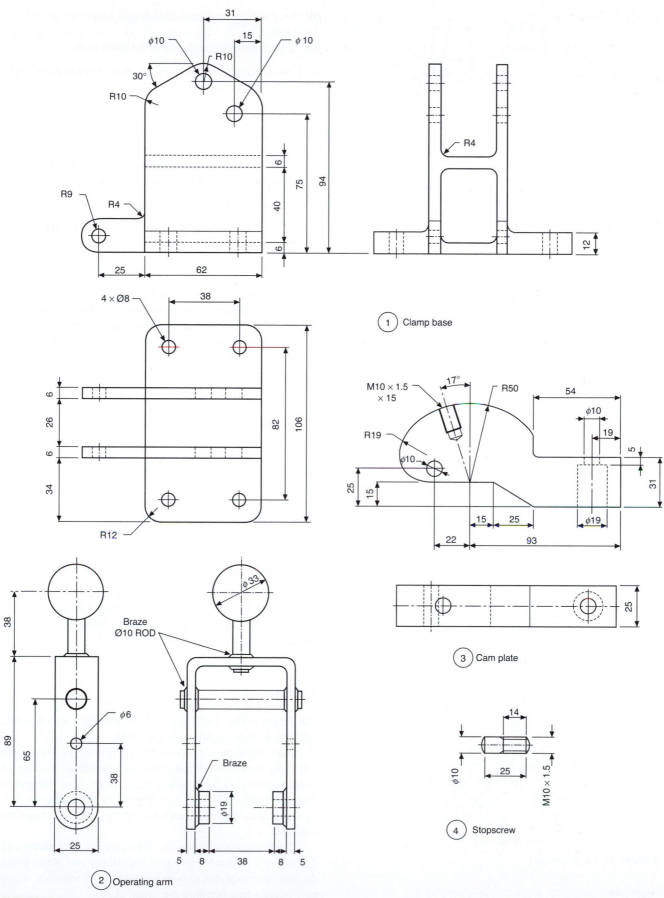

① Clamp base

③ Cam plate

④ Stopscrew

② Operating arm

FIGURE 18.3

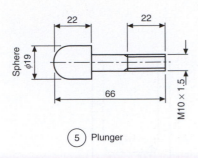

5 Plunger

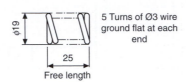

5 Turns of Ø3 wire ground flat at each end

25
Free length

6 Spring

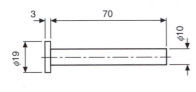

3

70

Ø19

Ø10

7 Arm pivot

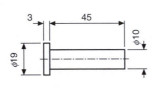

3

45

Ø19

Ø10

8 Cam pivot

Note: On assembly, fit a suitable washer over each pivot and drill for 2 mm Ø split pin

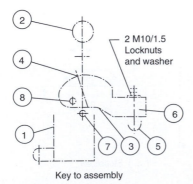

2

4

8

1

7 3 5

2 M10/1.5
Locknuts and washer

6

Key to assembly

FIGURE 18.3 (*Continued*)

(b) An end view projected on the right-hand side of the front view.

(c) A plan view drawn beneath the front view.

Add to your solution a title, scale, projection symbol, parts list and reference balloons.

Plug cock

Details of a plug cock are shown in Fig. 18.4. Draw the following two views of the assembled plug cock:

(a) A view in half section showing the outside view on the left of the vertical centre line and a section on the right-hand side, the position of the valve body being similar to that shown on the given detail.

(b) An outside-end view.

Add a title and a parts list. Hidden details are not required in either view. Include on your assembly suitable nuts and bolts and washers where applicable. Between the valve body and cover include a joint ring of thickness 2 mm. The gland should be positioned entering 10 mm into the valve cover. First or third angle projected views will be acceptable.

Air engine

The component parts of an oscillating air engine are detailed in Fig. 18.5. Draw in first angle projection and, at twice full size, the following views:

(a) A front view taken as a section through the engine cylinder and flywheel.

(b) A plan view in projection with the front view and drawn below the front view.

(c) An end view on the right-hand side and in projection with the front view.

Hidden detail is not required in any view. Add a parts list to your drawing and reference balloons to identify each of the components in the assembly.

Toolbox

The details in Fig. 18.6 show parts of a toolbox which can be made as a sheet–metal-work exercise.

A development is given of the end plates (two required) and these should be drilled at the corners before bending, to avoid possible cracking taking place. Note, from the photograph, that the ends have been pressed with grooves to provide additional stiffening and this can be done if facilities are available.

The bottom and sides are fabricated in one piece and drillings are shown for pop rivets. Drill the end plates using the sides as templates to take account of deformation which may occur while bending the metal plate.

FIGURE 18.4

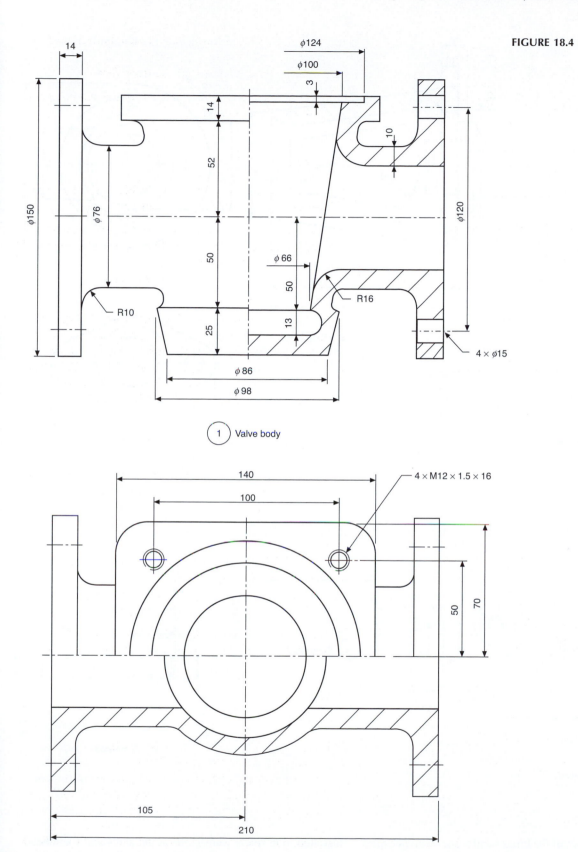

1 Valve body

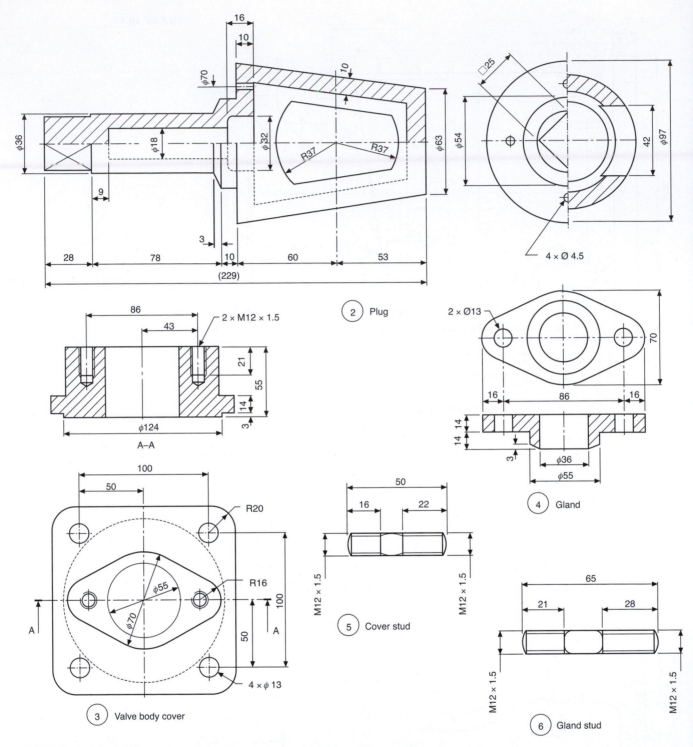

2 Plug

4 Gland

3 Valve body cover

5 Cover stud

6 Gland stud

FIGURE 18.4 (*Continued*)

Details are shown of the hinges and some clearance must be provided for their satisfactory operation. Before drilling the sides and lids for the hinges using the dimensions given, use the hinges as manufactured to check the hole centres for the rivets. One lid must overlap the other to close the box, so some adjustment here may be required. On the toolbox il-lustrated, pop rivets were used on the sides and roundhead rivets on the lid. Roundhead rivets were also used to assem-ble the handle clips. Check also the drill sizes for the rivets used as alternatives are available.

The top support member was pop riveted to the end plates after the catch had been welded on the underside. The slot in

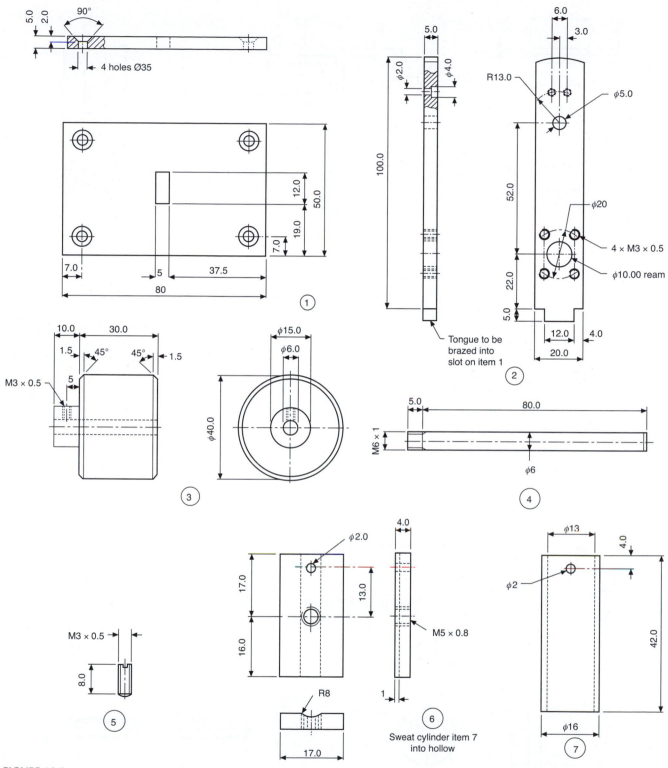

FIGURE 18.5

each lid slips over this catch to allow a padlock to secure the toolbox.

A tooltray may also be made if required and details are provided. The tray supports also offer added stability and were pop riveted on assembly.

Remove all sharp corners and edges before assembly of each part. The completed toolbox can be stove enamelled, painted or sprayed according to personal choice.

As an additional exercise, copy the given details and produce an assembly drawing complete with parts list. Also,

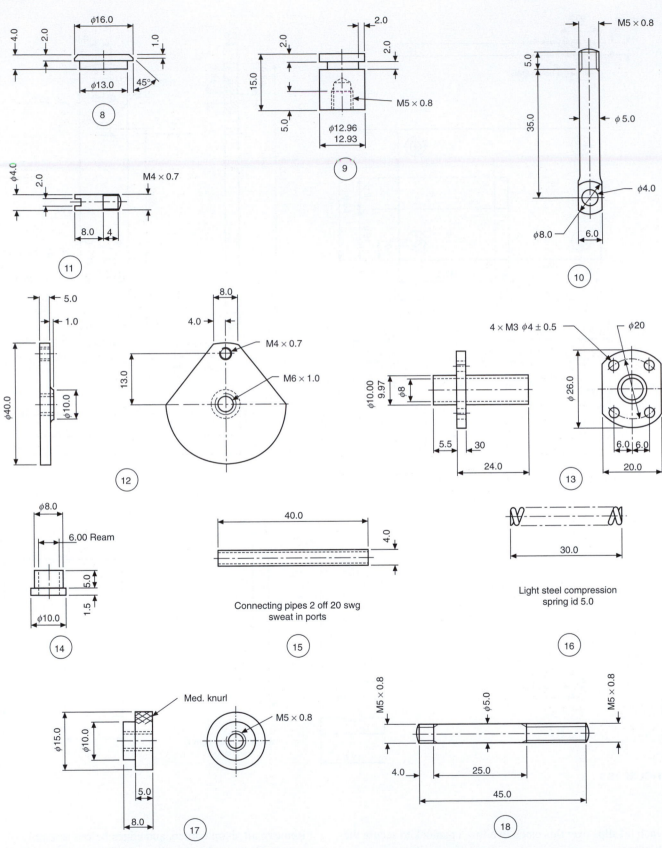

FIGURE 18.5 (Continued)

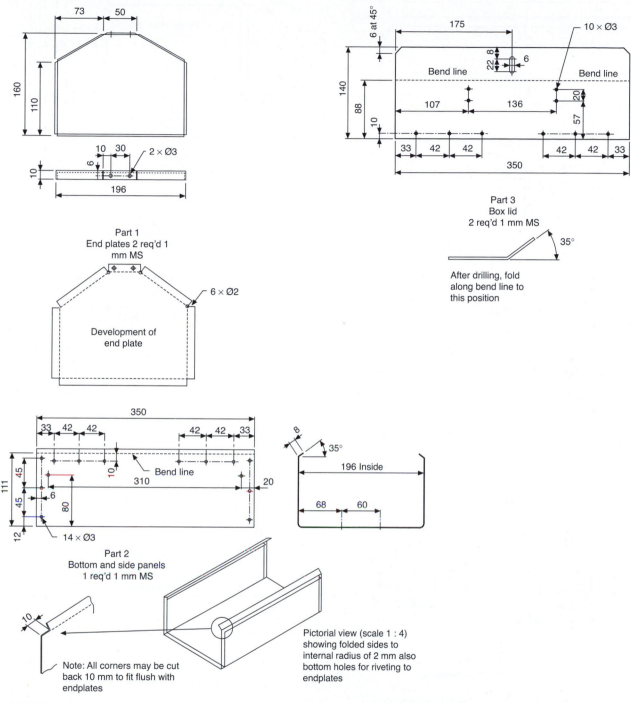

Part 1
End plates 2 req'd 1
mm MS

Development of
end plate

6 × Ø2

2 × Ø3

Part 3
Box lid
2 req'd 1 mm MS

35°

After drilling, fold
along bend line to
this position

10 × Ø3

Bend line

Bend line

Part 2
Bottom and side panels
1 req'd 1 mm MS

Bend line

14 × Ø3

196 Inside

35°

Pictorial view (scale 1 : 4)
showing folded sides to
internal radius of 2 mm also
bottom holes for riveting to
endplates

Note: All corners may be cut
back 10 mm to fit flush with
endplates

FIGURE 18.6

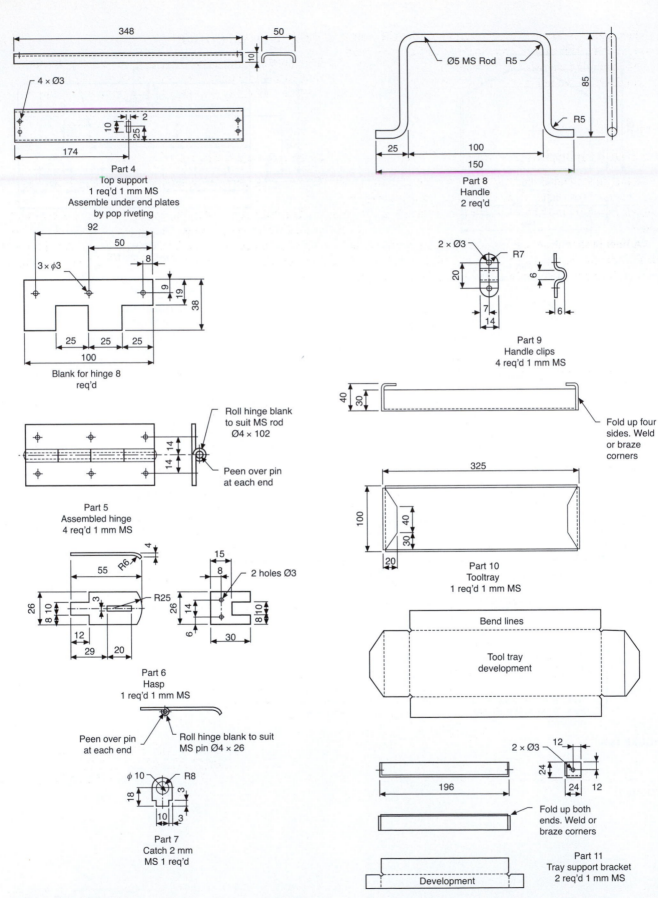

348

50

10

4 × Ø3

2

10

25

174

Part 4
Top support
1 req'd 1 mm MS
Assemble under end plates
by pop riveting

Ø5 MS Rod   R5

85

R5

25

100

150

Part 8
Handle
2 req'd

92

50

8

3 × φ3

9

19

38

25   25   25

100

Blank for hinge 8
req'd

2 × Ø3   R7

20

7

14

6

6

Part 9
Handle clips
4 req'd 1 mm MS

Roll hinge blank
to suit MS rod
Ø4 × 102

14

14

Peen over pin
at each end

Part 5
Assembled hinge
4 req'd 1 mm MS

40

30

Fold up four
sides. Weld
or braze
corners

325

100

40

30

20

Part 10
Tooltray
1 req'd 1 mm MS

4

55

R6

15

8

2 holes Ø3

26

8   10

3

R25

26

14

8   10

6

12

29   20

30

Part 6
Hasp
1 req'd 1 mm MS

Bend lines

Tool tray
development

Peen over pin
at each end

Roll hinge blank to suit
MS pin Ø4 × 26

φ 10   R8

18

3

10   3

Part 7
Catch 2 mm
MS 1 req'd

196

2 × Ø3   12

24

24   12

Fold up both
ends. Weld or
braze corners

Development

Part 11
Tray support bracket
2 req'd 1 mm MS

FIGURE 18.6 (*Continued*)

produce patterns to cut the components from sheet metal economically. This example provides useful practice in reading and applying engineering drawings.

Solution shown in Fig. 18.12 (and see Fig. 18.13).

Solution notes

Bushed bearing bracket (Fig. 18.7)

Note that in the end view there is an example of a thin web which is not cross hatched. The example shows three examples of counterbored holes used to contain heads of fixing screws.

A liner bush with an interference fit is also indicated, and since the bush is relatively thin compared with the main casting, its cross hatching lines are drawn closer together.

Drill table (Fig. 18.8)

The table is clamped to the drilling machine main vertical pillar using two bolts. The slot permits tightening.

Cross hatching is omitted on the left side of the front view.

Cam operated clamp (Fig. 18.9)

This example shows a typical assembly and includes a parts list.

The balloons containing the part numbers are equally spaced whenever possible. The leader lines to the components terminate in a dot. The leader line is also directed towards the centre of the balloon and touches the circumference.

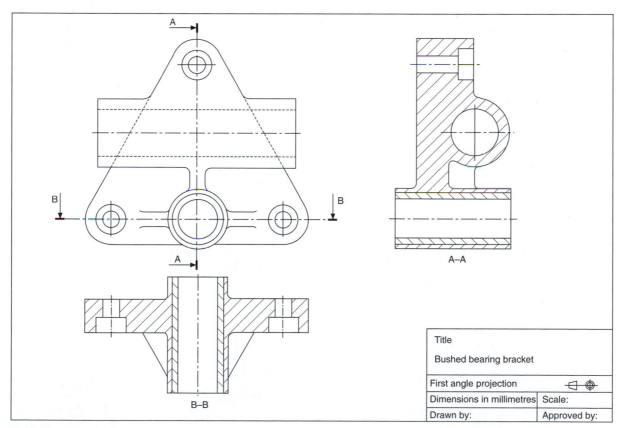

FIGURE 18.7

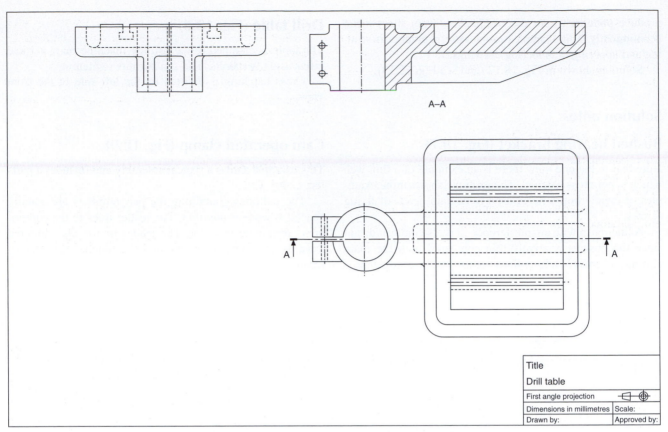

A–A

	Title		
	Drill table		
	First angle projection		
	Dimensions in millimetres	Scale:	
	Drawn by:	Approved by:	

FIGURE 18.8

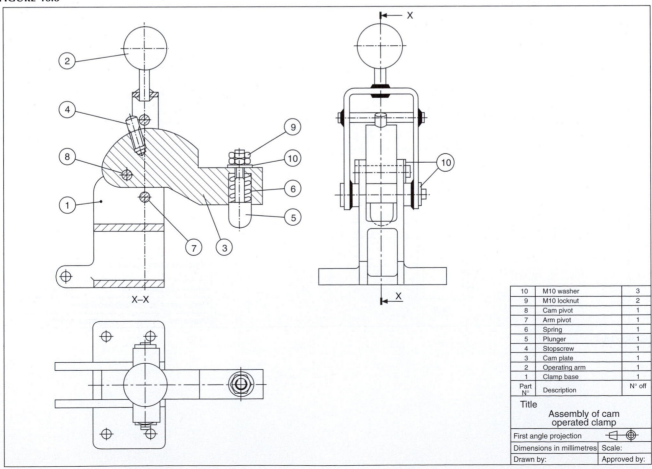

X–X

10	M10 washer	3
9	M10 locknut	2
8	Cam pivot	1
7	Arm pivot	1
6	Spring	1
5	Plunger	1
4	Stopscrew	1
3	Cam plate	1
2	Operating arm	1
1	Clamp base	1
Part N°	Description	N° off

Title		
Assembly of cam operated clamp		
First angle projection		
Dimensions in millimetres	Scale:	
Drawn by:	Approved by:	

FIGURE 18.9

It is considered good policy to position the parts list and commence writing in such a way that parts can be added on to the end of the list in the event of future modifications. The list can be drawn at the top or bottom of the drawing sheet.

Plug cock assembly (Fig. 18.10)

This example illustrates a typical industrial valve. The plug turns through 90° between the on and off positions. Spanner flats are provided and indicated by the diagonal lines. Gland packing (Part 10) is supplied in rings. These rings are contained by the body cover and on assembly are fed over the plug spindle. The gland is tightened and the com-

pressive force squeezes the packing to provide a seal, sufficient to prevent leaks, but enabling the spindle to be turned. The joint ring (Part 9) is too thin to be cross hatched and is shown filled in.

This is also an example of a symmetrical part where the half section gives an outside view and a sectional view to indicate the internal details.

Air engine assembly, illustrated below (Fig. 18.11)

The engine operates through compressed air entering the cylinder via one of the connecting pipes shown as item 15. The other pipe serves to exhaust the cylinder after the power

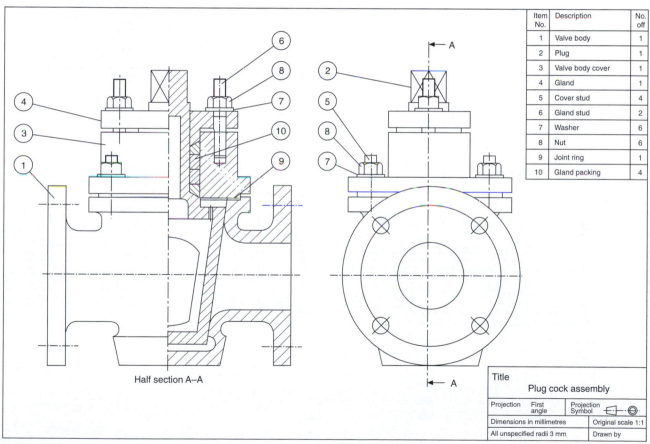

Item No.	Description	No. off
1	Valve body	1
2	Plug	1
3	Valve body cover	1
4	Gland	1
5	Cover stud	4
6	Gland stud	2
7	Washer	6
8	Nut	6
9	Joint ring	1
10	Gland packing	4

Half section A–A

Title	Plug cock assembly
Projection	First angle
	Projection Symbol
Dimensions in millimetres	Original scale 1:1
All unspecified radii 3 mm	Drawn by

FIGURE 18.10

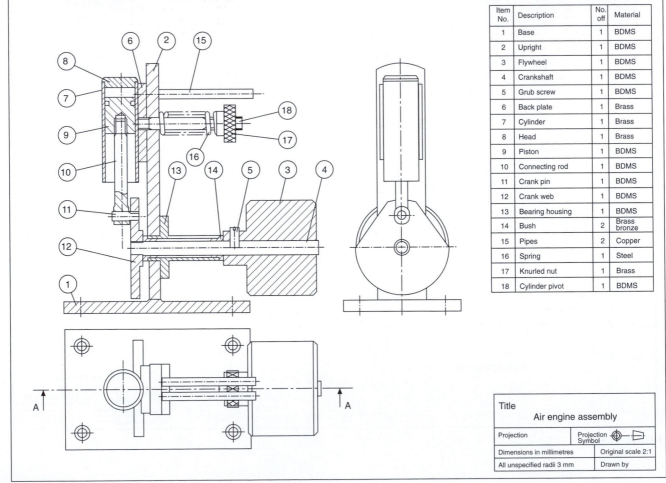

Item No.	Description	No. off	Material
1	Base	1	BDMS
2	Upright	1	BDMS
3	Flywheel	1	BDMS
4	Crankshaft	1	BDMS
5	Grub screw	1	BDMS
6	Back plate	1	Brass
7	Cylinder	1	Brass
8	Head	1	Brass
9	Piston	1	BDMS
10	Connecting rod	1	BDMS
11	Crank pin	1	BDMS
12	Crank web	1	BDMS
13	Bearing housing	1	BDMS
14	Bush	2	Brass bronze
15	Pipes	2	Copper
16	Spring	1	Steel
17	Knurled nut	1	Brass
18	Cylinder pivot	1	BDMS

Title	
	Air engine assembly
Projection	Projection Symbol
Dimensions in millimetres	Original scale 2:1
All unspecified radii 3 mm	Drawn by

FIGURE 18.11

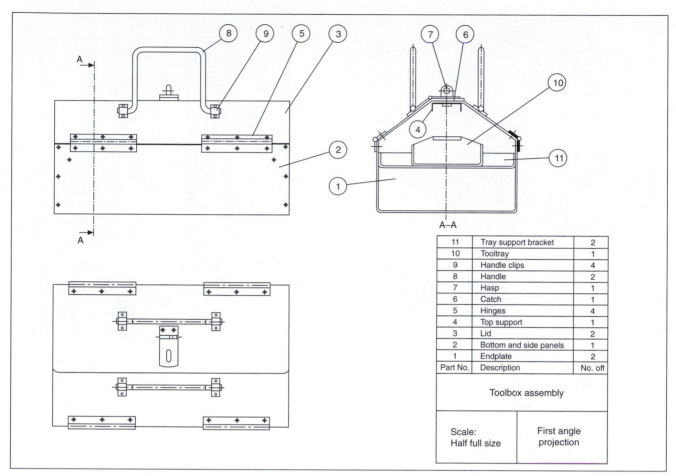

11	Tray support bracket	2
10	Tooltray	1
9	Handle clips	4
8	Handle	2
7	Hasp	1
6	Catch	1
5	Hinges	4
4	Top support	1
3	Lid	2
2	Bottom and side panels	1
1	Endplate	2
Part No.	Description	No. off

Toolbox assembly

Scale: Half full size	First angle projection

FIGURE 18.12

stroke. The cylinder oscillates in an arc and a hole through the cylinder wall lines up with the inlet and exhaust pipes at each 180° of rotation of the flywheel.

The spindles (Parts 4 and 18), grub screw (Part 5) and the pin (Part 11) would not normally be sectioned. A part section is illustrated at the bottom of the connecting rod in order to show its assembly with the crank pin (Part 11). The BS convention is shown for the spring (Part 16). The BS convention is also shown for cross knurling on the nut (Part 17) (Fig. 18.12).

Note: If the solutions to examples 3, 4, 5, and 6 had been required to be drawn in third angle projection, then the views would have been arranged as shown in Fig. 18.13.

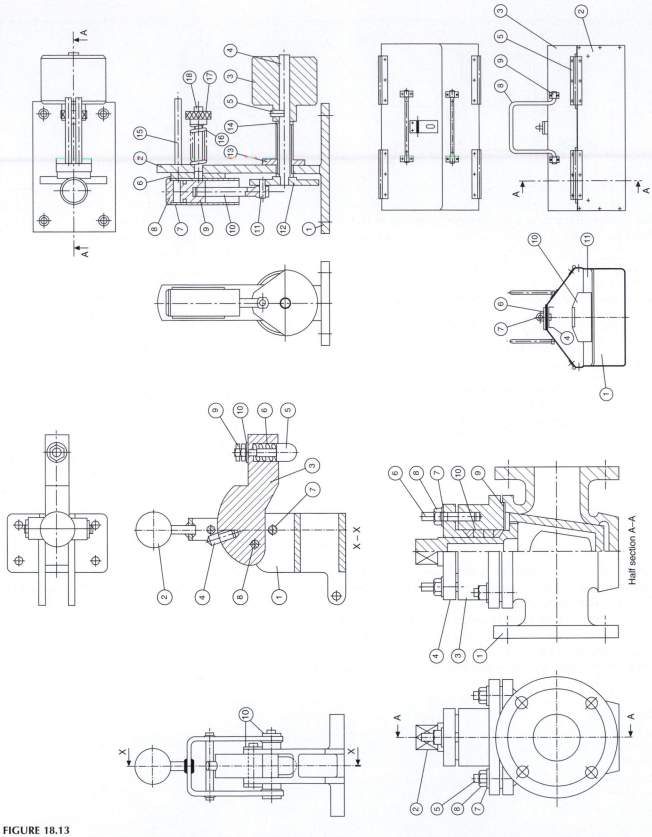

FIGURE 18.13

Chapter 19

Limits and fits

To ensure that an assembly will function correctly, its component parts must fit together in a predictable manner. Now, in practice, no component can be manufactured to an exact size, and one of the problems facing the designer is to decide the upper and lower limits of size which are acceptable for each of the dimensions used to define shape and form and which will ensure satisfactory operation in service. For example, a dimension of 10 0.02 means that a part will be acceptable if manufactured anywhere between the limits of size of 9.98 and 10.02 mm. The present system of manufacture of interchangeable parts was brought about by the advent of and the needs of mass production, and has the following advantages.

1. Instead of 'fitting' components together, which requires some adjustment of size and a high degree of skill, they can be 'assembled'.
2. An assembly can be serviced by replacing defective parts with components manufactured to within the same range of dimensions.
3. Parts can be produced in large quantities, in some cases with less demand on the skill of the operator. Invariably this requires the use of special-purpose machines, tools, jigs, fixtures, and gauges: but the final cost of each component will be far less than if made separately by a skilled craftsman.

It should be noted, however, that full interchangeability is not always necessary in practice; neither is it always feasible, especially when the dimensions are required to be controlled very closely in size. Many units used in the construction of motor vehicles are assembled after an elaborate inspection process has sorted the components into different groups according to size. Suppose, for example, that it was required to maintain the clearance between a piston and a cylinder to within 0.012 mm. To maintain full interchangeability would require both the piston and the cylinder bores to be finished to a tolerance of 0.006 mm, which would be difficult to maintain and also uneconomic to produce. In practice it is possible to manufacture both bores and pistons to within a tolerance of 0.06 mm and then divide them into groups for assembly; this involves the gauging of each component.

A designer should ensure that the drawing conveys clear instructions regarding the upper and lower limits of size for each dimension, and Figs. 19.1–19.4 show typical methods in common use.

The method shown in Fig. 19.1 is perhaps the clearest way of expressing limits of size on a drawing, since the upper and lower limits are quoted, and the machine operator is not involved in mental arithmetic. The dimensions are quoted in logical form, with the upper limit above the lower limit and both to the same number of decimal places.

$$\frac{60.05}{60.00}$$

FIGURE 19.1

As an alternative to the method above, the basic size may be quoted and the tolerance limits added as in Fig. 19.2. It is not necessary to express the nominal dimension to the same number of decimal places as the limits.

Fits can be taken directly from those tabulated in BS 4500, 'ISO limits and fits', and, in order to indicate the grade of fit, the following alternative methods of dimensioning a hole may be used:

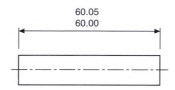

$$90\,H7\left(\begin{array}{c}90.035\\90.000\end{array}\right)\text{(first choice)}$$

$$\text{or }90\,H7\text{ or }90\,H7\left(\begin{array}{c}+0.035\\0\end{array}\right)$$

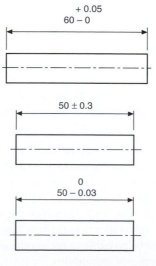

$$\frac{+\,0.05}{60-0}$$

$$50 \pm 0.3$$

$$\frac{0}{50-0.03}$$

FIGURE 19.2

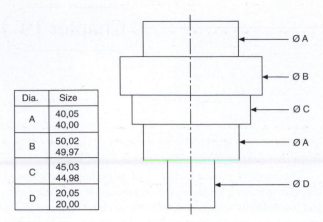

Dia.	Size
A	40,05 40,00
B	50,02 49,97
C	45,03 44,98
D	20,05 20,00

FIGURE 19.3

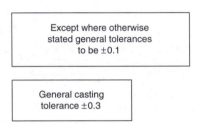

Except where otherwise
stated general tolerances
to be ±0.1

General casting
tolerance ±0.3

FIGURE 19.4

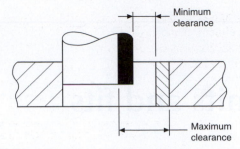

FIGURE 19.5   Clearance fits – allowance always positive.

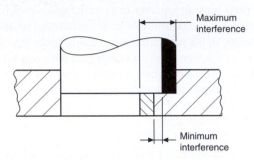

FIGURE 19.6   Interference fits – allowance always negative.

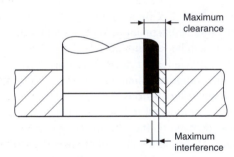

FIGURE 19.7   Transition fit – allowance may be positive or negative.

Similarly, a shaft may be dimensioned as follows:

$$90\,g6\left(\frac{89.988}{89.966}\right)(\text{first choice})$$

$$\text{or } 90\,g6 \text{ or } 90\,g6\left(\frac{-0.012}{-0.034}\right)$$

In cases where a large amount of repetition is involved, information can be given in tabulated form, and a typical component drawing is shown in Fig. 19.3.

In many cases, tolerances need be only of a general nature, and cover a wide range of dimensions. A box with a standard note is added to the drawing, and the typical examples in Fig. 19.4 are self-explanatory.

Engineering fits between two mating parts can be divided into three types:

1. a *clearance fit* (Fig. 19.5), in which the shaft is always smaller than the hole into which it fits;
2. an *interference fit* (Fig. 19.6), in which the shaft is always bigger than the hole into which it fits;
3. a *transition fit* (Fig. 19.7), in which the shaft may be either bigger or smaller than the hole into which it fits – it will therefore be possible to get interference or clearance fits in one group of assemblies.

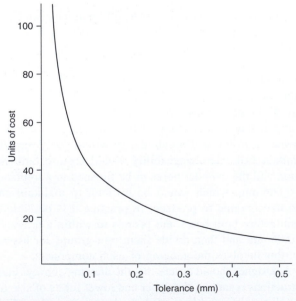

FIGURE 19.8   Approximate relationship between production cost and manufacturing tolerance.

It will be appreciated that, as the degree of accuracy required for each dimension increases, the cost of production to maintain this accuracy increases at a sharper rate.

Figure 19.8 shows the approximate relationship between cost and tolerance. For all applications, the manufacturing tolerance should be the largest possible which permits satisfactory operation.

Elements of interchangeable systems (Fig. 19.9)

Nominal size is the size by which a component is referred to as a matter of convenience, i.e. 25, 50 and 60 mm thread.

Actual size is the measured size.

Basic size is the size in relation to which all limits of size are fixed, and will be the same for both the male and female parts of the fit.

Limits of size – These are the maximum and minimum permissible sizes acceptable for a specific dimension.

Tolerance – This is the total permissible variation in the size of a dimension, and is the difference between the upper and lower acceptable dimensions.

Allowance – concerns mating parts, and is the difference between the high limit of size of the shaft and the low limit of size of its mating hole. An allowance may be positive or negative.

Grade – This is an indication of the tolerance magnitude: the lower the grade, the finer will be the tolerance.

Deviation – This is the difference between the maximum, minimum, or actual size of a shaft or hole and the basic size.

Maximum Material Condition (MMC) – This is the maximum limit of an external feature; for example, a shaft manufactured to its high limits would contain the maximum amount of material. It is also the minimum limit on an internal feature; for example, a component which has a hole produced to its lower limit of size would have the minimum of material removed and remain in its maximum metal condition.

Minimum or Least Material Condition (LMC) – This is the minimum limit of an external feature; for example a shaft manufactured to its low limits would contain the minimum amount of material. It is also the maximum limit on an internal feature; for example a hole produced to its maximum limit would have the maximum amount of material removed and remain in its minimum material condition.

Unilateral and bilateral limits

Figure 19.10 shows an example of unilateral limits, where the maximum and minimum limits of size are disposed on the same side of the basic size. This system is preferred since the basic size is used for the GO limit gauge; changes in the magnitude of the tolerance affect only the size of the other gauge dimension, the NOT GO gauge size.

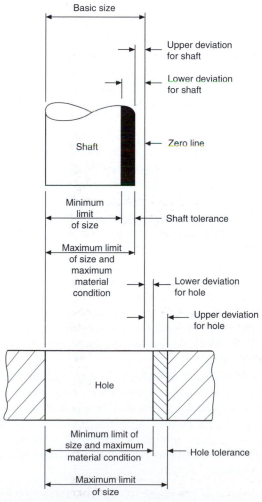

FIGURE 19.9 Elements of interchangeable systems.

FIGURE 19.10 Unilateral limits.

Figure 19.11 shows an example of bilateral limits, where the limits are disposed above and below the basic size.

FIGURE 19.11 Bilateral limits.

Bases of fits

The two bases of a system of limits and fits are:

(a) the hole basis,
(b) the shaft basis.

Hole basis (Fig. 19.12) – In this system, the basic diameter of the hole is constant while the shaft size varies according to the type of fit. This system leads to greater economy of production, as a single drill or reamer size can be used to

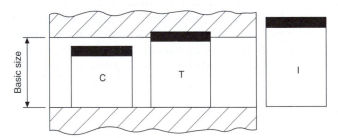

FIGURE 19.12 Hole-basis fits: C – clearance, T – transition, I – interference.

produce a variety of fits by merely altering the shaft limits. The shaft can be accurately produced to size by turning and grinding. Generally it is usual to recommend hole-base fits, except where temperature may have a detrimental effect on large sizes.

Shaft basis (Fig. 19.13) – Here the hole size is varied to produce the required class of fit with a basic-size shaft. A series of drills and reamers is required for this system, therefore it tends to be costly. It may, however, be necessary to use

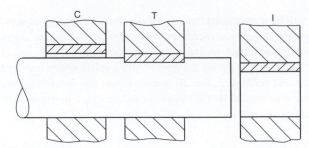

FIGURE 19.13 Shaft-basis fits: C – clearance, T – transition, I – interference.

it where different fits are required along a long shaft. This BSI data sheet 4500A gives a selection of ISO fits on the hole basis, and data sheet 4500B gives a selection of shaft-basis fits extracted from BS 4500, the current standard on limits and fits.

The ISO system contained in BS 4500 gives an extensive selection of hole and shaft tolerances to cover a wide range of applications. It has been found, however, that in the manufacture of many standard engineering components a limited selection of tolerances is adequate. These are provided on the data sheets referred to above. Obviously, by using only a selected range of fits, economic advantages are obtained from the reduced tooling and gauging facilities involved.

Selected ISO fits – hole basis (extracted from BS 4500)

BS EN 20286 Part 1, describes the ISO system of limits and fits, providing the basis of tolerances, deviations and fits. BS EN 20286 Part 2, provides tables of standard tolerance grades and limit deviations for holes and shafts. Both of these standards have replaced BS 4500 (which has been withdrawn by BSI). However, the information contained in BS 4500 is identical to that in the BS EN 20286 series and many engineers still use the BS 4500A (hole basis) and BS 4500B (shaft basis) data sheets because they are easier to read and include diagrams of the relationships between a hole and a shaft for commonly applied fits in the clearance, transition and interference classes. For this publication, reference is still made to BS 4500 as are the data sheets reproduced on the following pages.

The ISO system provides a great many hole and shaft tolerances so as to cater for a very wide range of conditions. However, experience shows that the majority of fit conditions required for normal engineering products can be provided by a quite limited selection of tolerances. The following selected hole and shaft tolerances have been found to be commonly applied:

● selected hole tolerances: H7 H8 H9 H11;
● selected shaft tolerances: c11 d10 e9 f7 g6 h6 k6 n6 p6 s6.

BRITISH STANDARD
SELECTED ISO FITS—HOLE BASIS

Extracted from
BS 4500 : 1969

Data Sheet
4500A
Issue 1. February 1970
confirmed August 1985

Diagram to scale for 25 mm diameter

Holes · Shafts

Clearance fits: H11/c11, H9/d10, H9/e9, H8/f7, H7/g6, H7/h6
Transition fits: H7/k6, H7/n6
Interference fits: H7/p6, H7/s6

All tolerance values in units of 0.001 mm. Nominal sizes in mm.

Over	To	H11	c11	H9	d10	H9	e9	H8	f7	H7	g6	H7	h6	H7	k6	H7	n6	H7	p6	H7	s6
—	3	+60 / 0	−60 / −120	+25 / 0	−20 / −60	+25 / 0	−14 / −39	+14 / 0	−6 / −16	+10 / 0	−2 / −8	+10 / 0	−6 / 0	+10 / 0	+6 / 0	+10 / 0	+10 / +4	+10 / 0	+12 / +6	+10 / 0	+20 / +14
3	6	+75 / 0	−70 / −145	+30 / 0	−30 / −78	+30 / 0	−20 / −50	+18 / 0	−10 / −28	+12 / 0	−4 / −12	+12 / 0	−8 / 0	+12 / 0	+9 / +1	+12 / 0	+16 / +8	+12 / 0	+20 / +12	+12 / 0	+27 / +19
6	10	+90 / 0	−80 / −170	+36 / 0	−40 / −98	+36 / 0	−25 / −61	+22 / 0	−13 / −28	+15 / 0	−5 / −14	+15 / 0	−9 / 0	+15 / 0	+10 / +1	+15 / 0	+19 / +10	+15 / 0	+24 / +15	+15 / 0	+32 / +23
10	18	+110 / 0	−95 / −205	+43 / 0	−50 / −120	+43 / 0	−32 / −75	+27 / 0	−16 / −34	+18 / 0	−6 / −17	+18 / 0	−11 / 0	+18 / 0	+12 / +1	+18 / 0	+23 / +12	+18 / 0	+29 / +18	+18 / 0	+39 / +28
18	30	+130 / 0	−110 / −240	+52 / 0	−65 / −149	+52 / 0	−40 / −92	+33 / 0	−20 / −41	+21 / 0	−7 / −20	+21 / 0	−13 / 0	+21 / 0	+15 / +2	+21 / 0	+28 / +15	+21 / 0	+35 / +22	+21 / 0	+48 / +35
30	40	+160 / 0	−120 / −280	+62 / 0	−80 / −180	+62 / 0	−50 / −112	+39 / 0	−25 / −50	+25 / 0	−9 / −25	+25 / 0	−16 / 0	+25 / 0	+18 / +2	+25 / 0	+33 / +17	+25 / 0	+42 / +26	+25 / 0	+59 / +43
40	50	+160 / 0	−130 / −290	+62 / 0	−80 / −180	+62 / 0	−50 / −112	+39 / 0	−25 / −50	+25 / 0	−9 / −25	+25 / 0	−16 / 0	+25 / 0	+18 / +2	+25 / 0	+33 / +17	+25 / 0	+42 / +26	+25 / 0	+59 / +43
50	65	+190 / 0	−140 / −330	+74 / 0	−100 / −220	+74 / 0	−60 / −134	+46 / 0	−30 / −60	+30 / 0	−10 / −29	+30 / 0	−19 / 0	+30 / 0	+21 / +2	+30 / 0	+39 / +20	+30 / 0	+51 / +32	+30 / 0	+72 / +53
65	80	+190 / 0	−150 / −340	+74 / 0	−100 / −220	+74 / 0	−60 / −134	+46 / 0	−30 / −60	+30 / 0	−10 / −29	+30 / 0	−19 / 0	+30 / 0	+21 / +2	+30 / 0	+39 / +20	+30 / 0	+51 / +32	+30 / 0	+78 / +59
80	100	+220 / 0	−170 / −390	+87 / 0	−120 / −260	+87 / 0	−72 / −159	+54 / 0	−36 / −71	+35 / 0	−12 / −34	+35 / 0	−22 / 0	+35 / 0	+25 / +3	+35 / 0	+45 / +23	+35 / 0	+59 / +37	+35 / 0	+93 / +71
100	120	+220 / 0	−180 / −400	+87 / 0	−120 / −260	+87 / 0	−72 / −159	+54 / 0	−36 / −71	+35 / 0	−12 / −34	+35 / 0	−22 / 0	+35 / 0	+25 / +3	+35 / 0	+45 / +23	+35 / 0	+59 / +37	+35 / 0	+101 / +79
120	140	+250 / 0	−200 / −450	+100 / 0	−145 / −305	+100 / 0	−84 / −185	+63 / 0	−43 / −83	+40 / 0	−14 / −39	+40 / 0	−25 / 0	+40 / 0	+28 / +3	+40 / 0	+52 / +27	+40 / 0	+68 / +43	+40 / 0	+117 / +92
140	160	+250 / 0	−210 / −460	+100 / 0	−145 / −305	+100 / 0	−84 / −185	+63 / 0	−43 / −83	+40 / 0	−14 / −39	+40 / 0	−25 / 0	+40 / 0	+28 / +3	+40 / 0	+52 / +27	+40 / 0	+68 / +43	+40 / 0	+125 / +100
160	180	+250 / 0	−230 / −480	+100 / 0	−145 / −305	+100 / 0	−84 / −185	+63 / 0	−43 / −83	+40 / 0	−14 / −39	+40 / 0	−25 / 0	+40 / 0	+28 / +3	+40 / 0	+52 / +27	+40 / 0	+68 / +43	+40 / 0	+133 / +108
180	200	+290 / 0	−240 / −530	+115 / 0	−170 / −355	+115 / 0	−100 / −215	+72 / 0	−50 / −96	+46 / 0	−15 / −44	+46 / 0	−29 / 0	+46 / 0	+33 / +4	+46 / 0	+60 / +31	+46 / 0	+79 / +50	+46 / 0	+151 / +122
200	225	+290 / 0	−260 / −550	+115 / 0	−170 / −355	+115 / 0	−100 / −215	+72 / 0	−50 / −96	+46 / 0	−15 / −44	+46 / 0	−29 / 0	+46 / 0	+33 / +4	+46 / 0	+60 / +31	+46 / 0	+79 / +50	+46 / 0	+159 / +130
225	250	+290 / 0	−280 / −570	+115 / 0	−170 / −355	+115 / 0	−100 / −215	+72 / 0	−50 / −96	+46 / 0	−15 / −44	+46 / 0	−29 / 0	+46 / 0	+33 / +4	+46 / 0	+60 / +31	+46 / 0	+79 / +50	+46 / 0	+169 / +140
250	280	+320 / 0	−300 / −620	+130 / 0	−190 / −400	+130 / 0	−110 / −240	+81 / 0	−56 / −108	+52 / 0	−17 / −49	+52 / 0	−32 / 0	+52 / 0	+36 / +4	+52 / 0	+66 / +34	+52 / 0	+88 / +56	+52 / 0	+190 / +158
280	315	+320 / 0	−330 / −650	+130 / 0	−190 / −400	+130 / 0	−110 / −240	+81 / 0	−56 / −108	+52 / 0	−17 / −49	+52 / 0	−32 / 0	+52 / 0	+36 / +4	+52 / 0	+66 / +34	+52 / 0	+88 / +56	+52 / 0	+202 / +170
315	355	+360 / 0	−360 / −720	+140 / 0	−210 / −440	+140 / 0	−125 / −265	+89 / 0	−62 / −119	+57 / 0	−18 / −54	+57 / 0	−36 / 0	+57 / 0	+40 / +4	+57 / 0	+73 / +37	+57 / 0	+98 / +62	+57 / 0	+226 / +190
355	400	+360 / 0	−400 / −760	+140 / 0	−210 / −440	+140 / 0	−125 / −265	+89 / 0	−62 / −119	+57 / 0	−18 / −54	+57 / 0	−36 / 0	+57 / 0	+40 / +4	+57 / 0	+73 / +37	+57 / 0	+98 / +62	+57 / 0	+244 / +208
400	450	+400 / 0	−440 / −840	+155 / 0	−230 / −480	+155 / 0	−135 / −290	+97 / 0	−68 / −131	+63 / 0	−20 / −60	+63 / 0	−40 / 0	+63 / 0	+45 / +5	+63 / 0	+80 / +40	+63 / 0	+108 / +68	+63 / 0	+272 / +232
450	500	+400 / 0	−480 / −880	+155 / 0	−230 / −480	+155 / 0	−135 / −290	+97 / 0	−68 / −131	+63 / 0	−20 / −60	+63 / 0	−40 / 0	+63 / 0	+45 / +5	+63 / 0	+80 / +40	+63 / 0	+108 / +68	+63 / 0	+292 / +252

BRITISH STANDARDS INSTITUTION, 2 Park Street, London, W1A 2BS
SBN: 580 05766 6

Extracted from
BS 4500 : 1969

BRITISH STANDARD

Data Sheet
4500B
Issue 1. February 1970

SELECTED ISO FITS—SHAFT BASIS

Diagram to scale for 25 mm. diameter

Holes / Shafts

Clearance fits: C 11 / h 11 — D 10 / h 9 — E 9 / h 9 — F 8 / h 7 — G 7 / h 6 — H 7 / h 6
Transition fits: K 7 / h 6 — N 7 / h 6
Interference fits: P 7 / h 6 — S 7 / h 6

All tolerance values in 0.001 mm.

Over	To	C11	h11	D10	h9	E9	h9	F8	h7	G7	h6	H7	h6	K7	h6	N7	h6	P7	h6	S7	h6
–	3	+120 / +60	0 / -60	+60 / +20	0 / -25	+39 / +14	0 / -25	+20 / +6	0 / -10	+12 / +2	0 / -6	+10 / 0	0 / -6	0 / -10	0 / -6	-4 / -14	0 / -6	-6 / -16	0 / -6	-14 / -24	0 / -6
3	6	+145 / +70	0 / -75	+78 / +30	0 / -30	+50 / +20	0 / -30	+28 / +10	0 / -12	+16 / +4	0 / -8	+12 / 0	0 / -8	+3 / -9	0 / -8	-4 / -16	0 / -8	-8 / -20	0 / -8	-15 / -27	0 / -8
6	10	+170 / +80	0 / -90	+98 / +40	0 / -36	+61 / +25	0 / -36	+35 / +13	0 / -15	+20 / +5	0 / -9	+15 / 0	0 / -9	+5 / -10	0 / -9	-4 / -19	0 / -9	-9 / -24	0 / -9	-17 / -32	0 / -9
10	18	+205 / +95	0 / -110	+120 / +50	0 / -43	+75 / +32	0 / -43	+43 / +16	0 / -18	+24 / +6	0 / -11	+18 / 0	0 / -11	+6 / -12	0 / -11	-5 / -23	0 / -11	-11 / -29	0 / -11	-21 / -39	0 / -11
18	30	+240 / +110	0 / -130	+149 / +65	0 / -52	+92 / +40	0 / -52	+53 / +20	0 / -21	+28 / +7	0 / -13	+21 / 0	0 / -13	+6 / -15	0 / -13	-7 / -28	0 / -13	-14 / -35	0 / -13	-27 / -48	0 / -13
30	40	+280 / +120	0 / -160	+180 / +80	0 / -62	+112 / +50	0 / -62	+64 / +25	0 / -25	+34 / +9	0 / -16	+25 / 0	0 / -16	+7 / -18	0 / -16	-8 / -33	0 / -16	-17 / -42	0 / -16	-34 / -59	0 / -16
40	50	+290 / +130	0 / -160	+180 / +80	0 / -62	+112 / +50	0 / -62	+64 / +25	0 / -25	+34 / +9	0 / -16	+25 / 0	0 / -16	+7 / -18	0 / -16	-8 / -33	0 / -16	-17 / -42	0 / -16	-42 / -72	0 / -16
50	65	+330 / +140	0 / -190	+220 / +100	0 / -74	+134 / +60	0 / -74	+76 / +30	0 / -30	+40 / +10	0 / -19	+30 / 0	0 / -19	+9 / -21	0 / -19	-9 / -39	0 / -19	-21 / -51	0 / -19	-48 / -78	0 / -19
65	80	+340 / +150	0 / -190	+220 / +100	0 / -74	+134 / +60	0 / -74	+76 / +30	0 / -30	+40 / +10	0 / -19	+30 / 0	0 / -19	+9 / -21	0 / -19	-9 / -39	0 / -19	-21 / -51	0 / -19	-58 / -93	0 / -19
80	100	+390 / +170	0 / -220	+260 / +120	0 / -87	+159 / +72	0 / -87	+90 / +36	0 / -35	+47 / +12	0 / -22	+35 / 0	0 / -22	+10 / -25	0 / -22	-10 / -45	0 / -22	-24 / -59	0 / -22	-66 / -101	0 / -22
100	120	+400 / +180	0 / -220	+260 / +120	0 / -87	+159 / +72	0 / -87	+90 / +36	0 / -35	+47 / +12	0 / -22	+35 / 0	0 / -22	+10 / -25	0 / -22	-10 / -45	0 / -22	-24 / -59	0 / -22	-77 / -117	0 / -22
120	140	+450 / +200	0 / -250	+305 / +145	0 / -100	+185 / +85	0 / -100	+106 / +43	0 / -40	+54 / +14	0 / -25	+40 / 0	0 / -25	+12 / -28	0 / -25	-12 / -52	0 / -25	-28 / -68	0 / -25	-85 / -125	0 / -25
140	160	+460 / +210	0 / -250	+305 / +145	0 / -100	+185 / +85	0 / -100	+106 / +43	0 / -40	+54 / +14	0 / -25	+40 / 0	0 / -25	+12 / -28	0 / -25	-12 / -52	0 / -25	-28 / -68	0 / -25	-93 / -133	0 / -25
160	180	+480 / +230	0 / -250	+305 / +145	0 / -100	+185 / +85	0 / -100	+106 / +43	0 / -40	+54 / +14	0 / -25	+40 / 0	0 / -25	+12 / -28	0 / -25	-12 / -52	0 / -25	-28 / -68	0 / -25	-105 / -151	0 / -25
180	200	+530 / +240	0 / -290	+355 / +170	0 / -115	+215 / +100	0 / -115	+122 / +50	0 / -46	+61 / +15	0 / -29	+46 / 0	0 / -29	+13 / -33	0 / -29	-14 / -60	0 / -29	-33 / -79	0 / -29	-113 / -159	0 / -29
200	225	+550 / +260	0 / -290	+355 / +170	0 / -115	+215 / +100	0 / -115	+122 / +50	0 / -46	+61 / +15	0 / -29	+46 / 0	0 / -29	+13 / -33	0 / -29	-14 / -60	0 / -29	-33 / -79	0 / -29	-123 / -169	0 / -29
225	250	+570 / +280	0 / -290	+355 / +170	0 / -115	+215 / +100	0 / -115	+122 / +50	0 / -46	+61 / +15	0 / -29	+46 / 0	0 / -29	+13 / -33	0 / -29	-14 / -60	0 / -29	-33 / -79	0 / -29	-138 / -190	0 / -29
250	280	+620 / +300	0 / -320	+400 / +190	0 / -130	+240 / +110	0 / -130	+137 / +56	0 / -52	+62 / +17	0 / -32	+52 / 0	0 / -32	+16 / -36	0 / -32	-14 / -66	0 / -32	-36 / -88	0 / -32	-150 / -202	0 / -32
280	315	+650 / +330	0 / -320	+400 / +190	0 / -130	+240 / +110	0 / -130	+137 / +56	0 / -52	+62 / +17	0 / -32	+52 / 0	0 / -32	+16 / -36	0 / -32	-14 / -66	0 / -32	-36 / -88	0 / -32	-169 / -226	0 / -32
315	355	+720 / +360	0 / -360	+440 / +210	0 / -140	+265 / +125	0 / -140	+151 / +62	0 / -57	+75 / +18	0 / -36	+57 / 0	0 / -36	+17 / -40	0 / -36	-16 / -73	0 / -36	-41 / -98	0 / -36	-187 / -244	0 / -36
355	400	+760 / +400	0 / -360	+440 / +210	0 / -140	+265 / +125	0 / -140	+151 / +62	0 / -57	+75 / +18	0 / -36	+57 / 0	0 / -36	+17 / -40	0 / -36	-16 / -73	0 / -36	-41 / -98	0 / -36	-209 / -272	0 / -36
400	450	+840 / +440	0 / -400	+480 / +230	0 / -155	+290 / +135	0 / -155	+165 / +68	0 / -63	+83 / +20	0 / -40	+63 / 0	0 / -40	+18 / -45	0 / -40	-17 / -80	0 / -40	-45 / -108	0 / -40	-229 / -292	0 / -40
450	500	+880 / +480	0 / -400	+480 / +230	0 / -155	+290 / +135	0 / -155	+165 / +68	0 / -63	+83 / +20	0 / -40	+63 / 0	0 / -40	+18 / -45	0 / -40	-17 / -80	0 / -40	-45 / -108	0 / -40		0 / -40

BRITISH STANDARDS INSTITUTION, 2 Park Street, London, W1Y 4AA
SBN: 580 05567 1

Table 19.1 shows a range of fits derived from these selected hole and shaft tolerances. As will be seen, it covers fits from loose clearance to heavy interference, and it may therefore be found to be suitable for most normal requirements. Many users may in fact find that their needs are met by a further selection within this selected range.

It should be noted, however, that this table is offered only as an example of how a restricted selection of fits can be made. It is clearly impossible to recommend selections of fits which are appropriate to all sections of industry, but it must be emphasized that a user who decides upon a selected range will always enjoy the economic advantages this conveys. Once he has installed the necessary tooling and gauging facilities, he can combine his selected hole and shaft tolerances in different ways without any additional investment in tools and equipment.

For example, if it is assumed that the range of fits shown in the table has been adopted but that, for a particular application the fit H8-f7 is appropriate but provides rather too much variation, the hole tolerance H7 could equally well be associated with the shaft f7 and may provide exactly what is required without necessitating any additional tooling.

For most general applications, it is usual to recommend hole-basis fits, as, except in the realm of very large sizes where the effects of temperature play a large part, it is usually considered easier to manufacture and measure the male member of a fit, and it is thus desirable to be able to allocate the larger part of the tolerance available to the hole and adjust the shaft to suit.

In some circumstances, however, it may in fact be preferable to employ a shaft basis. For example, in the case of driving shafts where a single shaft may have to accommodate a variety of accessories such as couplings, bearings, collars, etc., it is preferable to maintain a constant diameter for the permanent member, which is the shaft, and vary the bore of the accessories. For use in applications of this kind, a selection of shaft basis fits is provided in data sheet BS 4500B.

Note: Data sheet 4500A (p. 157) refers to hole basis fits. Data sheet 4500B (p. 158) refers to shaft basis fits.

Interpretations of limits of size in relation to form

There are two ways of interpreting the limits of size of an individual feature, which are known by:

1. The Principle of Independency, where the limits of size apply to local two point measurements of a feature regardless of form.
2. The Envelope Requirement, also known as The Taylor Principle, where the limits of size of an individual feature are intended to have a mutual dependency of size and form.

BRITISH STANDARD
SELECTED ISO FITS – HOLES BASIS
Figure 19.14 a, b and c illustrate the Principle of Independency.

Figure 19.15 a, b, c, d and e illustrate the Envelope Requirement.

The drawing indication in Fig. 19.15a shows a linear tolerance followed by the symbol Ⓔ. Two functional requirements are implied by the use of the symbol.

1. That the surface of the cylindrical feature is contained within an envelope of perfect form at maximum material size of Ø120.
2. That no actual local size shall be less than Ø119,96. An exaggerated view of the feature in Fig. 19.15b, shows that each actual local diameter of the shaft must remain within the size tolerance of 0.04 and may vary between Ø120 and Ø119,96.

In the examples which follow, the entire shaft must remain within the boundary of the Ø120 envelope cylinder of perfect form.

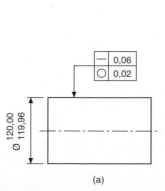
(a)

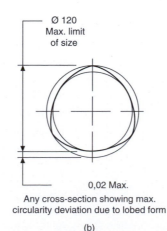
0,02 Max.
Any cross-section showing max. circularity deviation due to lobed form
(b)

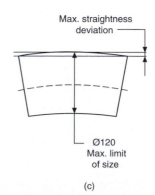
(c)

FIGURE 19.14

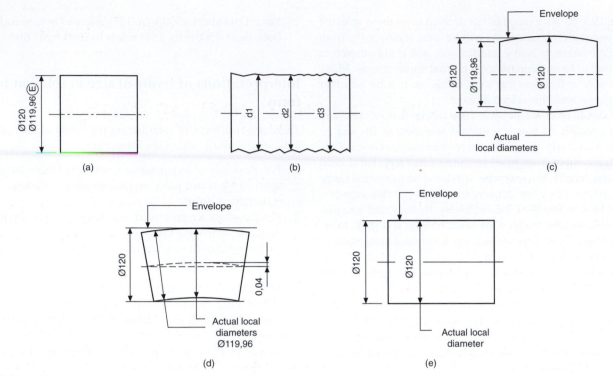

FIGURE 19.15

It follows therefore that the shaft will be perfectly cylindrical when all actual local diameters are at the maximum material size of Ø120.

Maximum material condition. For further reading see ISO 2692 which states that: if for functional and economic reasons there is a requirement for the mutual dependency of the size and orientation or location of the feature(s), then the maximum material principle ⓂM may be applied.

Chapter 20

Geometrical tolerancing and datums

Geometrical tolerances

The object of this section is to illustrate and interpret in simple terms the advantages of calling for geometrical tolerances on engineering drawings, and also to show that, when correctly used, they ensure that communications between the drawing office and the workshop are complete and incapable of mis-interpretation, regardless of any language barrier.

Applications

Geometrical tolerances are applied over and above normal dimensional tolerances when it is necessary to control more precisely the form or shape of some feature of a manufactured part, because of the particular duty that the part has to perform. In the past, the desired qualities would have been obtained by adding to drawings such expressions as 'surfaces to be true with one another', 'surfaces to be square with one another', 'surfaces to be flat and parallel', etc., and leaving it to workshop tradition to provide a satisfactory interpretation of the requirements.

Advantages

Geometrical tolerances are used to convey in a brief and precise manner complete geometrical requirements on engineering drawings. They should always be considered for surfaces which come into contact with other parts, especially when close tolerances are applied to the features concerned.

No language barrier exists, as the symbols used are in agreement with published recommendations of the International Organization for Standardization (ISO) and have been internationally agreed. BS 8888 incorporates these symbols.

Caution – It must be emphasized that geometrical tolerances should be applied only when real advantages result, when normal methods of dimensioning are considered inadequate to ensure that the design function is kept, especially where repeatability must be guaranteed. Indiscriminate use of geometrical tolerances could increase costs in manufacture and inspection. Tolerances should be as wide as possible, as the satisfactory design function permits.

General rules

The symbols relating to geometrical characteristics are shown in Fig. 20.1 with additional symbols used in tolerancing in Fig. 20.1A. Examination of the various terms – flatness, straightness, concentricity, etc. – used to describe the geometrical characteristics shows that one type of geometrical tolerance can control another form of geometrical error.

For example, a positional tolerance can control perpendicularity and straightness; parallelism, perpendicularity, and angularity tolerances can control flatness.

The use of geometrical tolerances does not involve or imply any particular method of manufacture or inspection. Geometrical tolerances shown in this book, in keeping with international conventions, must be met regardless of feature size unless modified by one of the following conditions:

(a) Maximum material condition, denoted by the symbol Ⓜ describes a part, which contains the maximum amount of material, i.e. the minimum size hole or the maximum size shaft.
(b) Least material condition, denoted by the symbol Ⓛ describes a part, which contains the minimum amount of material, i.e. the maximum size hole or the minimum size shaft.

Theoretically exact dimensions (TED's)

These dimensions are identified by enclosure in a rectangular box, e.g. $\boxed{50}$ $\boxed{60°}$ $\boxed{Ø30}$ and are commonly known as 'Boxed dimensions' or 'True Position' dimensions. They define the true position of a hole, slot, boss profile, etc.

TED's are never individually toleranced but are always accompanied by a positional or zone tolerance specified within the tolerance frame referring to the feature (see Fig. 20.2).

Note: If two or more groups of features are shown on the same axis, they shall be considered to be a single pattern when not related to a datum.

Type of tolerance	Characteristics to be toleranced	Symbol	Datum needed	Applications
Form	Straightness	—	No	A straight line. The edge or axis of a feature.
	Flatness	▱	No	A plane surface.
	Roundness	○	No	The periphery of a circle. Cross-section of a bore, cylinder, cone or sphere.
	Cylindricity	⌭	No	The combination of circularity, straightness and parallelism of cylindrical surfaces. Mating bores and plungers.
	Profile of a line	⌒	No	The profile of a straight or irregular line.
	Profile of a surface	⌓	No	The profile of a straight or irregular surface.
Orientation	Parallelism	//	Yes	Parallelism of a feature related to a datum. Can control flatness when related to a datum.
	Perpendicularity	⊥	Yes	Surfaces, axes, or lines positioned at right angles to each other.
	Angularity	∠	Yes	The angular displacement of surfaces, axes, or lines from a datum.
	Profile of a line	⌒	Yes	The profile of a straight or irregular line positioned by theoretical exact dimensions with respect to datum plane(s).
	Profile of a surface	⌓	Yes	The profile of a straight or irregular surface positioned by theoretical exact dimensions with respect to datum plane(s).
Location	Position	⌖	See note below	The deviation of a feature from a true position.
	Concentricity and coaxiality	◎	Yes	The relationship between two circles having a common centre or two cylinders having a common axis.
	Symmetry	═	Yes	The symmetrical position of a feature related to a datum.
	Profile of a line	⌒	Yes	The profile of a straight or irregular line positioned by theoretical exact dimensions with respect to datum plane(s).
	Profile of a surface	⌓	Yes	The profile of a straight or irregular surface positioned by theoretical exact dimensions with respect to datum plane(s).
Runout	Circular runout	↗	Yes	The position of a point fixed on a surface of a part which is rotated 360° about its datum axis.
	Total runout	↗↗	Yes	The relative position of a point when traversed along a surface rotating about its datum axis.

FIGURE 20.1

Additional symbols

Description	Symbols
Toleranced feature indication	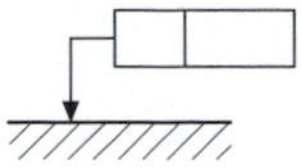
Datum feature indication	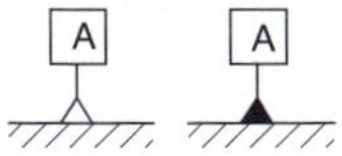 A A
Datum target indication	 Ø 2 / A1
Theoretically exact dimension	50
Projected tolerance zone	Ⓟ
Maximum material requirement	Ⓜ
Least material requirement	Ⓛ
Free state condition (non-rigid parts)	Ⓕ
All around profile	 ⌀
Envelope requirement	Ⓔ
Common zone	CZ
Minor diameter	LD
Major diameter	MD
Pitch diameter	PD
Line element	LE
Not convex	NC
Any cross-section	ACS

FIGURE 20.1A

Minor diameter	LD
Major diameter	MD
Pitch diameter	PD
Line element	LE
Not convex	NC
Any cross section	ACS
Unilateral or unequally disposed tolerance	UZ
Median feature	Ⓐ
All around (profile)	 ⌀
Between (two points)	◄──►
From (two points)	──►
Counterbore or Spotface	⊔
Countersink	⌄
Deep / Depth	⤓
Intersection plane	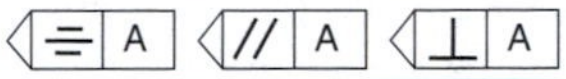 ⟨= A⟩ ⟨// A⟩ ⟨⊥ A⟩
Orientation plane	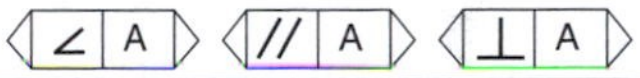 ⟨∠ A⟩ ⟨// A⟩ ⟨⊥ A⟩

FIGURE 20.1B

Definitions

Limits: The maximum and minimum dimensions for a given feature are known as the 'limits'. For example, 20 0.1.

The upper and lower limits of size are 20.1 and 19.9 mm, respectively.

Tolerance: The algebraic difference between the upper and lower limit of size is known as the 'tolerance'. In the example above, the tolerance is 0.2 mm. The tolerance is the amount of variation permitted.

Nominal dimension:Limits and tolerances are based on 'nominal dimensions' which are target dimensions.

In practice there is no such thing as a nominal dimension, since no part can be manufactured to a theoretical exact size.

The limits referred to above can be set in two ways:

(a) *unilateral* limits – limits set wholly above or below the nominal size;
(b) *bilateral* limits – limits set partly above and partly below the nominal size.

Geometrical tolerance: These tolerances specify the maximum error of a component's geometrical characteristic, over its whole dimensioned length or surface. Defining a zone in which the feature may lie does this.

Tolerance zone: A tolerance zone is the space in which any deviation of the feature must be contained
  e.g. – the space within a circle;

the space between two concentric circles;
the space between two equidistant lines or two parallel straight lines;
the space within a cylinder;
the space between two coaxial cylinders;
the space between two equidistant surfaces or two parallel planes;
the space within a sphere.

The tolerance applies to the whole extent of the considered feature unless otherwise specified.

FIGURE 20.2

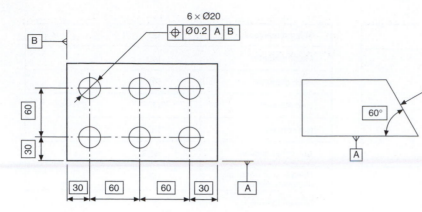

Method of indicating geometrical tolerances on drawings

Geometrical tolerances are indicated by stating the following details in compartments in a rectangular frame:

(a) the characteristic symbol, for single or related features;
(b) the tolerance value;
    (i) preceded by Ø if the zone is circular or cylindrical,
    (ii) preceded by SØ if the zone is spherical;
(c) letter or letters identifying the datum or datum systems.

    Figure 20.3 shows examples.

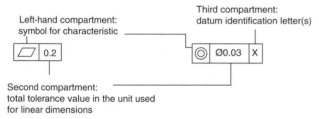

FIGURE 20.3

Methods of applying the tolerance frame to the toleranced feature

Figures 20.4 and 20.5 illustrate alternative methods of referring the tolerance to the surface or the plane itself. Note that

in Fig. 20.5 the dimension line and frame leader line are offset.

    The tolerance frame as shown in Fig. 20.6 refers to the axis or median plane only of the dimensioned feature.

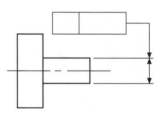

FIGURE 20.6

    Figure 20.7a illustrates the method of referring the tolerance to the axis or median plane. Note that the dimension line and frame leader line are drawn in line.

    Figure 20.7b illustrates an alternative way of referring to an axis or median feature, this has been introduced to aid the annotation of 3D models (see Chapter 25). In this method the tolerance frame is connected to the feature by a leader line terminating with an arrowhead pointing directly at the surface, but with the addition of the modifier symbol Ⓐ (median feature) placed to the right-hand end of the second compartment of the tolerance frame.

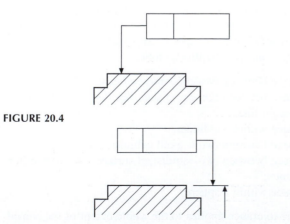

FIGURE 20.4

FIGURE 20.5

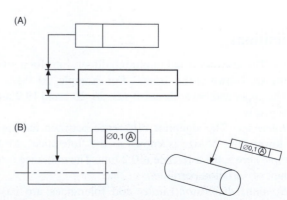

FIGURE 20.7

Procedure for positioning remarks which are related to tolerance

Remarks related to the tolerance for example '6x' should be written above the frame (see Figs. 20.8 and 20.9).

4 × Ø10

| ⊕ | Ø 0.01 |

FIGURE 20.8

6 ×

| ⊕ | Ø0.1 |

FIGURE 20.9

Indications qualifying the feature within the tolerance zone should be written under the tolerance frame (see Fig. 20.10).

| ▱ | 0.3 |

NC

FIGURE 20.10

If it is necessary to specify more than one tolerance characteristic for a feature, the tolerance specification should be given in tolerance frames positioned one under the other as shown in Fig. 20.11.

| ○ | 0.01 | |
| // | 0.06 | B |

FIGURE 20.11

The application of tolerances to a restricted length of a feature

Figure 20.12 shows the method of applying a tolerance to only a particular part of a feature.

The tolerance frame in Fig. 20.13 shows the method of applying another tolerance, similar in type but smaller in magnitude, on a shorter length. In this case, the whole flat surface must lie between parallel planes 0.2 apart, but over any length of 180 mm, in any direction, the surface must be within 0.05.

Figure 20.14 shows the method used to apply a tolerance over a given length; it allows the tolerance to accumulate over a longer length. In this case, straightness tolerance of

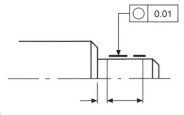

FIGURE 20.12

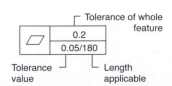

FIGURE 20.13

0.02 is applicable over a length of 100 mm. If the total length of the feature was 800 mm, then the total permitted tolerance would accumulate to 0.16.

FIGURE 20.14

Tolerance zones

The width of the tolerance zone is in the direction of the leader line arrow joining the symbol frame to the toleranced feature unless the tolerance zone is preceded by the symbol Ø. An example is given in Fig. 20.15a.

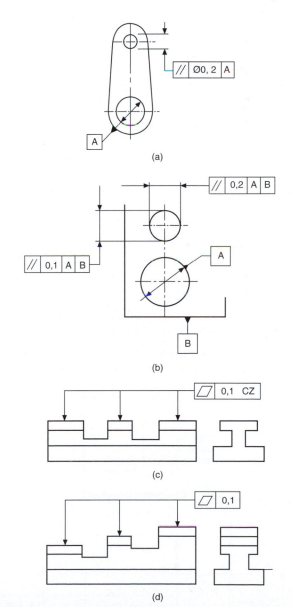

FIGURE 20.15

If two tolerances are given, then they are considered to be perpendicular to each other, unless otherwise stated. Figure 20.15b shows an example.

Figure 20.15c gives an example where a single tolerance zone is applied to several separate features. In the tolerance frame the symbol 'CZ' is added. CZ is the standard abbreviation for 'Common Zone'.

Figure 20.15d gives an example where individual tolerance zones of the same valve are applied to several separate features.

Projected toleranced zone

Figure 20.16 shows a part section through a flange where it is required to limit the variation in perpendicularity of each hole axis. The method used is to apply a tolerance to a projected zone. The enlargement shows a possible position for the axis through one hole. Each hole axis must lie somewhere within a projected cylinder of Ø0.02 and 30 deep.

As an alternative to indicating the projected tolerance zone using supplemental geometry, the length of projection can also be specified indirectly by adding the value, after the ⓟ symbol, in the tolerance frame (Fig. 20.16b). This method of indication is only applicable to blind holes.

Note: Projected tolerance zones are indicated by the symbol ⓟ.

Note that the zone is projected from the specified datum.

Datums

A datum surface on a component should be accurately finished, since other locations or surfaces are established by measuring from the datum. Figure 20.17 shows a datum surface indicated by the letter A.

In this example, the datum edge is subject to a straightness tolerance of 0.05, shown in the tolerance frame.

FIGURE 20.16

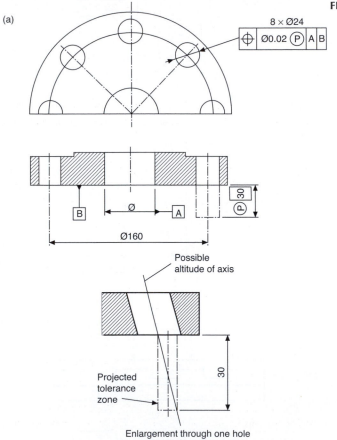

(a)

$8 \times Ø24$

⊕ | Ø0.02 | ⓟ | A | B

Ø160

Possible altitude of axis

Projected tolerance zone

30

Enlargement through one hole

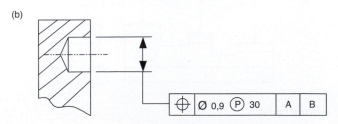

(b)

⊕ | Ø 0,9 | ⓟ | 30 | A | B

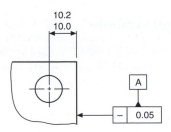

FIGURE 20.17

Methods of specifying datum features

A datum is designated by a capital letter enclosed by a datum box. The box is connected to a solid or a blank datum triangle. There is no difference in understanding between solid or blank datum triangles. Figures 20.18 and 20.19 show alternative methods of designating a flat surface as Datum A.

FIGURE 20.18

FIGURE 20.19

Figure 20.20 illustrates alternative positioning of datum boxes. Datum A is designating the main outline of the feature. The shorter stepped portion Datum B is positioned on an extension line, which is clearly separated from the dimension line.

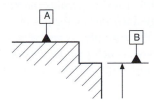

FIGURE 20.20

Figure 20.21 shows the datum triangle placed on a leader line pointing to a flat surface.

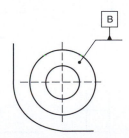

FIGURE 20.21

Figures 20.22, 20.23 and 20.24 illustrate the positioning of a datum box on an extension of the dimension line, when the datum is the axis or median plane or a point defined by the dimensioned feature.

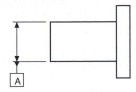

FIGURE 20.22

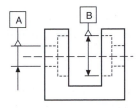

FIGURE 20.23

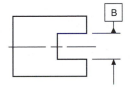

FIGURE 20.24

Note: If there is insufficient space for two arrowheads, one of them may be replaced by the datum triangle, as shown in Fig. 20.24.

Multiple datums: Figure 20.25 illustrates two datum features of equal status used to establish a single datum plane. The reference letters are placed in the third compartment of the tolerance frame, and have a hyphen separating them.

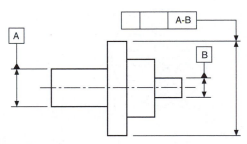

FIGURE 20.25

Figure 20.26 shows the drawing instruction necessary when the application of the datum features is required in a particular order of priority.

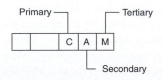

FIGURE 20.26

Figure 20.27 shows an application where a geometrical tolerance is related to two separate datum surfaces indicated in order of priority.

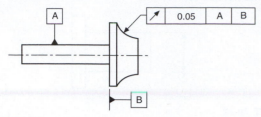

FIGURE 20.27

If a positional tolerance is required as shown in Fig. 20.28 and no particular datum is specified, then the individual feature to which the geometrical-tolerance frame is connected is chosen as the datum.

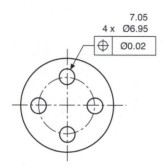

FIGURE 20.28

Figure 20.29 illustrates a further positional-tolerance application where the geometrical requirement is related to another feature indicated as the datum. In the given example, this implies that the pitch circle and the datum circle must be coaxial, i.e. they have common axes.

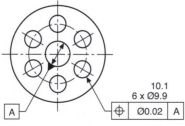

FIGURE 20.29

Datum targets

Surfaces produced by forging casting or sheet metal may be subject to bowing, warping or twisting; and not necessarily be flat. It is therefore impractical to designate an entire surface as a functional datum because accurate and repeatable measurements cannot be made from the entire surface.

In order to define a practical datum plane, selected points or areas are indicated on the drawing. Manufacturing processes and inspection utilizes these combined points or areas as datums.

Datum target symbols

The symbol for a datum target is a circle divided by a horizontal line (see Fig. 20.30). The lower part identifies the datum target. The upper area may be used only for information relating to datum target.

FIGURE 20.30

Indication of datum targets

If the datum target is:

(a) a point: it is indicated by a cross . . . . X.
(b) a line: it is indicated by two crosses, connected by a thin line . . . . X ————X.
(c) an area: it is indicated by a hatched area surrounded by a thin double dashed chain.

All symbols appear on the drawing view which most clearly shows the relevant surface.

Practical application of datum targets

Interpretation – In Figure 20.32, it is understood in that:

(a) Datum targets A1, A2 and A3 establish Datum A.
(b) Datum targets B1 and B2 establish Datum B.
(c) Datum target C1 establishes Datum C.

Dimensioning and tolerancing non-rigid parts

The basic consideration is that distortion of a non-rigid part must not exceed that which permits the part to be brought within the specified tolerances for positioning, at assembly and verification. For example, by applying pressure or forces not exceeding those which may be expected under normal assembly conditions (Figs. 20.31 and 20.32).

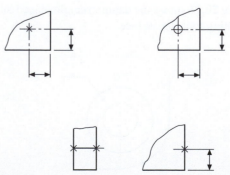

FIGURE 20.31

FIGURE 20.32

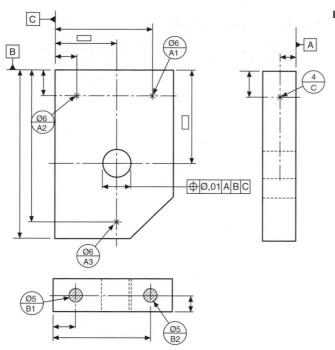

Definitions

(a) A non-rigid part relates to the condition of that part which deforms in its free state to an extent beyond the dimensional and geometrical tolerances on the drawing.

(b) Free-state relates to the condition of a part when subjected only to the force of gravity.

(c) The symbol used is Ⓕ.

Figure 20.33 shows a typical application of a buffer detail drawing. In its restrained condition, Datum's A and B position the buffer.

Interpretation – The geometrical tolerance followed by symbol Ⓕ is maintained in its free state. Other geometrical tolerances apply in its assembled situation.

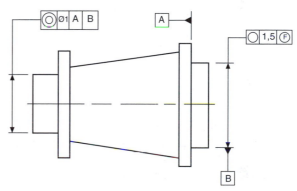

FIGURE 20.33

Chapter 21

Application of geometrical tolerances

In this chapter, examples are given of the application of tolerances to each of the characteristics on engineering drawings by providing a typical specification for the product and the appropriate note which must be added to the drawing. In every example, the tolerance values given are only typical figures: the product designer would normally be responsible for the selection of tolerance values for individual cases.

Straightness

A straight line is the shortest distance between two points. A straightness tolerance controls:

1. the straightness of a line on a surface,
2. the straightness of a line in a single plane,
3. the straightness of an axis.

Case 1

Product requirement

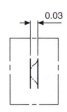

The specified line shown on the surface must lie between two parallel straight lines 0.03 apart.

Drawing instruction

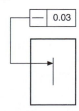

A typical application for this type of tolerance could be a graduation line on an engraved scale.

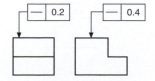

In the application shown at left column, tolerances are given controlling the straightness of two lines at right angles to one another. In the left-hand view the straightness control is 0.2, and in the right-hand view 0.4. As in the previous example, the position of the graduation marks would be required to be detailed on a plan view.

Case 2

Product requirement

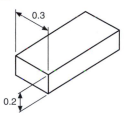

The axis of the whole part must lie in a boxed zone of 0.3 × 0.2 over its length.

Drawing instruction

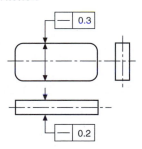

As indicated, the straightness of the axis is controlled by the dimensions of the box, and could be applied to a long-rectangular key.

Case 3

Product requirement

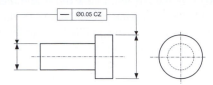

The axis of the whole part must lie within the cylindrical-tolerance zone of 0.05.

Drawing instruction

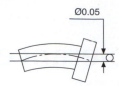

Case 4

Product requirement

The geometrical tolerance may be required to control only part of the component. In this example the axis of the dimensioned portion of the feature must lie within the cylindrical-tolerance zone of 0.1 diameter.

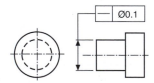

Drawing instruction

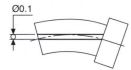

Flatness

Flatness tolerances control the divergence or departure of a surface from a true plane.

The tolerance of flatness is the specified zone between two parallel planes. It does not control the squareness or parallelism of the surface in relation to other features, and it can be called for independently of any size tolerance.

Case 1

Product requirement

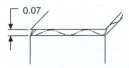

The surface must be contained between two parallel planes 0.07 apart.

Drawing instruction

Case 2

Product requirement

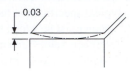

The surface must be contained between two parallel planes 0.03 apart, but must not be convex.

Drawing instruction

Note that these instructions could be arranged to avoid a concave condition.

Circularity (roundness)

Circularity is a condition where any point of a feature's continuous curved surface is equidistant from its centre, which lies in the same plane.

The tolerance of circularity controls the divergence of the feature, and the annular space between the two co-planar concentric circles defines the tolerance zone, the magnitude being the algebraic difference of the radii of the circles.

Case 1

Product requirement

The circumference of the bar must lie between two co-planar concentric circles 0.5 apart.

Drawing instruction

Note that, at any particular section, a circle may not be concentric with its axis but may still satisfy a circularity tolerance. The following diagram shows a typical condition.

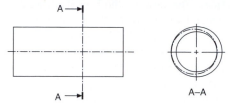

Case 2

Product requirement
   The circumference at any cross-section must lie between two co-planar concentric circles 0.02 apart.

Drawing instruction

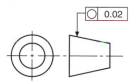

Case 3

Product requirement
   The periphery of any section of maximum diameter of the sphere must lie between concentric circles a radial distance 0.001 apart in the plane of the section.

Drawing instruction

Cylindricity

The combination of parallelism, circularity and straightness defines cylindricity when applied to the surface of a cylinder, and is controlled by a tolerance of cylindricity. The tolerance zone is the annular space between two coaxial cylinders, the radial difference being the tolerance value to be specified.

   It should be mentioned that, due to difficulties in checking the combined effects of parallelism, circularity and

straightness, it is recommended that each of these characteristics is toleranced and inspected separately.

   Product requirement
   The whole curved surface of the feature must lie between an annular-tolerance zone 0.04 wide formed by two cylindrical surfaces coaxial with each other.

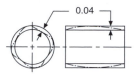

Drawing instruction

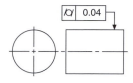

Profile tolerance of a line

Profile tolerance of a line is used to control the ideal contour of a feature. The contour is defined by theoretically exact boxed dimensions and must be accompanied by a relative-tolerance. This tolerance zone, unless otherwise stated, is taken to be equally disposed about the true form, and the tolerance value is equal to the diameter of circles whose centres lie on the true form. If it is required to call for the tolerance zone to be positioned on one side of the true form (i.e. unilaterally), the circumferences of the tolerance-zone circles must touch the theoretical contour of the profile.\

Case 1

Product requirement
   The profile is required to be contained within the bilateral-tolerance zone.

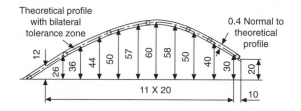

Drawing instruction

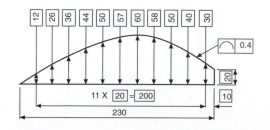

Case 2

Product requirement

The profile is required to be contained within the unilateral or unequal-tolerance zone.

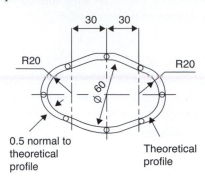

Drawing instruction (traditional 2D method)

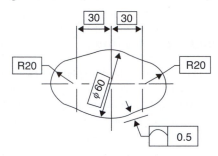

p0540

Drawing instruction (new 2D and 3D method)

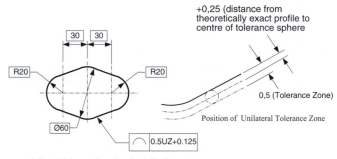

Unilateral Tolerance Zone Drawing Indication

+0,25 (distance from theoretically exact profile to centre of tolerance sphere

0,5 (Tolerance Zone)

Position of Unilateral Tolerance Zone

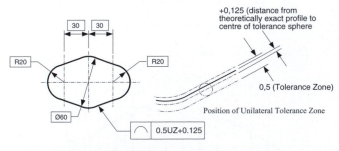

Unequal Tolerance Zone Drawing Indication

+0,125 (distance from theoretically exact profile to centre of tolerance sphere

0,5 (Tolerance Zone)

Position of Unilateral Tolerance Zone

Drawing instruction (new 2D and 3D method)

Prompted by the increase in 3D specification and the publication of BS ISO 16792 – Digital product definition data practices; the new method as illustrated above has been introduced to eliminate the need to define the position of the tolerance zone by using supplemental geometry. No graphical method of specifying unilateral or unequal-tolerance zones existed prior to this new method. In this case a modifier UZ is placed in the tolerance frame after the tolerance value to signify a unilateral or unequal condition. A value as to the location of the centre of the tolerance zone in relation to the theoretical profile is then placed after the UZ modifier. A plus (+) value indicates the tolerance zone is positioned outwards from the theoretical profile and a minus (−) value indicates the tolerance zone is positioned inwards from the theoretical profile. As the case illustrated is a unilateral tolerance, the modifying value is half of the total tolerance.

This method can be used for both profile of a line and profile of a surface.

The figure below shows an example where the toleranced profile of a feature has a sharp corner. The inner-tolerance zone is considered to end at the intersection of the inner boundary lines, and the outer-tolerance zone is considered to extend to the outer boundary-line intersections. Sharp corners such as these could allow considerable rounding; if this is desirable, then the design requirement must be clearly defined on the drawing by specifying a radius or adding a general note such as 'ALL CORNERS 0.5 MAX'. It should be noted that such radii apply regardless of the profile tolerance.

In the example given, the product is required to have a sharp corner.

Product requirement

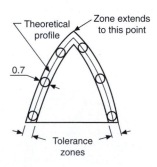

Drawing instruction

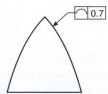

Profile tolerance of a surface

Profile tolerance of a surface is used to control the ideal form of a surface, which is defined by theoretically exact boxed dimensions and must be accompanied by a relative-tolerance zone. The profile-tolerance zone, unless otherwise stated, is taken to be bilateral and equally disposed about its true-form surface. The tolerance value is equal to the diameter of spheres whose centre lines lie on the true form of the surface. The zone is formed by surfaces which touch the circumferences of the spheres on either side of the ideal form.

If it is required to call for a unilateral-tolerance zone, then the circumferences of the tolerance-zone spheres must touch the theoretical contour of the surface.

Product requirement

SØ0.3

The tolerance zone is to be contained by upper and lower surfaces which touch the circumference of spheres 0.3 diameter whose centres lie on the theoretical form of the surface.

Drawing instruction

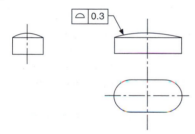

Parallelism

Two parallel lines or surfaces are always separated by a uniform distance. Lines or surfaces may be required to be parallel with datum planes or axes. Tolerance zones may be the space between two parallel lines or surfaces, or the space contained within a cylinder positioned parallel to its datum. The magnitude of the tolerance value is the distance between the parallel lines or surfaces, or the cylinder diameter.

Case 1

Product requirement

The axis of the hole on the left-hand side must be contained between two straight lines 0.2 apart, parallel to the datum axis A and lying in the same vertical plane.

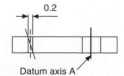

Datum axis A

Drawing instruction

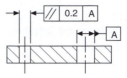

Case 2

Product requirement

The axis of the upper hole must be contained between two straight lines 0.3 apart which are parallel to and symmetrically disposed about the datum axis A and lie in the same horizontal plane.

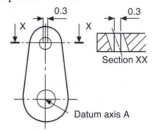

Section XX

Datum axis A

Drawing instruction

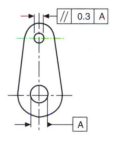

Case 3

Product requirement

The upper hole axis must be contained in a cylindrical zone 0.4 diameter, with its axis parallel to the datum axis A.

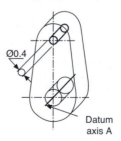

Ø0.4

Datum axis A

Drawing instruction

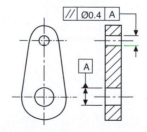

Case 4

Product requirement

The axis of the hole on the left-hand side must be contained in a tolerance box $0.5 \times 0.2 \times$ width, as shown, with its sides parallel to the datum axis A and in the same horizontal plane.

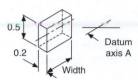

Drawing instruction

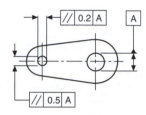

Case 5

Product requirement

The axis of the hole must be contained between two planes 0.06 apart parallel to the datum surface A.

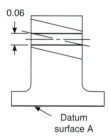

Drawing instruction

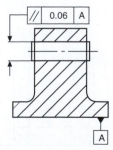

Case 6

Product requirement

The top surface of the component must be contained between two planes 0.7 apart and parallel to the datum surface A.

Drawing instruction

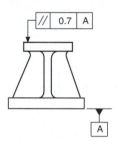

Perpendicularity (squareness)

Perpendicularity is the condition when a line, plane or surface is at right angles to a datum feature.

The tolerance zone is the space between two parallel lines or surfaces; it can also be the space contained within a cylinder. All tolerance zones are perpendicular to the datum feature.

The magnitude of the tolerance value is the specified distance between these parallel lines or surfaces, or the diameter of the cylinder.

Case 1

Product requirement

The axis of the vertical hole must be contained between two planes 0.1 apart which are perpendicular to the datum axis.

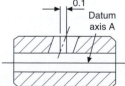

Drawing instruction

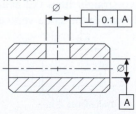

Case 2

Product requirement

The axis of the upright must be contained between two straight lines 0.2 apart which are perpendicular to the datum. Squareness is controlled here in one plane only.

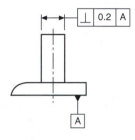

Drawing instruction

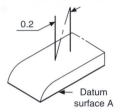

Case 3

Product requirement

The axis of the column must be contained in a cylindrical-tolerance zone 0.3 diameter, the axis of which is perpendicular to the datum surface A. Squareness is controlled in more than one plane by this method.

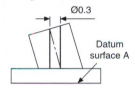

Drawing instruction

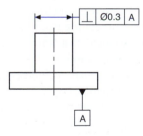

Case 4

Product requirement

The axis of the column must be contained in a tolerance zone box 0.2 × 0.4 which is perpendicular to the datum surface A.

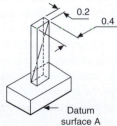

Drawing instruction

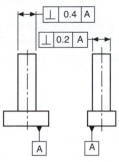

Case 5

Product requirement

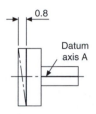

The left-hand end face of the part must be contained between two parallel planes 0.8 apart and perpendicular to the datum axis A.

Drawing instruction

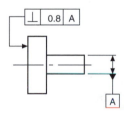

Case 6

Product requirement

The left-hand surface must be contained between two parallel planes 0.7 apart and perpendicular to the datum surface A.

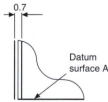

Drawing instruction

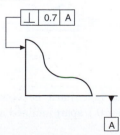

Angularity

Angularity defines a condition between two related planes, surfaces, or lines which are not perpendicular or parallel to one another. Angularity tolerances control this relationship.

The specified angle is a basic dimension, and is defined by a theoretically exact boxed dimension and must be accompanied by a tolerance zone. This zone is the area between two parallel lines inclined at the specified angle to the datum line, plane, or axis. The tolerance zone may also be the space within a cylinder, the tolerance value being equal to the cylinder diameter. In this case, symbol ø precedes tolerance value in the tolerance frame.

Case 1

Product requirement

The inclined surface must be contained within two parallel planes 0.2 apart which are at an angle of 42° to the datum surface.

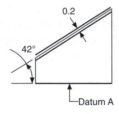

Drawing instruction

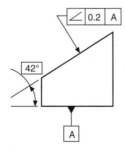

Case 2

Product requirement

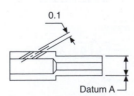

The axis of the hole must be contained within two parallel straight lines 0.1 apart inclined at 28° to the datum axis.

Drawing instruction

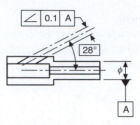

Case 3

Product requirement

The inclined surface must be contained within two parallel planes 0.5 apart which are inclined at 100° to the datum axis.

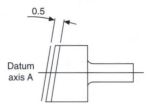

Drawing instruction

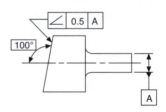

Circular run-out

Circular run-out is a unique geometrical tolerance. It can be a composite form control relating two or more characteristics, and it requires a practical test where the part is rotated through 360° about its own axis.

The results of this test may include errors of other characteristics such as circularity, concentricity, perpendicularity, or flatness, but the test cannot discriminate between them. It should therefore not be called for where the design function of the part necessitates that the other characteristics are to be controlled separately. The sum of any of these errors will be contained within the specified circular run-out tolerance value. The tolerance value is equal to the full indicator movement of a fixed point measured during one revolution of the component about its axis, without axial movement. Circular run-out is measured in the direction specified by the arrow at the end of the leader line which points to the toleranced feature. It must always be measured regardless of feature size, and it is convenient for practical purposes to establish the datum as the diameter(s) from which measurement is to be made, if true centres have not been utilized in manufacturing the part.

Case 1

Product requirement

The circular radial run-out must not exceed 0.4 at any point along the cylinder, measured perpendicular to the datum axis without axial movement.

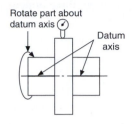

Drawing instruction

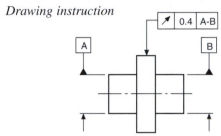

Case 2

Product requirement

Circular run-out must not exceed 0.2 measured at any point normal to the surface, without axial movement.

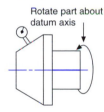

Drawing instruction

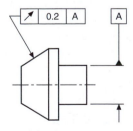

Case 3

Product requirement

At any radius, the circular run-out must not exceed 0.06 measured parallel to the datum axis.

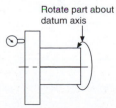

Drawing instruction

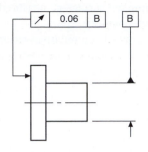

Case 4

Product requirement

The component is required to be rotated about datum axis C, with datum face B set to ensure no axial movement.

The circular radial run-out on the cylindrical portion must not exceed 0.05 at any point measured perpendicular to the datum axis.

The circular run-out on the tapered portion must not exceed 0.07 at any point measured normal to its surface.

The circular run-out on the curved portion must not exceed 0.04 at any point measured normal to its surface.

The axial run-out of the end face must not exceed 0.1 at any point measured parallel to the datum axis of rotation.

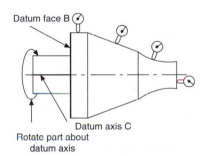

Drawing instruction

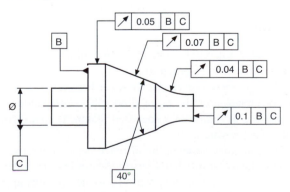

Circular run-out provides composite control of circular elements of a surface.

Total run-out provides composite control of all surface elements. The complete surface is measured, and not single points, as in circular run-out.

Total run-out controls cumulative variations of perpendicularity which can detect wobble, also flatness which can detect concavity and convexity.

Total run-out

Case 1

Product requirement

The total run-out must not exceed 0.06 at any point measured across the entire surface parallel to the datum axis.

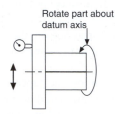

Drawing instruction

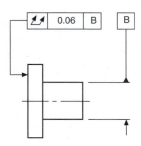

Note: The symbol means that the measuring instrument is guided across a theoretically exact surface true to the datum axis.

Position

A positional tolerance controls the location of one feature from another feature or datum.

The tolerance zone can be the space between two parallel lines or planes, a circle, or a cylinder. The zone defines the permissible deviation of a specified feature from a theoretically exact position.

The tolerance value is the distance between the parallel lines or planes, or the diameters of the circle or cylinder.

The theoretically exact position also incorporates squareness and parallelism of the tolerance zones with the plane of the drawing.

Case 1

Product requirement

The point must be contained within a circle of 0.1 diameter in the plane of the surface. The circle has its centre at the intersection of the two theoretically exact dimensions. If the point were to be located by three dimensions, the tolerance zone would be a sphere.

Drawing instruction

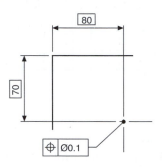

Case 2

Product requirement

The axis of the hole must be contained in a cylindrical-tolerance zone of 0.3 diameter with its axis coincident with the theoretically exact position of the hole axis.

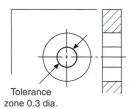

Drawing instruction

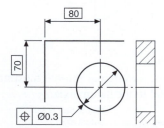

Case 3

Product requirement

Each line must be contained between two parallel straight lines on the surface, 0.2 apart, which are symmetrical with the theoretically exact positions of the required lines.

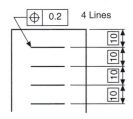

Drawing instruction

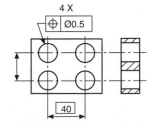

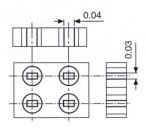

Drawing instruction

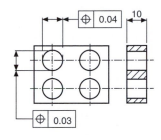

Case 4

Product requirement

The axes of each of the four holes must be contained in a cylindrical-tolerance zone of 0.5 diameter, with its own axis coincident with the theoretically exact position of each hole.

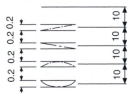

Drawing instruction

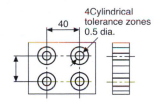

Case 5

Product requirement

The axes of each of the four holes must be contained in a boxed zone of 0.04 × 0.03 × 10, symmetrically disposed about the theoretically exact position of each hole axis.

Case 6

Product requirement

The angled surface must be contained between two parallel planes 0.7 apart, which are symmetrically disposed about the theoretically exact position of the surface relative to datum axis *X* and datum face *Y*.

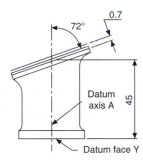

Drawing instruction

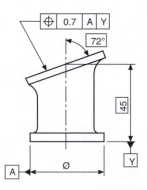

Case 7

Product requirement

The axes of the two holes must be contained in cylindrical-tolerance zones of 0.01 diameter, with their own axes coincident with the theoretically exact hole positions related to datum face *X* and the datum centre-line axis *Y*.

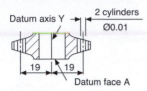

Drawing instruction

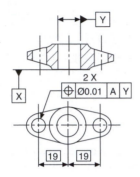

Concentricity and coaxiality

Two circles are said to be concentric when their centres are coincident.

Two cylinders are said to be coaxial when their axes are coincident.

The deviation from the true centre or datum axis is controlled by the magnitude of the tolerance zone.

Case 1 (Concentricity)

Product requirement

To contain the centre of the large circle within a circular-tolerance zone of 0.001 diameter which has its centre coincident with the datum-circle centre.

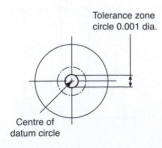

Drawing instruction

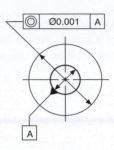

Case 2 (Coaxiality)

Product requirement

To contain the axis of the right-hand cylinder within a cylindrical-tolerance zone which is coaxial with the axis of the datum cylinder.

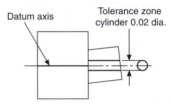

Drawing instruction

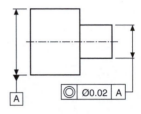

Case 3 (Coaxiality)

Product requirement

To contain the axes of both the left- and right-hand cylinders within a cylindrical-tolerance zone.

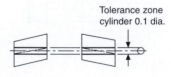

Drawing instruction

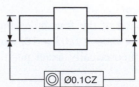

Case 4 (Coaxiality)

Product requirement

   To contain the axis of the central cylinder within a cylindrical-tolerance zone which is coaxial with the mean axes of the left- and right-hand cylinders.

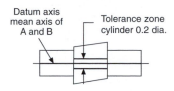

Datum axis
mean axis of
A and B

Tolerance zone
cylinder 0.2 dia.

Drawing instruction

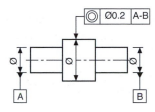

Symmetry

Symmetry involves the division of spacing of features so that they are positioned equally in relation to a datum which may be a line or plane. The tolerance zone is the space between two parallel lines or planes, parallel to, and positioned symmetrically with the datum. The tolerance magnitude is the distance between these two parallel lines or planes.

   Symmetry also implies perpendicularity with the plane of the drawing where depth is involved.

Case 1

Product requirement

   The specified line XX must lie in a tolerance zone formed by two parallel straight lines 0.01 apart and disposed symmetrically between datums *A* and *B*.

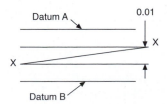

Datum A

0.01

X

X

Datum B

Drawing instruction

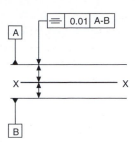

Case 2

Product requirement

   The axis of a hole in a plate must lie in a rectangular-box tolerance zone $0.03 \times 0.06 \times$ depth of the plate, parallel with and symmetrically disposed about the common median planes formed by slots AC and BD.

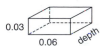

0.03

0.06

depth

Drawing instruction

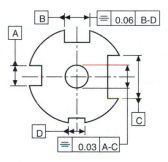

Chapter 22

Maximum material and least material principles

Maximum material condition (MMC)

MMC is that condition of a part or feature which contains the maximum amount of material, e.g. minimum-size hole, or a maximum-size shaft. In certain cases its use allows an increase in the specified tolerance if it is indicated that the tolerance applies to the feature at its maximum material condition.

The *maximum material principle* takes into account the mutual dependence of tolerances of size, form, orientation and/or location and permits additional tolerance as the considered feature departs from its maximum material condition.

The free assembly of components depends on the combined effect of the actual finished sizes and the errors of form or position of the parts.

Any errors of form or position between two mating parts have the effect of virtually altering their respective sizes. The tightest condition of assembly between two mating parts occurs when each feature is at the MMC, plus the maximum errors permitted by any required geometrical tolerance.

Assembly clearance is increased if the actual sizes of the mating features are finished away from their MMC, and if any errors of form or position are less than that called for by any geometrical control. Also, if either part is finished away from its MMC, the clearance gained could allow for an increased error of form or position to be accepted. Any increase of tolerance gained this way, provided it is functionally acceptable for the design, is advantageous during the process of manufacture (Fig. 22.1).

The symbol for maximum material condition is the letter M enclosed by a circle, Ⓜ. The symbol is positioned in the tolerance frame as follows:

(a) | ⊕ | Ø 0.05 Ⓜ | A | Refers to the tolerance only.

(b) | ⊕ | Ø 0.05 | A Ⓜ | Refers to the datum only.

(c) | ⊕ | Ø 0.05 Ⓜ | A Ⓜ | Refers to both tolerance and datum.

Least material condition (LMC)

LMC is that condition of a part or feature which contains the minimum amount of material, e.g. maximum-size hole or a minimum-size shaft.

Circumstances do arise where, for example, a designer would require to limit the minimum wall thickness between a hole and the side of a component. In such a case, we need to control the least material condition where a part contains the minimum amount of material.

The appropriate tolerance would then be quoted, followed by the letter L inside a circle Ⓛ.

Generally such examples are very few. The applications which follow cover the more widely found conditions of MMC.

Maximum material condition related to geometrical form

The limit of size, together with geometrical form or position of a feature, are factors of the maximum material principle, and its application is restricted to those features whose size is specified by toleranced dimensions incorporating an axis or median plane. It can never be applied to a plane, surface, or line on a surface.

The characteristics to which it can be applied are as follows:

- straightness,
- parallelism,
- squareness,
- angularity,

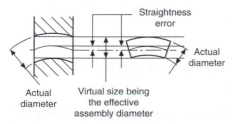

FIGURE 22.1

- position,
- concentricity,
- symmetry.

The characteristics to which the maximum material condition concept cannot be applied are as follows:

- flatness,
- roundness,
- cylindricity,
- profile of a line,
- profile of a surface,
- run-out.

Maximum material condition applied to straightness

Figure 22.2 shows a typical drawing instruction where limits of size are applied to a pin, and in addition a straightness tolerance of 0.2 is applicable at the maximum material condition.

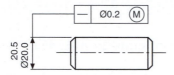

FIGURE 22.2

Figure 22.3 shows the condition where the pin is finished at the maximum material condition with the maximum straightness error. The effective assembly diameter will be equal to the sum of the upper limit of size and the straightness tolerance.

The straightness error is contained within a cylindrical tolerance zone of ⌀ 0.2.

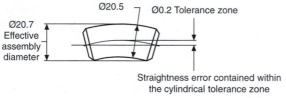

FIGURE 22.3

To provide the same assembly diameter of 20.7 as shown in Fig. 22.4 when the pin is finished at its low limit of size of 20.0, it follows that a straightness error of 0.7 could be acceptable. This increase may in some cases have no serious effect on the function of the component, and can be permitted.

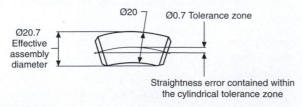

FIGURE 22.4

Maximum material condition applied to squareness

Figure 22.5 shows a typical drawing instruction where limits of size are applied to a pin, and in addition a squareness tolerance of 0.3 is applicable at the maximum material condition.

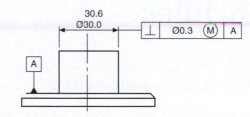

FIGURE 22.5

Figure 22.6 shows the condition where the pin is finished at the maximum material condition with the maximum squareness error of 0.3. The effective assembly diameter will be the sum of the upper limit of size and the squareness error. The squareness error will be contained within a cylindrical tolerance zone of ⌀ 0.3.

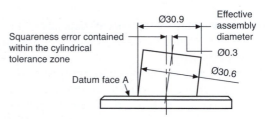

FIGURE 22.6

To provide the same assembly diameter of 30.9, as shown in Fig. 22.7, when the pin is finished at its low limit of size of 30.0, it follows that the squareness error could increase from 0.3 to 0.9. This permitted increase should be checked for acceptability.

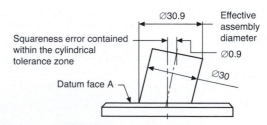

FIGURE 22.7

Maximum material condition applied to position

A typical drawing instruction is given in Fig. 22.8, and the following illustrations show the various extreme dimensions which can possibly arise.

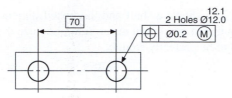

FIGURE 22.8

Condition A (Fig. 22.9)
Minimum distance between hole centres and the maximum material condition of holes.

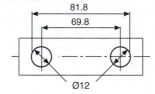

FIGURE 22.9

Condition B (Fig. 22.10)
Maximum distance between hole centres and maximum material condition of holes.

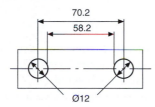

FIGURE 22.10

Condition C (Fig. 22.11)
To give the same assembly condition as in A, the minimum distance between hole centres is reduced when the holes are finished away from the maximum material condition.

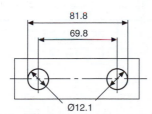

FIGURE 22.11

Condition D (Fig. 22.12)
To give the same assembly condition as in B, the maximum distance between hole centres is increased when the holes are finished away from the maximum material condition.

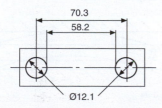

FIGURE 22.12

Note that the total tolerance zone is 0.2 + 0.1 = 0.3, and therefore the positional tolerance can be increased where the two holes have a finished size away from the maximum material condition.

Maximum material condition applied to coaxiality

In the previous examples, the geometrical tolerance has been related to a feature at its maximum material condition, and, provided the design function permits, the tolerance has increased when the feature has been finished away from the maximum material condition. Now the geometrical tolerance can also be specified in relation to a datum feature, and Fig. 22.13 shows a typical application and drawing instruction of a shoulder on a shaft. The shoulder is required to be coaxial with the shaft, which acts as the datum. Again, provided the design function permits, further relaxation of the quoted geometrical control can be achieved by applying the maximum material condition to the datum itself.

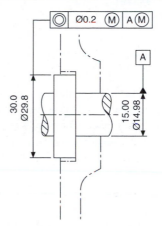

FIGURE 22.13

Various extreme combinations of size for the shoulder and shaft can arise, and these are given in the drawings below. Note that the increase in coaxiality error which could be permitted in these circumstances is equal to the total amount that the part is finished away from its maximum material condition, i.e. the shoulder tolerance plus the shaft tolerance.

Condition A (Fig. 22.14)

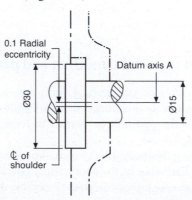

FIGURE 22.14

Shoulder and shaft at maximum material condition; shoulder at maximum permissible eccentricity to the shaft datum axis *X*.

Condition B (Fig. 22.15)

Shoulder at minimum material condition and shaft at maximum material condition. Total coaxiality tolerance = specified coaxiality tolerance + limit of size tolerance of shoulder = 0.2 + 0.2 = 0.4 diameter. This gives a maximum eccentricity of 0.2.

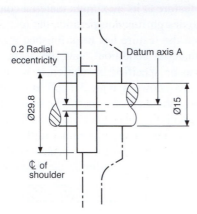

FIGURE 22.15

Condition C (Fig. 22.16)

Shows the situation where the smallest size shoulder is associated with the datum shaft at its low limit of size. Here, the total coaxiality tolerance which may be permitted is the sum of the specified coaxiality tolerance + limit of size tolerance for the shoulder + tolerance on the shaft = 0.2 + 0.2 + 0.02 = 0.42 diameter.

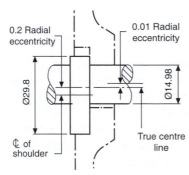

FIGURE 22.16

Maximum material condition and perfect form

When any errors of geometrical form are required to be contained within the maximum material limits of size, it is assumed that the part will be perfect in form at the upper limit of size.

In applying this principle, two conditions are possible.

Case 1: The value of the geometrical tolerance can progressively increase provided that the assembly diameter does not increase above the maximum material limit of size.

Figure 22.17 shows a shaft and the boxed dimension, and indicates that at maximum material limit of size the shaft is required to be perfectly straight.

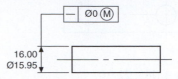

FIGURE 22.17

Figure 22.18 shows the shaft manufactured to its lower limit of size, where the permitted error in straightness can be 0.05, since the assembly diameter will be maintained at 16.00. Similarly, a shaft manufactured to, say, 15.97 can have a permitted straightness error of 0.03.

FIGURE 22.18

Case 2: The geometrical tolerance can also be limited to a certain amount where it would be undesirable for the part to be used in service too much out of line.

Figure 22.19 shows a shaft, with a tolerance frame indication that at the maximum material limit of size the shaft is required to be perfectly straight. Also, the upper part of the box indicates that a maximum geometrical tolerance error of 0.02 can exist, provided that for assembly purposes the assembly diameter does not exceed 14.00.

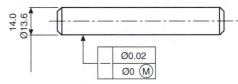

FIGURE 22.19

Figure 22.20 shows the largest diameter shaft acceptable, assuming that it has the full geometrical error of 0.02. Note that a shaft finished at 13.99 would be permitted a maximum straightness error of only 0.01 to conform with the drawing specification.

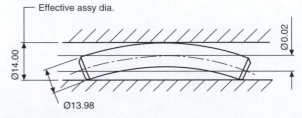

FIGURE 22.20

Figure 22.21 shows the smallest diameter shaft acceptable, and the effect of the full geometrical error of straightness.

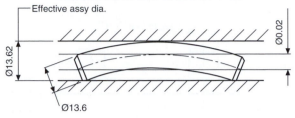

FIGURE 22.21

The application of maximum material condition and its relationship with perfect form and squareness

A typical drawing instruction is shown in Fig. 22.22.

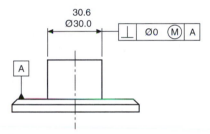

FIGURE 22.22

Condition A (Fig. 22.23)
Maximum size of feature: zero geometrical tolerance.

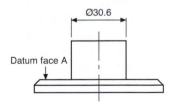

FIGURE 22.23

Condition B (Fig. 22.24)
Minimum size of feature; Permitted geometrical error = 0.6.

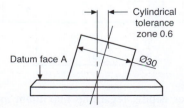

FIGURE 22.24

Note that between these extremes the geometrical tolerance will progressively increase; i.e. when the shaft diameter is 30.3, then the cylindrical tolerance error permitted will be 0.3.

The application of maximum material condition and its relationship with perfect form and coaxiality

A typical drawing instruction is shown in Fig. 22.25.

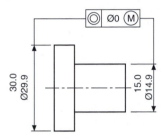

FIGURE 22.25

Condition A (Fig. 22.26)
Head and shank at maximum material condition. No geometrical error is permitted, and the two parts of the component are coaxial.

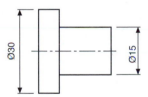

FIGURE 22.26

Condition B (Fig. 22.27)
Head at maximum material condition; shank at minimum material condition. The permitted geometrical error is equal to the tolerance on the shank size. This gives a tolerance zone of 0.1 diameter.

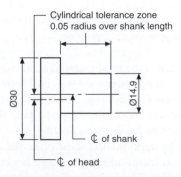

FIGURE 22.27

Condition C (Fig. 22.28)

Shank at maximum material condition; head at minimum material condition. The permitted geometrical error is equal to the tolerance on the head size. This gives a tolerance zone of 0.1 diameter.

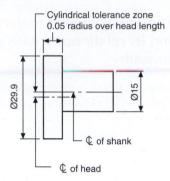

FIGURE 22.28

Condition D (Fig. 22.29)

Both shank and shaft are finished at their low limits of size; hence the permitted geometrical error will be the sum of the two manufacturing tolerances, namely 0.2 diameter.

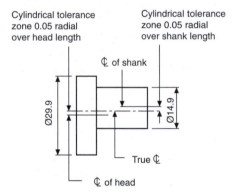

FIGURE 22.29

The application of maximum material condition to two mating components

Figure 22.30 shows a male and female component dimensioned with a linear tolerance between centres, and which will assemble together under the most adverse conditions allowed by the specified tolerances. The male component has centre distance and diameters of pins at maximum condition. The female component has centre distance and diameter of holes at minimum condition.

The tolerance diagram, Fig. 22.31, shows that, when the pin diameters are at the least material condition, their centre distance may vary between 74.9 − 14.7 = 60.2, or 45.1 + 14.7 = 59.8. Now this increase in tolerance can be used to advantage, and can be obtained by applying the maximum material concept to the drawing detail.

Similarly, by applying the same principle to the female component, a corresponding advantage is obtainable. The lower part of Fig. 22.31 shows the female component in

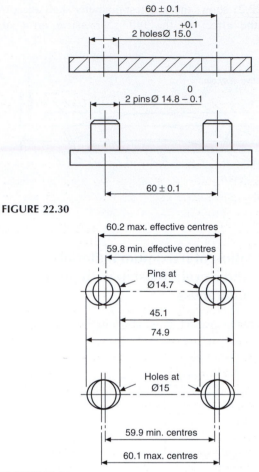

FIGURE 22.30

FIGURE 22.31

its maximum material condition. Assembly with the male component will be possible if the dimension over the pins does not exceed 74.9 and the dimension between the pins is no less than 45.1.

Figure 22.32 shows the method of dimensioning the female component with holes controlled by a positional tolerance, and modified by maximum material condition. This ensures assembly with the male component, whose pins are manufactured regardless of feature size.

When the maximum condition is applied to these features, any errors of form or position can be checked by using suitable gauges.

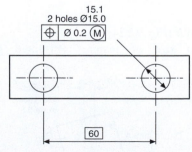

FIGURE 22.32

For further details regarding maximum and minimum condition refer to BS EN ISO 2692.

Chapter 23

Positional tolerancing

The essential requirement is to be able to define the limits for location of actual features, e.g., axes, points, median surfaces and nominally plane surfaces, relative to each other or in relation to one or more datum.

To accurately achieve this aim, it is essential that the primary constituents, theoretically exact dimensions, tolerance zones, and datums are utilized. The tolerance zone is symmetrically disposed about its theoretically exact location.

Utilizing these primary constituents ensures positional tolerances do not accumulate when dimensions are arranged in a chain, as would be the case if the feature pattern location were to be specified by coordinate tolerances.

Note: The practice of locating groups of features by positional tolerancing and their pattern location by coordinate tolerances, is no longer recommended by BS 8888 and BS EN ISO 5458.

Figure 23.1 illustrates the advantage of specifying a circular tolerance zone to a feature located by positional tolerancing. Note that the shaded tolerance area represents an increase of more than 57%.

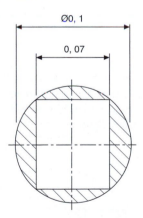

FIGURE 23.1

Theoretically exact dimensioning (TED) (true-position)

True-position dimensioning defines the exact location on a component of features such as holes, slots, keyways, etc., and also differentiates between 'ideal' and other toleranced dimensions. True-position dimensions are always shown 'boxed' on engineering drawings; they are never individually

toleranced, and must always be accompanied by a positional or zone tolerance for the feature to which they are applied.

The positional tolerance is the permitted deviation of a feature from a true position.

The positional-tolerance zone defines the region which contains the extreme limits of position and can be, for example, rectangular, circular, cylindrical, etc.

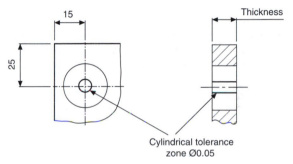

FIGURE 23.2   Product requirement.

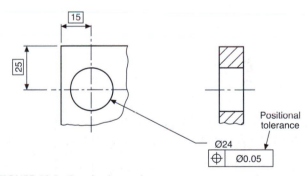

FIGURE 23.3   Drawing instruction.

Typical product requirement

In the examples shown in Figs. 23.2 and 23.3 respectively, the hole axis must lie within the cylindrical tolerance zone fixed by the true-position dimensions.

Some advantages of using this method are:

1. interpretation is easier, since true boxed dimensions fix the exact positions of details;
2. there are no cumulative tolerances;
3. it permits the use of functional gauges to match the mating part;

4. it can ensure interchangeability without resorting to small position tolerances, required by the coordinate tolerancing system;
5. the tolerancing of complicated components is simplified;
6. positional-tolerance zones can control squareness and parallelism.

The following examples show some typical cases where positional tolerances are applied to engineering drawings.

Case 1 (Figs. 23.4 and 23.5)

The axes of the four fixing holes must be contained within cylindrical tolerance zones 0.03 diameter.

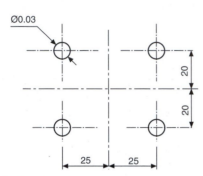

FIGURE 23.4 Case 1: Product requirement.

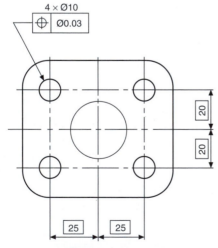

FIGURE 23.5 Case 1: Drawing instruction.

Case 2 (Figs. 23.6 and 23.7)

The axes of the four fixing holes must be contained within rectangular tolerance zones 0.04 × 0.02.

In cases 3 and 4, the perpendicularity and co-axial symbols shown, are constituents of the position characteristic, and could have been indicated with the position symbol equally as well.

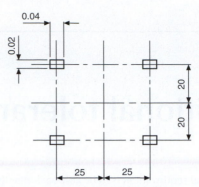

FIGURE 23.6 Case 2: Product requirement.

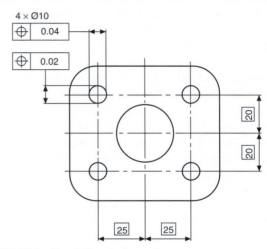

FIGURE 23.7 Case 2: Drawing instruction.

Case 3

Figure 23.8 shows a component where the outside diameter at the upper end is required to be square and coaxial within a combined tolerance zone with face A and diameter B as the primary and secondary datums.

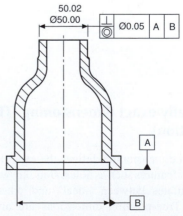

FIGURE 23.8 Case 3.

Case 4

In the component illustrated in Fig. 23.9, the three-dimensional features are required to be perfectly square to the datum face A, and also truly coaxial with each other in the maximum material condition.

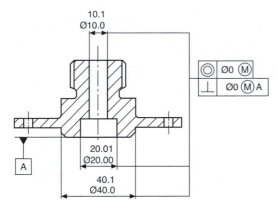

FIGURE 23.9 Case 4.

Case 5 (Figs. 23.10 and 23.11)

The six boltholes on the flange in Fig. 23.10 must have their centres positioned within six tolerance zones of $\emptyset = 0.25$ when the boltholes are at their maximum material condition (i.e., minimum limit of size).

Note in Fig. 23.11 that all the features in the group have the same positional tolerance in relation to each other. This method also limits in all directions the relative displacement of each of the features to each other.

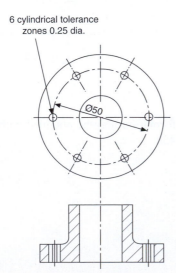

FIGURE 23.10 Case 5: Product requirement.

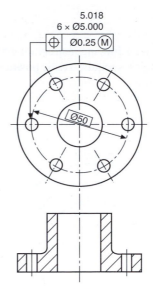

FIGURE 23.11 Case 5: Drawing instruction.

Case 6 (Figs. 23.12 and 23.13)

The group of holes in Fig. 23.12, dimensioned with a positional tolerance, is also required to be positioned with respect to the datum spigot and the face of the flange.

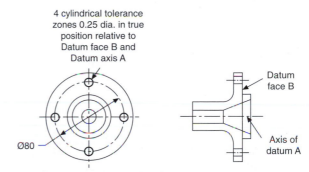

FIGURE 23.12 Case 6: Product requirement.

Note in Fig. 23.13 that the four holes and the spigot are dimensioned at the maximum material condition. It follows that, if any hole is larger than 12.00, it will have the effect of increasing the positional tolerance for that hole. If the spigot is machined to less than 50.05, then the positional tolerance for the four holes as a group will also increase.

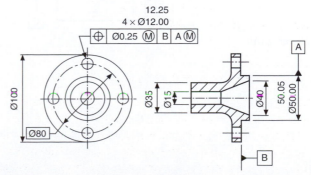

FIGURE 23.13 Case 6: Drawing instruction.

Case 7

Figure 23.14 shows a drawing instruction where the group of equally spaced holes is required to be positioned relative to a coaxial datum bore.

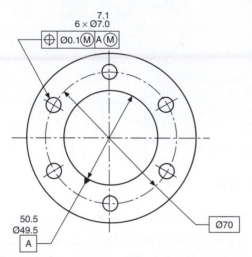

FIGURE 23.14 Case 7.

Case 8

Figure 23.15 shows a drawing instruction where a pattern of features is located by positional tolerancing. Each specific

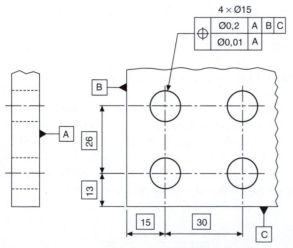

FIGURE 23.15

requirement is met independently. The product requirement in Fig. 23.16 shows that the axis of each of the four holes is required to lie within a cylindrical tolerance of $\emptyset = 0.01$. The positional tolerance zones are located in their theoretically exact positions to each other and perpendicular to datum A.

In Fig. 23.17, the axis of each of the four holes must lie within the cylindrical tolerance zone of $\emptyset = 0.2$ and the cylindrical tolerance must lie perpendicular to datum A and also located in their theoretical exact positions to each other and to datums B and C.

Note that in product requirement drawings, Figs. 23.16 and 23.17, simulated datums A, B and C are numbered 1, 2 and 3.

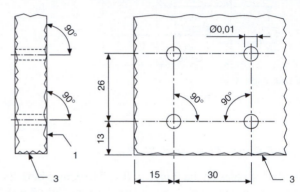

FIGURE 23.16

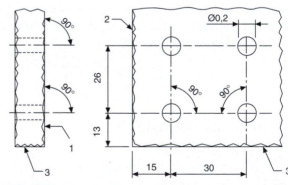

FIGURE 23.17

Further information may be obtained with reference to BS EN ISO 5458.

Chapter 24

Surface texture

Graphical symbols to indicate surface texture

The quality and type of surface texture has a direct connection with the manufacturing cost, function and wear of a component. Each of the symbols shown below has their own special interpretation. Individual surface texture values and text may be added to the symbols. The basic graphical symbol is shown in Fig. 24.1. The centre line between the lines of unequal length is positioned square to the considered surface.

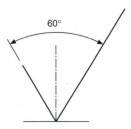

FIGURE 24.1

The symbol should not be indicated alone, without complementary information. It may, however be used for collective indication.

Expanded graphical symbols

Figure 24.2 shows the symbol indicating that removal of material is required. Figure 24.3 shows the symbol indicating that removal of material is not permitted.

FIGURE 24.2

FIGURE 24.3

Complete graphical symbols

Note. If complementary requirements for surface texture characteristics are required, then a line is added to the longer arm of the symbols, as shown below. Any manufacturing process permitted, in Fig. 24.4. Material shall be removed, in Fig. 24.5. Material shall not be removed, in Fig. 24.6.

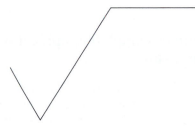

FIGURE 24.4

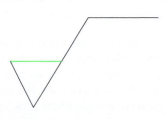

FIGURE 24.5

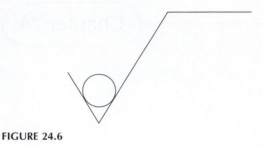

FIGURE 24.6

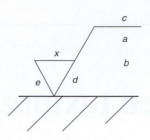

FIGURE 24.8

'All surfaces around a workpiece' graphical symbol

When the same surface texture is required on every surface around a workpiece, then a small circle is added to the symbol as shown in Fig. 24.7. The texture applies to all eight sides. If ambiguity is considered likely, then each surface may have its own independent symbol.

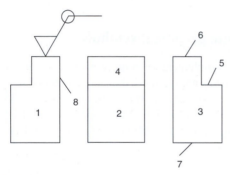

FIGURE 24.7

Composition of complete graphical symbols for surface texture

To avoid ambiguity, it may also be necessary to add additional requirements to the surface texture symbol and its numerical value, i.e. machining allowances, manufacturing process, sampling length and surface lay.

Mandatory positions for complementary requirements

Figure 24.8 shows mandatory positions of associated surface texture requirements. Note that in accordance with BS EN ISO 1302: 2002 the position 'X' is no longer used. Only positions a, b, c, d and e should be used. Position 'a' indicates one single surface texture requirement. Positions 'a' and 'b' indicate two or more surface texture requirements. Position

'c' indicates manufacturing method, treatment, coating or other process. Position 'd' indicates surface lay. Position 'e' shows machining allowance.

Three principal groups of surface texture parameters have been standardized in connection with the complete symbol and are defined as R, W and P profiles. The R profile series relates to roughness parameters. The W profile series relates to waviness parameters. The P profile series relates to structure parameters. The Ra value is the most commonly specified value throughout the world and examples follow in this chapter. Figure 24.9 illustrates an application with a single Ra requirement. Fig. 24.10 shows an application with an upper and lower call out requirement. Figure 24.11 shows examples of position and orientation on various surfaces. Note that the general rule when reading drawings is to read from the bottom or the right-hand side of the drawing. Figure 24.12 is a typical application in connection with features of size of mating parts. Figures 24.13 and 24.14 show examples of typical tolerance frames for geometrical tolerance applications.

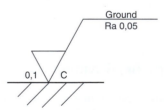

FIGURE 24.9

FIGURE 24.10

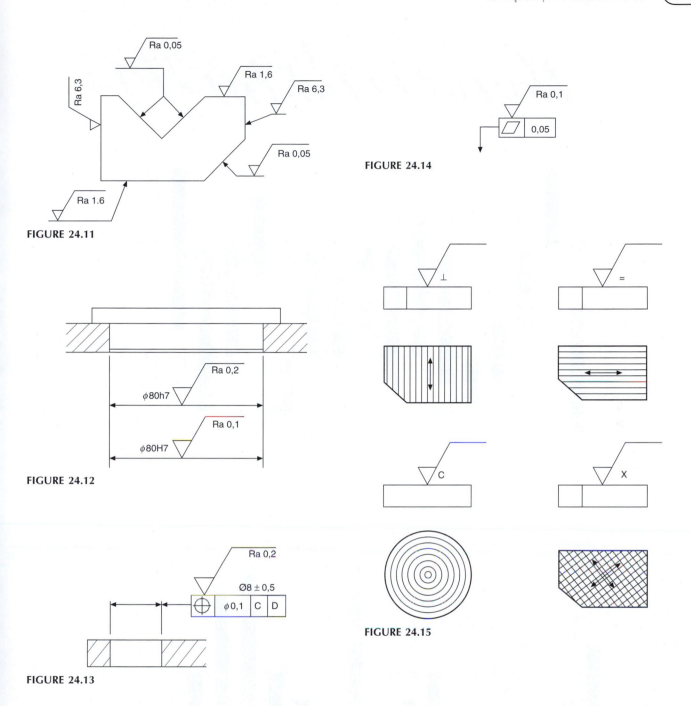

FIGURE 24.11

FIGURE 24.14

FIGURE 24.12

FIGURE 24.13

FIGURE 24.15

Further information on this specialized area of Metrology is given in BS EN ISO 1302 – 2002. Indication of Surface Texture in Product Documentation. The examples in Fig. 24.15 show methods of indicating four typical examples of surface lay.

(a) The lay is parallel to the plane of projection of the view where the symbol is drawn. The 'equals' symbol is added to the graphical symbol in the position shown. An arrow is indicated on the workpiece for clarification.

(b) The lay is perpendicular to the plane of projection and the symbol drawn is an inverted letter 'T'.

(c) The lay crosses the workpiece in two oblique directions indicated by the letter 'X'.

(d) The lay consists of concentric circles and the symbol used is the letter 'C'.

The standard also states that where a surface pattern is required which is not covered by BS 8888, then an explanatory note shall be added to the drawing.

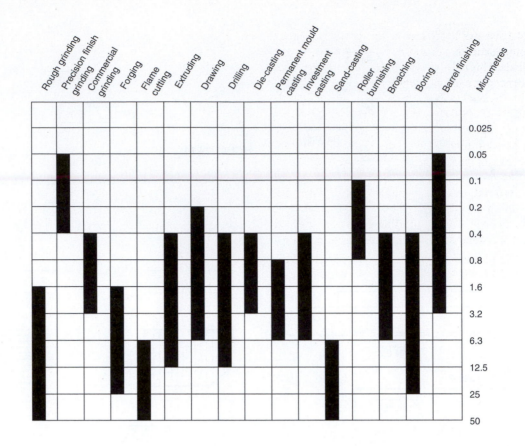

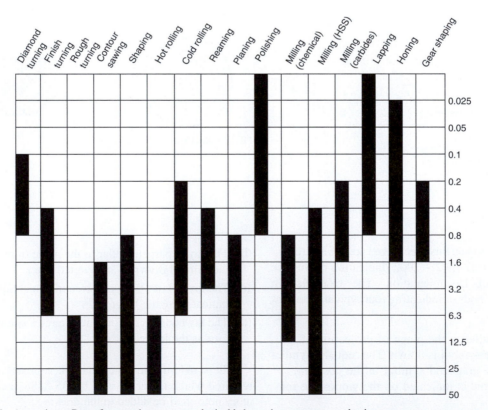

FIGURE 24.16   Approximate Ra surface roughness ranges obtainable by various common production processes.

Figure 24.16 shows approximate Ra surface roughness ranges for components manufactured by some common production methods. This information can only be approximate, since finish depends on many factors, such as the skill of the machinist, the accuracy and condition of the machine, the speeds and feeds selected for the operation, and the quality and condition of the cutting tools.

The approximate relationship between surface roughness and the cost of producing such a finish is shown in Fig. 24.17. The cost of rough machining can be considered as the zero datum on the *y* axis of the graph, and other processes can be compared with it. For example, a finish of 6.3 μm produced by grinding may well cost four times as much as rough machining. Many factors contribute towards production costs, and this information again can be only approximate.

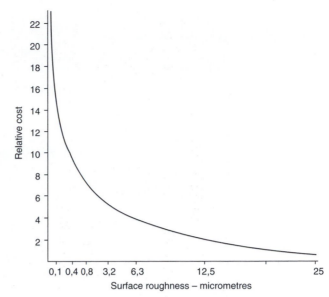

FIGURE 24.17 Approximate relationship between surface roughness and cost.

Chapter 25

3D annotation

Three dimensional (3D) or 'Solid' modelling is now used worldwide by industry, as the tool of choice for engineering design. Driven by the race to shorten 'time to market' full exploitation of the benefits of using the 3D model as a master which can be used to enable the production of product data specifications (drawings), rapid prototyping, computer aided manufacture (CAM), computer aided inspection or verification (CAI) and finite-element analysis (FEA).

There are many examples of 'paperless manufacture', i.e., a product being manufactured from a model, by CAM, without the aid of drawings or any supplementary specification. In this application, the accuracy and uncertainty of the manufacturing process is known and is acceptable for the finished product. To ensure repeatability the manufacturing process must be tightly controlled as any change could impact the product. Unfortunately, this method does not lend itself readily to changes in manufacturing environments such as differing machine tools or global manufacturing, so it is important that the full design intent can be added to the model.

Despite the common use of 3D modelling, two-dimensional (2D) drawings are still widely used to enable manufacture. The introduction of geometrical product specification (GPS) has given the designer the tools to convey full design intent using unambiguous 2D drawings. By using annotated 3D models the requirement to produce 2D drawings has been reduced.

Two standards relating to 3D Annotation exist; ASME Y14.41 Technical product documentation—Digital product definition data practices and ISO 16792—Technical Product Documentation—Digital Product Definition Data Practices (which is closely aligned to ASME 14-41 but adapted to suit the ISO system). Additional methods of 3D specification are being developed by ISO and are, or will be, included in the standards such as ISO 1101—Geometrical Product Specification—Geometrical Tolerancing—Tolerances of Form, Orientation, Location and Run-out, and ISO 5459—Geometrical Product Specification—Geometrical Tolerancing Datums and Datum Systems.

These new standards have been introduced to standardize 3D annotation. For example; 3D CAD systems on the market today differ slightly in the way of constructing models, assigning attributes, determining associativity and specifying tolerances; 2D drawings contain tolerances that are view or direction-dependent, e.g., straightness (see Fig. 25.1) the limit of the tolerances (see Fig. 25.3) and, as 3D models are not confined to orthographic views,

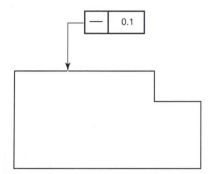

FIGURE 25.1 In 2D the choice of view determines the location and orientation of the feature.

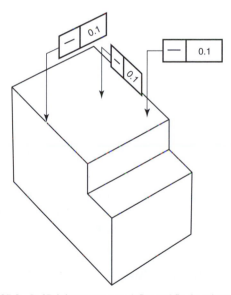

FIGURE 25.2 In 3D it is necessary to define and fix the orientation of the tolerance set (tolerance frame, leader line and arrow) to have the same expression as in Fig. 25.1.

the positioning of tolerance indicators often require controlling (see Figs. 25.2 and 25.4). These examples highlight the need for a consistent approach towards 3D annotation which ensures that any rules introduced into new and existing standards are equally applicable to 2D and 3D specification or application, see Figs. 25.16 and 25.17 for typical examples.

The following are examples of alternative methods of specifying tolerances within 2D and 3D applications together with the conventional method. Both methods are equally applicable and the designer should decide which method best

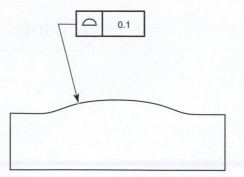

FIGURE 25.3 In 2D, the choice of view determines the limits of the toleranced feature.

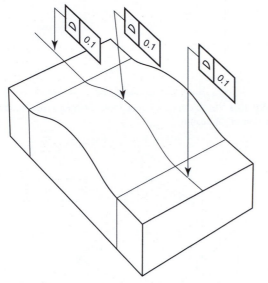

FIGURE 25.4 In 3D, the software generally defines several geometrical features for the same surface identified in Fig. 25.3.

suits their requirements. It is strongly recommended that where a product is specified in both 2D and 3D formats the same method of specification is used on both.

Axis or median feature

Figure 25.5 illustrates an alternative way of referring the tolerance to an axis or median feature. In this method the

tolerance frame is connected to the feature by a leader line terminating with an arrowhead pointing directly at the surface, but with the addition of the modifier symbol Ⓐ (median feature) placed to the right hand end of the second compartment of the tolerance frame.

Figure 25.6 illustrates the conventional method of referring the tolerance to an axis or median feature.

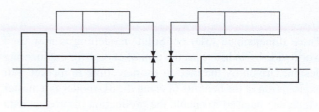

FIGURE 25.6 3D annotated model.

Projected tolerance zone

Figure 25.7 illustrates an alternative to indicating the projected tolerance zone without using supplemental geometry, the length of projection can also be specified indirectly by adding the value, after the Ⓟ symbol, in the tolerance frame. This method of indication only applies to blind holes.

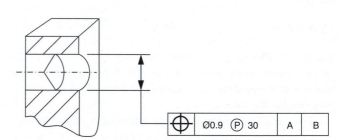

FIGURE 25.7 Alternative method.

Figure 25.8 illustrates the conventional method of specifying projected tolerance zone.

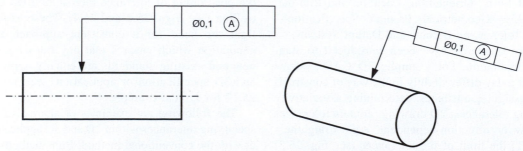

FIGURE 25.5 Alternative method.

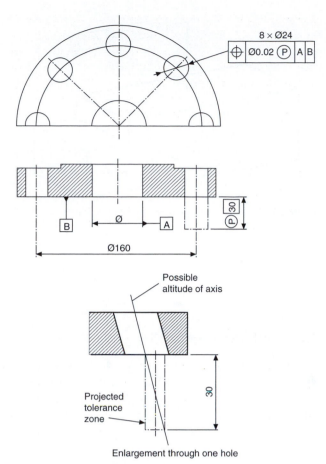

8 × Ø24

⊕ | Ø0.02 Ⓟ | A | B

B   Ø   A   Ⓟ 30

Ø160

Possible altitude of axis

Projected tolerance zone

30

Enlargement through one hole

FIGURE 25.8 Conventional method.

Indicating a tolerance zone between two points

During the development of 3D annotation it was established that there was some ambiguity in identifying the extent (start and finish) of tolerance zones, such as profile of a surface. Simplistically, where a surface is the product of a couple of features such as a curve and a flat, it was practice to assume that both elements were one surface. When 3D modelling it becomes more obvious that each feature can be a separate entity and as such any tolerance indicated by a leader line could only apply to that entity. To give the designer more flexibility and to eliminate multi tolerance indications the 'Between' symbol has been introduced. This introduction has aligned the ISO (ISO 1101) with ASME (Y14-5) where the 'between' symbol has been widely used.

The between symbol '↔' is used between two letters identifying the start and the end of the considered toleranced zone. This zone includes all segments or areas between the start and the end of the identified features.

To clearly identify the tolerance zone, the tolerance frame is connected to the compound toleranced feature by a leader line and terminating with an arrowhead on the outline of the compound toleranced feature (see example of Fig. 25.9). Note that the tolerance does not apply to surfaces a, b, c and d.

If the tolerance zone applies to part of a feature or surface it can be positioned by using theoretical exact dimensions.

Unilateral and unequal profile tolerance

To reduce the need to use supplemental geometry, new symbology has been introduced to indicate unilateral and unequal tolerance zones. See Chapter 21 for rules of application (Figs. 25.10 and 25.11).

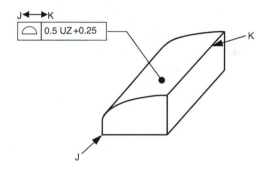

FIGURE 25.10 Application of unilateral profile tolerance.

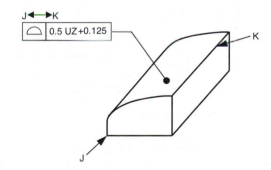

FIGURE 25.11 Application of unequal profile tolerance.

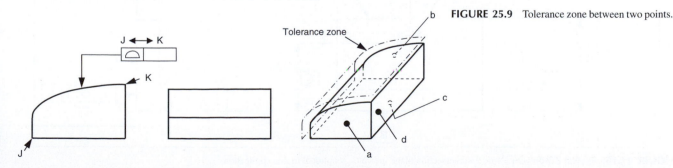

FIGURE 25.9 Tolerance zone between two points.

Indicating the direction of tolerance zones

Tolerances specified on 2D drawings are often associated to the orthographic view to which they are applied to, i.e., view dependent. 3D models are not confined to orthographic projection and as such it is vital that the application of tolerances is unambiguous. If the designer feels there is any ambiguity in specifying the direction of a tolerance zone the indicators shown in Figs. 25.12 and 25.13 may be applied after the tolerance frame, see Figs. 25.14 and 25.15. In Fig. 25.14 the straightness tolerance applies parallel to datum face A. In Fig. 25.15 the tolerance is applied at the specified angle to datum A.

For intersection planes, the symbol symmetrical, parallel, or perpendicular, defining how the intersection plane is derived from the datum is placed in the first compartment of the intersection plane indicator.

For orientation planes, the symbol perpendicular, parallel, or angular, defining how the intersection plane is derived from the datum is placed in the first compartment of the intersection plane indicator (Figs. 25.12 and 25.13).

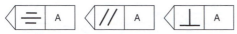

FIGURE 25.12   Intersection plane indications.

FIGURE 25.13   Orientation plane indicators.

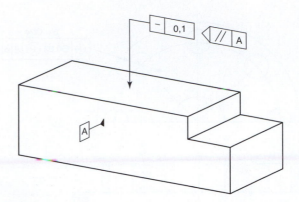

FIGURE 25.14   Application of intersection plane indicators.

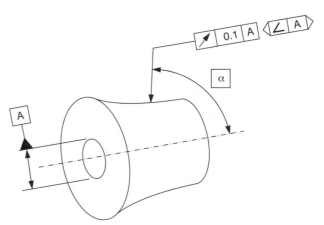

FIGURE 25.15   Application of orientation plane indicators.

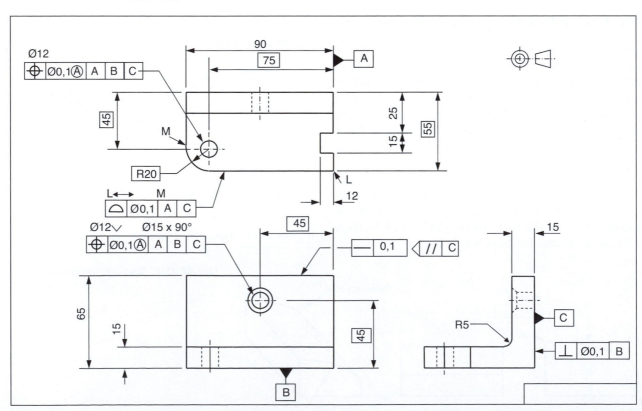

FIGURE 25.16   Typical ZD drawing. Note the intersection plane indicator is optional in this application.

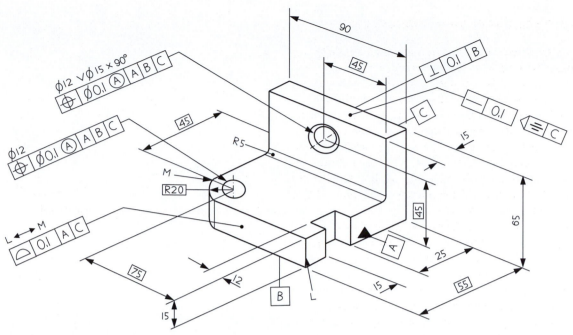

FIGURE 25.17 Annotated model.

Chapter 26

The Duality Principle—the essential link between the design intent and the verification of the end product

Introduction

In today's modern world of mass production and the sophisticated techniques available to the 'manufacturing and verification fields of operation' it is essential that the communication between all factions involved in the production of a workpiece, from it's functional concept through to the end product, must be as complete as possible, without ambiguity in order to minimize any unaccountable uncertainties that may inadvertently be found, see Fig. 26.1. Compliance with the Duality Principle greatly assists in this aim.

Design specification and verification

When a workpiece is inspected or verified a number of stage processes take place, see Fig. 26.2. Because much of these stage processes are common practices, they are not

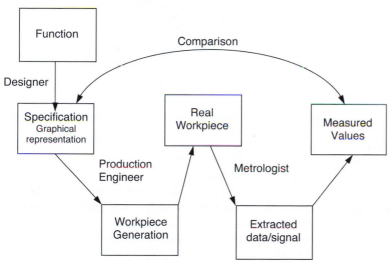

FIGURE 26.1 Relationship between design and the verification of an actual finished part.

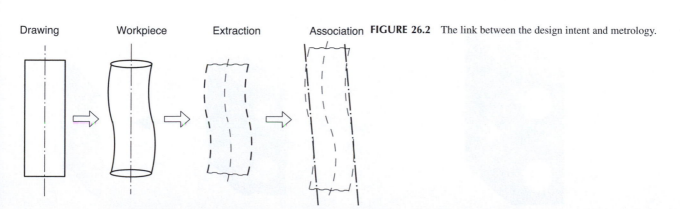

FIGURE 26.2 The link between the design intent and metrology.

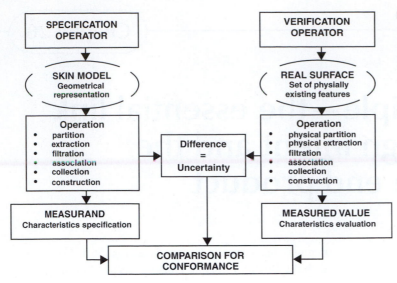

FIGURE 26.3 The Duality Principle.

consciously classified and considered just the systematic way we already do things. Now, by identifying and classifying each of these individual stage operations, a common language is created for use between the designer and verification engineer, which must greatly assist and improve the communication and understanding between them. The classifications given to these stage operations are partition, extraction, filtration, association, collection and construction.

Figure 26.3 shows these processes mirrored to each other at each operation between the design intent and the verification stage process, in order to achieve an acceptable end product. This essential link is known as the Duality Principle.

Detailed illustrations and explanations of each stage operation are shown in Figs. 26.4–26.11 depicting the Duality Principle.

Figure 26.4 illustrates the simplistic design intent, i.e. the nominal model as conceived by the designer delineated in perfect geometry.

Figure 26.5 illustrates the realistic design intent, i.e. the skin model representing the imperfect geometry of the inter-

FIGURE 26.5 Skin model.

face of the workpiece with its surrounding medium, as perceived by the designer.

Figure 26.6 illustrates the term Partition, the operation of when a feature or features such as flat surfaces, cylindrical surfaces, curved surfaces, etc., are partitioned to obtain from the skin model or real surface(s) the non-ideal surfaces corresponding to the nominal features.

FIGURE 26.4 Nominal model.

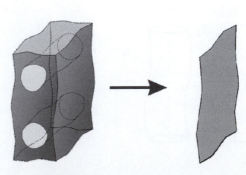

FIGURE 26.6 Partition.

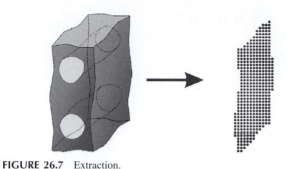

FIGURE 26.7 Extraction.

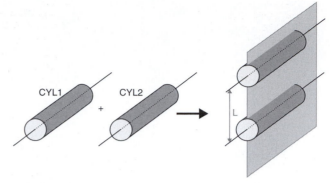

FIGURE 26.10 Collection.

Figure 26.7 illustrates the term known as Extraction, the operation that results in the representation of the surface (skin model or real surface), obtained by extracting a finite number of points from a partitioned surface.

Figure 26.8—in practice it is generally found that in addition to the extraction operation, some filtering or smoothing of the extracted data is necessary, in order to remove any unwanted detail. This operation is known as filtering.

than one feature together, i.e., such as the collection of data, relating to a group or pattern of holes.

Figure 26.11 illustrates the term known as construction, this being the operation used to determine the toleranced feature, when tolerances are applied to other features which are dependent on, or resultants of other features. Hence the operation used to build ideal features from other ideal features with constraints is known by the term 'construction'.

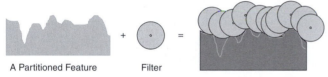

A Partitioned Feature   Filter

FIGURE 26.8 Filtration.

Figure 26.9 illustrates the term known as association, this being the operation resulting in a perfect form associated to the extracted surface in accordance with specified convention, which may include the method of least squares, minimum zone, maximum inscribing and minimum circumscribing.

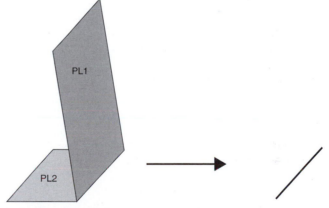

FIGURE 26.11 Construction.

Advantages of the Duality Principle

It is a critical factor of the synergy between the disciplines of the design, manufacturing, and verification.

It provides a unified system, a common language, and a common understanding.

It removes the option of informal communication and understanding that so often exists between the design, manufacture and verification factions, in house and/or between different manufacturing plants, and sub-contractors.

FIGURE 26.9 Association.

Figure 26.10 illustrates the term known as collection, this being the operation when there is a need to consider more

Differences between American ASME Y 14.5M Geometric dimensioning and tolerancing (GD & T) and ISO/BS 8888 geometrical tolerancing, standards

This chapter aims to highlight, identify and analyse, the differences between the ASME and ISO/BS 8888 systems. The Authors make no claim as to which might be the better and/or accommodate in any specific application. Some of these differences are of a smaller nature, and as such are self-evident, while others are not, and involve indications which are interpreted differently by users of the two different systems, giving rise to significant differences of the intended design specification. All Geometric controls included throughout this manual are to the ISO/BS 8888 standards.

There are a few differences in terminology as detailed in Table 27.1.

TABLE 27.1 Comparison of ASME Y 14.5M and ISO terminologies

ASME Y 14.5M	ISO
Basic dimension	Theoretical exact dimension (TED)
Feature control frame	Tolerance frame
Variation	Deviation
True position (TP)	Theoretical exact position
Reference dimension	Auxiliary dimension

Applicability of standards

The rules to which any drawing is produced must be indicated within the framework of the drawing, i.e. ISO or ASME Y 14.5M.

If a provision from ASME Y 14.5M were needed to be invoked on a drawing conforming to ISO rules, the relevant ASME Y 14.5M cross-reference must be specifically identified at the point of application.

Symbology

ASME Y14.5 specifies in addition to or deviating from ISO 1101 the symbols shown in Table 27.2.

Specification of datums

Whilst the general understanding that a datum is basically a 'good starting surface or point' is not wrong. The advancement and availability of today's manufacturing technology, has created many more options of specific applications of datums that may be required. ISO 5459 and ASME Y 14.5 M standards comprehensively define these conceptual options together by their respective rules. These two set of rules can lead to substantially different conclusions. A brief explanation appears below.

ISO standards keeping in line with these developments of modern techniques, give a wide range of different, sophisticated, practical conceptual terms and procedures for the various types of datums which may be specified on a drawing. Basically, where a specified datum feature has a form that allows the work piece to 'rock' within itself, the ISO rule is to 'equalize' the rock, in order to establish an 'average' position and orientation, to be used as the intended datum.

ASME Y 14.5M specifies the concept of 'candidate datums' which allows every position that an unstable datum can rock to (with some limitations) is a valid 'candidate datum'. A set of candidate datum reference frames can be derived for each set of requirements that are referenced to the same datum system, using the same precedence and the same material conditions. These sets of requirements are, by default, evaluated simultaneously to each candidate datum reference frame. If there is a candidate datum reference frame where all the requirements are fulfilled exists, the workpiece is acceptable with regard to the requirements.

In general, the ASME Y 14.5M system accepts more workpieces as the form error of the datum feature increases. However, some workpieces accepted under the applied ISO rules can be rejected upon application of the ASME Y 14.5M rules, so assumptions should not be made.

TABLE 27.2 Additional symbols found in ASME Y 14.5M

Symbol	Designation	Interpretation
Ⓣ	Tangent	Symbol placed within the tolerance frame indicating a tolerance applies to the contacting tangential element
CR	Controlled radius	Symbol placed before the toleranced radius dimension. The tolerance zone is defined by two arcs (the minimum and maximum radii) that are tangent to the adjacent surfaces. The part contour within the crescent-shaped tolerance zone must be a fair curve without reversals with all points on the radii being within the tolerance zone
⟨ST⟩	Statistical tolerancing	Symbol placed after a toleranced dimension indicating the assigning of tolerances to related components of an assembly on the basis of sound statistics (such as the assembly tolerance is equal to the square root of the sum of the squares of the individual tolerances)
⊔	Counterbore or spotface	Symbol indicating a flat bottom hole, presented before the associated dimension
∨	Countersink	Symbol indicating a countersink, presented before the associated dimension
⤓	Deep/depth	Symbol indicating depth of a feature, presented before the dimension

Exclusion of surface texture

The ISO standards do not currently state whether surface texture should be included or excluded within a specified geometric control, when evaluated. (However, the application of BS 8888 requires that surface texture be excluded by the use of appropriate filtering techniques.)

ASME Y 14.5M states that 'all requirements apply after application of the smoothing functions'. In other words surface texture shall be disregarded when evaluating workpieces using 'ASME Y 14.5M' and thus is similar to the dictates of BS 8888.

Tolerancing principle

The ASME Y 14.5M interprets size tolerances using the 'Principle of Dependency of size and form' in the same terms as the ISO envelope principle (Taylor principle), i.e. when only a size tolerance is quoted, the form of a workpiece is always within its maximum size when at maximum material condition (MMC). This is known as 'Rule 1' and stated in the ASME Y 14.5M standard. It was realized that this overall rule was not practical in all cases, and some exceptions to this rule are as follows:

(a) It does not apply to stock materials (bar stock, sheet, tubing, etc.).
(b) It does not apply to flexible parts, subject to free-state variation in the unstrained condition.
(c) It does not apply to features of size which have a straightness tolerance applied to their axes or median plane.

(d) It may be overruled where a feature of size has a specified relationship between size and a geometric control for example the use of Ⓜ or Ⓛ in the tolerance frame.
(e) It may be overruled with a statement such as 'PERFECT FORMAT MMC NOT REQUIRED' placed by a feature of size tolerance.

ISO promotes (ISO 8015) the 'The Principle of Independency' which states: 'Each specified dimensional or geometrical requirement on a drawing shall be met independently, unless a particular relationship is specified.' i.e. Maximum or Minimum Material Condition Ⓜ, Ⓛ or the envelope principle (the Taylor principle) Ⓔ.

This means that local two point measurements control the linear dimensional tolerances only, and not the form deviations of the feature.

Features-of-size

The following table lists features of size recognized by each standard (Table 27.3).

TABLE 27.3 Feature-of-size

ISO	ASME Y 14.5M
Cylindrical surfaces	Cylindrical surfaces
Spherical surfaces	Spherical surfaces
Two parallel, opposed surfaces	Two parallel, opposed surfaces
A cone	Two opposed elements (such as the radiused ends of a slot)
A wedge	

Tolerance characteristics (Table 27.4)

TABLE 27.4 Comparison of tolerance characteristics

Tolerance	BS 8888 and ISOs	ASME Y 14.5M:1994
Positional	Positional tolerance can be used to control the location of features-of-size and also points, lines and flat planes	The positional tolerance is *only* used with features of size ASME Y 14.5M recommends the use of Profile of a Surface to control a flat planar surface
Concentricity coaxiality	These characteristics have the same symbol even though they distinctively relate to different characteristics, with the term concentricity frequently and mistakenly confused with coaxially, and visa versa	Known only as 'Concentricity' tolerance. It is defined as the condition whereby the median points of all diametrically opposed elements of a figure of revolution are congruent with the axis or centre point of a datum feature
	The ISO definition describes concentricity as the situation whereby the centre point of a feature is located on a datum point or axis Coaxially is described as the situation where an axis of a feature is aligned to a datum axis. Concentricity/coaxially tolerances can be replaced by using a positional tolerance to provide an identical control Both these characteristics like the positional tolerance can be used with the maximum and minimum material condition modifiers	The standard states that concentricity cannot be used with the maximum and minimum material modifiers
Symmetry	ISO considers this as a special case of the positional tolerance, which can be used to control the location of an axis or median plane of a feature of size in relation to a datum axis	Symmetry is defined as the condition where the median points of all opposed or corresponding located elements of two or more feature surfaces are congruent with the axis or centre plane of a datum feature It is also stated that symmetry cannot be used with the maximum or minimum condition modifiers
Profile of a line and surface	These tolerance zones are generated by placing a theoretical circle or sphere, with a diameter corresponding to the size of the tolerance, on every point of the theoretically exact profile (or surface) to generate the boundary limits.	These tolerance zones are generated by a vector offset from the theoretically exact profile (or surface) to generate the boundary limits
	Where the theoretically exact profile (or surface) contains sharp corners (or edges) the tolerance zone boundary external to the corners (or edges) is radiused	Where the theoretically exact profile (or surface) contains sharp corners or (edges) the tolerance zone boundary is extended to give a sharp corner (or edge)
Roundness	ISO uses the term 'Roundness' for this form of tolerance	ASME uses the term 'Circularity' for this form of tolerance

Chapter 28

Cams and gears

A cam is generally a disc or a cylinder mounted on a rotating shaft, and it gives a special motion to a *follower*, by direct contact. The cam profile is determined by the required follower motion and the design of the type of follower.

The motions of cams can be considered to some extent as alternatives to motions obtained from linkages, but they are generally easier to design, and the resulting actions can be accurately predicted. If, for example, a follower is required to remain stationary, then this is achieved by a concentric circular arc on the cam. For a specified velocity or acceleration, the displacement of the follower can easily be calculated, but these motions are very difficult to arrange precisely with linkages.

Specialist cam-manufacturers computerize design data and, for a given requirement, would provide a read-out with cam dimensions for each degree, minute, and second of camshaft rotation.

When used in high-speed machinery, cams may require to be balanced, and this becomes easier to perform if the cam is basically as small as possible. A well-designed cam system will involve not only consideration of velocity and acceleration but also the effects of out-of-balance forces, and vibrations. Suitable materials must be selected to withstand wear and the effect of surface stresses.

Probably the most widely used cam is the *plate cam*, with its contour around the circumference. The line of action of the follower is usually either vertical or parallel to the camshaft, and Fig. 28.1 shows several examples.

Examples are given later of a *cylindrical* or *drum cam*, where the cam groove is machined around the circumference, and also a *face cam*, where the cam groove is machined on a flat surface.

Cam followers

Various types of cam followers are shown in Fig. 28.1. Knife-edge followers are restricted to use with slow-moving mechanisms, due to their rapid rates of wear. Improved stability can be obtained from the roller follower, and increased surface area in contact with the cam can be obtained from the flat and mushroom types of follower. The roller follower is the most expensive type, but is ideally suited to high speeds and applications where heat and wear are factors.

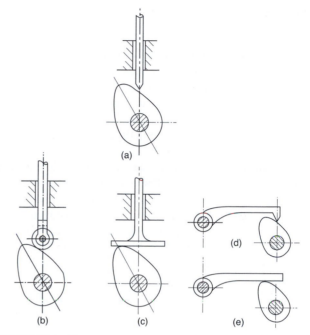

FIGURE 28.1 Plate cams. (a) Plate cam with knife-edge follower. (b) Plate cam with roller follower. (c) Plate cam with flat follower. (d) Plate cam with oscillating knife-edge follower. (e) Plate cam with oscillating flat follower.

Cam follower motions

1. *Uniform velocity*: This motion is used where the follower is required to rise or fall at a constant speed, and is often referred to as straight-line motion. Part of a uniform-velocity cam graph is shown in Fig. 28.2.

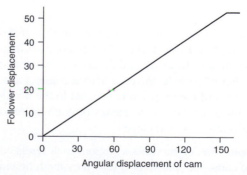

FIGURE 28.2

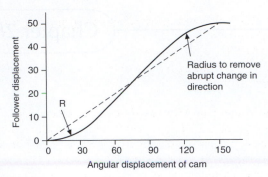

FIGURE 28.3

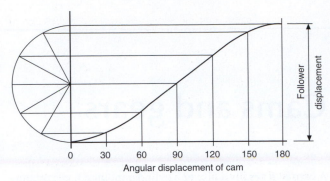

FIGURE 28.5

Abrupt changes in velocity with high-speed cams result in large accelerations and cause the followers to jerk or chatter. To reduce the shock on the follower, the cam graph can be modified as indicated in Fig. 28.3 by adding radii to remove the sharp corners. However, this action results in an increase in the average rate of rise or fall of the follower.

2. *Uniform acceleration and retardation motion* is shown in Fig. 28.4. The graphs for both parts of the motion are parabolic. The construction for the parabola involves dividing the cam-displacement angle into a convenient number of parts, and the follower displacement into the same number of parts. Radial lines are drawn from the start position to each of the follower division lines, and the parabola is obtained by drawing a line through successive intersections. The uniform-retardation parabola is constructed in a similar manner, but in the reverse position.

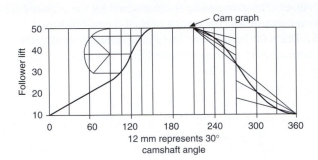

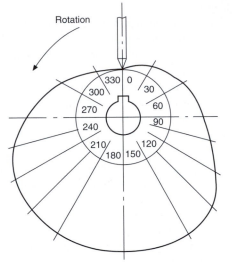

FIGURE 28.6

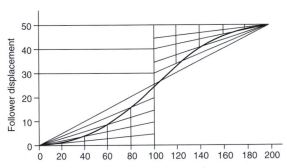

FIGURE 28.4

3. *Simple harmonic motion* is shown in Fig. 28.5 where the graph is a sine curve. The construction involves drawing a semi-circle and dividing it into the same number of parts as the cam-displacement angle. The diameter of the semi-circle is equal to the rise or fall of the follower. The graph passes through successive intersections as indicated.

The application of the various motions to different combinations of cams and followers is shown by the following practical example.

Case 1 (Fig. 28.6)

Cam specification:

Plate cam, rotating anticlockwise. Point follower.

Least radius of cam, 30 mm. Camshaft diameter, 20 mm.

0°–90°, follower rises 20 mm with uniform velocity.

90°–150°, follower rises 30 mm with simple harmonic motion.

150°–210°, dwell period.

210°–270°, follower falls 20 mm with uniform acceleration.

270°–360°, follower falls 30 mm with uniform retardation.

1. Draw the graph as shown. Exact dimensions are used for the *Y* axis, where the follower lift is plotted. The *X* axis

has been drawn to scale, where 12 mm represents 30° of shaft rotation.

2. To plot the cam, draw a 20 mm diameter circle to represent the bore for the camshaft, and another circle 30 mm radius to represent the base circle, or the least radius of the cam, i.e. the nearest the follower approaches to the centre of rotation.

3. Draw radial lines 30° apart from the cam centre, and number them in the reverse direction to the cam rotation.

4. Plot the Y ordinates from the cam graph along each of the radial lines in turn, measuring from the base circle. Where rapid changes in direction occur, or where there is uncertainty regarding the position of the profile, more points can be plotted at 10° or 15° intervals.

5. Draw the best curve through the points to give the required cam profile.

Note: The user will require to know where the cam program commences, and the zero can be conveniently established on the same centre line as the shaft keyway. Alternatively, a timing hole can be drilled on the plate, or a mark may be engraved on the plate surface. In cases where the cam can be fitted back to front, the direction or rotation should also be clearly marked.

Case 2 (Fig. 28.7)

Cam specification:

Plate cam, rotating anticlockwise. Flat follower. Least distance from follower to cam centre, 30 mm. Camshaft diameter, 20 mm.

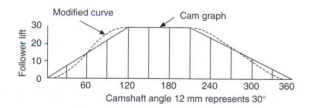

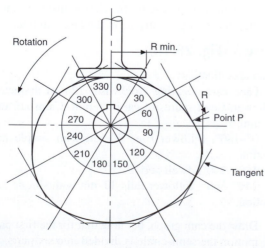

FIGURE 28.7

0°–120°, follower rises 30 mm with uniform velocity (modified).

120°–210°, dwell period.

220°–360°, follower falls 30 mm with uniform velocity (modified).

1. Draw the cam graph as shown, and modify the curve to remove the sharp corners. Note that in practice the size of the radius frequently used here varies between one-third and the full lift of the follower for the uniform-velocity part of the graph; the actual value depends on the rate of velocity and the speed of rotation. This type of motion is not desirable for high speeds.

2. Draw the base circle as before 30 mm radius, divide it into 30° intervals, and number them in the reverse order to the direction of rotation.

3. Plot the Y ordinates from the graph, radially from the base circle along each 30° interval line. Draw a tangent at each of the plotted points, as shown, and draw the best curve to touch the tangents. The tangents represent the face of the flat follower in each position.

4. Check the point of contact between the curve and each tangent and its distance from the radial line. Mark the position of the widest point of contact.

In the illustration given, point *P* appears to be the greatest distance, and hence the follower will require to be at least *R* in radius to keep in contact with the cam profile at this point. Note also that a flat follower can be used only where the cam profile is always convex.

Although the axis of the follower and the face are at 90° in this example, other angles are in common use.

Case 3 (Fig. 28.8)

Cam specification:

Plate cam, rotating clockwise. 20 mm diameter roller follower.

30 mm diameter camshaft. Least radius of cam, 35 mm.

0°–180°, rise 64 mm with simple harmonic motion.

180°–240°, dwell period.

240°–360°, fall 64 mm with uniform velocity.

1. Draw the cam graph as shown.

2. Draw a circle (shown as RAD Q) equal to the least radius of the cam plus the radius of the roller, and divide it into 30° divisions. Mark the camshaft angles in the anticlockwise direction.

3. Along each radial line plot the Y ordinates from the graph, and at each point draw a 20 mm circle to represent the roller.

4. Draw the best profile for the cam so that the cam touches the rollers tangentially, as shown.

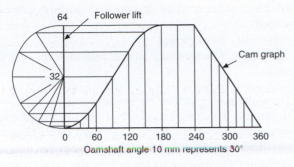

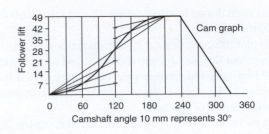

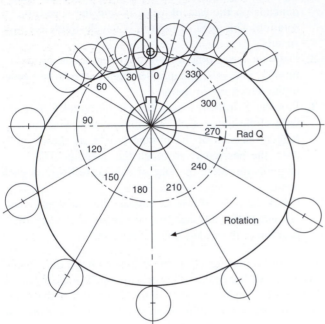

FIGURE 28.8

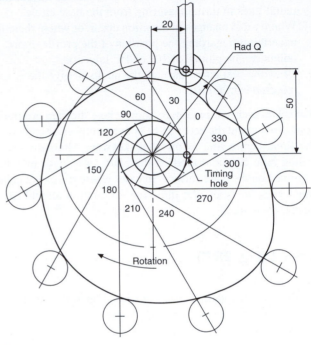

FIGURE 28.9

Case 4 (Fig. 28.9)

Cam specification:

Plate cam, rotating clockwise. 20 mm diameter roller follower set 20 mm to the right of the centre line for the camshaft. Least distance from the roller centre to the camshaft centre line, 50 mm. 25 mm diameter camshaft.

0°–120°, follower rises 28 mm with uniform acceleration.

120°–210°, follower rises 21 mm with uniform retardation.

210°–240°, dwell period.

240°–330°, follower falls 49 mm with uniform velocity.

330°–360°, dwell period.

1. Draw the cam graph as shown.
2. Draw a 20 mm radius circle, and divide it into 30° divisions as shown.
3. Where the 30° lines touch the circumference of the 20 mm circle, draw tangents at these points.
4. Draw a circle of radius Q, as shown, from the centre of the camshaft to the centre of the roller follower. This circle is the base circle.

5. From the base circle, mark lengths equal to the lengths of the Y ordinates from the graph, and at each position draw a 20 mm diameter circle for the roller follower.
6. Draw the best profile for the cam so that the cam touches the rollers tangentially, as in the last example.

Case 5 (Fig. 28.10)

Cam specification:

Face cam, rotating clockwise. 12 mm diameter roller follower. Least radius of cam, 26 mm. Camshaft diameter, 30 mm.

0°–180°, follower rises 30 mm with simple harmonic motion.

180°–240°, dwell period.

240°–360°, follower falls 30 mm with simple harmonic motion.

1. Draw the cam graph, but note that for the first part of the motion the semi-circle is divided into six parts, and that for the second part it is divided into four parts.

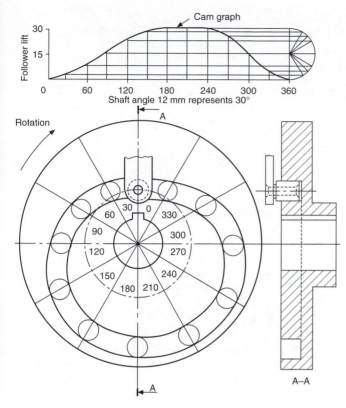

FIGURE 28.10

3. Underneath the front elevation, draw a development of the cylindrical cam surface, and on this surface draw the cam graph.
4. Using the cam graph as the centre line for each position of the roller, draw 14 mm diameter circles as shown.
5. Draw the cam track with the sides tangential to the rollers.
6. Plot the track on the surface of the cylinder by projecting the sides of the track in the plan view up to the front elevation. Note that the projection lines for this operation do not come from the circles in the plan, except at each end of the track.
7. The dotted line in the end elevation indicates the depth of the track.
8. Plot the depth of the track in the front elevation from the end elevation, as shown. Join the plotted points to complete the front elevation.

Note that, although the roller shown is parallel, tapered rollers are often used, since points on the outside of the cylinder travel at a greater linear speed than points at the bottom of the groove, and a parallel roller follower tends to skid.

2. Draw a base circle 32 mm radius, and divide into 30° intervals.
3. From each of the base-circle points, plot the lengths of the *Y* ordinates. Draw a circle at each point for the roller follower.
4. Draw a curve on the inside and the outside, tangentially touching the rollers, for the cam track.

The drawing shows the completed cam together with a section through the vertical centre line.

Note that the follower runs in a track in this example. In the previous cases, a spring or some resistance is required to keep the follower in contact with the cam profile at all times.

Case 6 (Fig. 28.11)

Cam specification:

Cylindrical cam, rotating anticlockwise, operating a roller follower, 14 mm diameter. Cam cylinder, 60 mm diameter. Depth of groove, 7 mm.

0°–180°, follower moves 70 mm to the right with simple harmonic motion.

180°–360°, follower moves 70 mm to the left with simple harmonic motion.

1. Set out the cylinder blank and the end elevation as shown.
2. Divide the end elevation into 30° divisions.

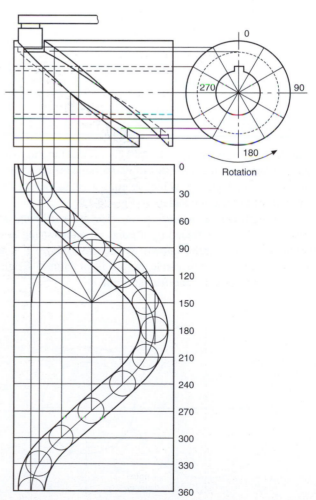

FIGURE 28.11

Dimensioning cams

Figure 28.12 shows a cam in contact with a roller follower; note that the point of contact between the cam and the roller is at A, on a line which joins the centres of the two arcs. To dimension a cam, the easiest method of presenting the data is in tabular form which relates the angular displacement 0 of the cam with the radial displacement R of the follower. The cam could then be cut on a milling machine using these point settings. For accurate master cams, these settings may be required at half- or one-degree intervals.

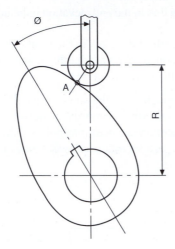

FIGURE 28.12

Spur gears

The characteristics feature of spur gears is that their axes are parallel. The gear teeth are positioned around the circumference of the pitch circles which are equivalent to the circumferences of the friction rollers in Fig. 28.13.

The teeth are of *involute* form, the involute being described as the locus traced by a point on a taut string as it unwinds from a circle, known as the base circle. For an involute rack, the base circle radius is of infinite length, and the tooth flank is therefore straight.

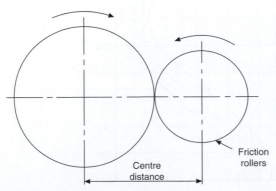

FIGURE 28.13

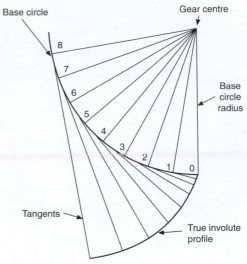

FIGURE 28.14 Involute construction. The distance along the tangent from each point is equal to the distance around the circumference from point 0.

The construction for the involute profile is shown in Fig. 28.14. The application of this profile to an engineering drawing of a gear tooth can be rather a tedious exercise, and approximate methods are used, as described later.

Spur-gear terms (Fig. 28.15)

The *gear ratio* is the ratio of the number of teeth in the gear to the number of teeth in the pinion, the pinion being the smaller of the two gears in mesh.

The *pitch-circle diameters* of a pair of gears are the diameters of cylinders co-axial with the gears which will roll together without slip. The pitch circles are imaginary friction discs, and they touch at the *pitch point*.

The *base circle* is the circle from which the involute is generated.

The *root diameter* is the diameter at the base of the tooth.

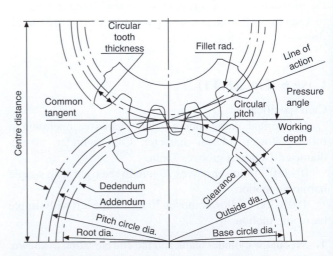

FIGURE 28.15 Spur-gear terms.

The *centre distance* is the sum of the pitch-circle radii of the two gears in mesh.

The *addendum* is the radial depth of the tooth from the pitch circle to the tooth tip.

The *dedendum* is the radial depth of the tooth from the pitch circle to the root of the tooth.

The *clearance* is the algebraic difference between the addendum and the dedendum.

The *whole depth* of the tooth is the sum of the addendum and the dedendum.

The *circular pitch* is the distance from a point on one tooth to the corresponding point on the next tooth, measured round the pitch-circle circumference.

The *tooth width* is the length of arc from one side of the tooth to the other, measured round the pitch-circle circumference.

The *module* is the pitch-circle diameter divided by the number of teeth.

The *diametral pitch* is the reciprocal of the module, i.e. the number of teeth divided by the pitch-circle diameter.

The *line of action* is the common tangent to the base circles, and the *path of contact* is that part of the line of action where contact takes place between the teeth.

The *pressure angle* is the angle formed between the common tangent and the line of action.

The fillet is the rounded portion at the bottom of the tooth space.

The various terms are illustrated in Fig. 28.15.

Involute gear teeth proportions and relationships

$$\text{Module} = \frac{\text{pitch} - \text{circle diameter, PCD}}{\text{number of teeth, } T}$$

$$\text{Circular pitch} = \pi \times \text{module}$$

$$\text{Tooth thickness} = \frac{\text{circular pitch}}{2}$$

Addendum = Module
Clearance = 0.25 × module
Dedendum = addendum + clearance

Involute gears having the same pressure angle and module will mesh together. The British Standard recommendation for pressure angle is 20°.

The conventional representation of gears shown in Fig. 28.16 is limited to drawing the pitch circles and outside diameters in each case. In the sectional end elevation, a section through a tooth space is taken as indicated. This convention is common practice with other types of gears and worms.

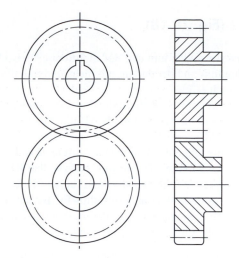

FIGURE 28.16   Gear conventions.

Typical example using Professor Unwin's approximate construction

Gear data:

Pressure angle, 20°. Module, 12 mm: Number of teeth, 25.
Gear calculations:

$$\text{Pitch} - \text{circle diameter} = \text{module} \times \text{no. of teeth}$$
$$= 12 \times 25 = 300 \text{ mm}$$
$$\text{Addendum} = \text{module} = 12 \text{ mm}$$
$$\text{Clearance} = 0.25 \times \text{module} = 0.25 \times 12 = 3 \text{ mm}$$
$$\text{Dedendum} = \text{addendum} + \text{clearance} = 12 + 3 = 15 \text{ mm}$$
$$\text{Circular pitch} = \pi \times \text{module} = \pi \times 12 = 37.68 \text{ mm}$$
$$\text{Tooth thickness} = 1/2(\text{circular pitch}) = 18.84 \text{ mm}$$

Stage 1 (Fig. 28.17)

(a) Draw the pitch circle and the common tangent.
(b) Mark out the pressure angle and the normal to the line of action.
(c) Draw the base circle. Note that the length of the normal is the base-circle radius.

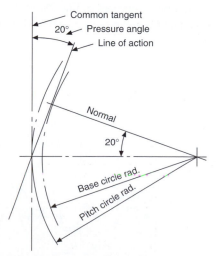

FIGURE 28.17   Unwin's construction-stage 1.

Stage 2 (Fig. 28.18)

(a) Draw the addendum and dedendum circles. Both addendum and dedendum are measured radially from the pitch circle.
(b) Mark out point A on the addendum circle and point B on the dedendum circles. Divide AB into three parts so that CB = 2AC.
(c) Draw the tangent CD to the base circle. D is the point of tangency. Divide CD into four parts so that CE = 3DE.
(d) Draw a circle with centre O and radius OE. Use this circle for centres of arcs of radius EC for the flanks of the teeth after marking out the tooth widths and spaces around the pitch-circle circumference.

Note that it may be more convenient to establish the length of the radius CE by drawing this part of the construction further round the pitch circle, in a vacant space, if the flank of one tooth, i.e. the pitch point, is required to lie on the line AO.

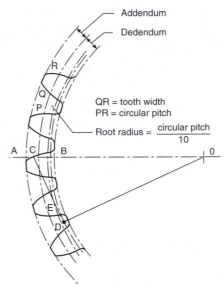

FIGURE 28.18 Unwin's construction-stage 2.

The construction is repeated in Fig. 28.19 to illustrate an application with a rack and pinion. The pitch line of the rack touches the pitch circle of the gear, and the values of the addendum and dedendum for the rack are the same as those for the meshing gear.

If it is required to use the involute profile instead of the approximate construction, then the involute must be constructed from the base circle as shown in Fig. 28.14. Complete stage 1 and stage 2(a) as already described, and mark off the tooth widths around the pitch circle, commencing at the pitch point. Take a tracing of the involute in soft pencil on transparent tracing paper, together with part of the base circle in order to get the profile correctly oriented on the required drawing. Using a French curve, mark the profile in pencil on either side of the tracing paper, so that, whichever side is used, a pencil impression can be obtained. With care, the profile can now be traced onto the required layout, lining up the base circle and ensuring that the profile of the tooth flank passes through the tooth widths previously marked out on the pitch circle. The flanks of each tooth will be traced from either side of the drawing paper. Finish off each tooth by adding the root radius.

Helical gears

Helical gears have been developed from spur gears, and their teeth lie at an angle to the axis of the shaft. Contact between the teeth in mesh acts along the diagonal face flanks in a progressive manner; at no time is the full length of any one tooth completely engaged. Before contact ceases between one pair of teeth, engagement commences between the following pair. Engagement is therefore continuous, and this fact results in a reduction of the shock which occurs when straight teeth operate under heavy loads. Helical teeth give a smooth, quiet action under heavy loads; backlash is considerably reduced; and, due to the increase in length of the tooth, for the same thickness of gear wheel, the tooth strength is improved.

Figure 28.20 illustrates the lead and helix angle applied to a helical gear. For single helical gears, the helix angle is generally 12°–20°.

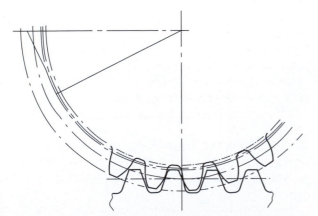

FIGURE 28.19 Unwin's construction applied to a rack and pinion.

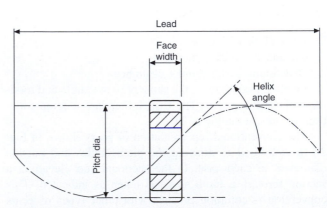

FIGURE 28.20 Lead and helix angle for a helical gear.

Since the teeth lie at an angle, a side or end thrust occurs when two gears are engaged, and this tends to separate the gears. Figure 28.21 shows two gears on parallel shafts and the position of suitable thrust bearings. Note that the position of the thrust bearings varies with the direction of shaft rotation and the 'hand' of the helix.

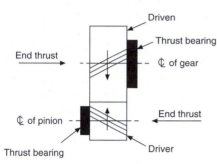

FIGURE 28.21

In order to eliminate the serious effect of end thrust, pairs of gears may be arranged as shown in Fig. 28.22 where a double helical gear utilizes a left- and a right-hand helix. Instead of using two gears, the two helices may be cut on the same gear blank.

Where shafts lie parallel to each other, the helix angle is generally 15°–30°. Note that a right-hand helix engages with a left-hand helix, and the hand of the helix must be correctly stated on the drawing. On both gears the helix angle will be the same.

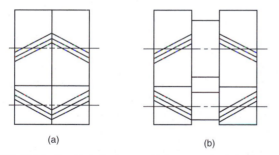

(a)   (b)

FIGURE 28.22 Double helical gears. (a) On same wheel. (b) On separate wheels.

For shafts lying at 90° to each other, both gears will have the same hand of helix, Fig. 28.23.

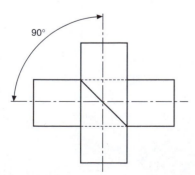

FIGURE 28.23

Helical gears can be used for shafts which lie at an angle less than 90°, but the hand of helix should be checked with a specialist gear manufacturer. The hand of a helix depends on the helix angle used and the shaft angles required.

Bevel gears

If the action of spur and helical gears can be related to that of rolling cylinders, then the action of bevel gears can be compared to a friction cone drive. Bevel gears are used to connect shafts which lie in the same plane and whose axes intersect. The size of the tooth decreases as it passes from the back edge towards the apex of the pitch cone, hence the cross-section varies along the whole length of the tooth. When viewed on the curved surface which forms part of the back cone, the teeth normally have the same profiles as spur gears. The addendum and dedendum have the same proportions as a spur gear, being measured radially from the pitch circle, parallel to the pitch-cone generator.

Data relating to bevel gear teeth is shown in Fig. 28.24. Note that the crown wheel is a bevel gear where the pitch angle is 90°. Mitre gears are bevel gears where the pitch-cone angle is 45°.

The teeth on a bevel gear may be produced in several different ways, e.g. straight, spiral, helical, or spiraloid. The advantages of spiral bevels over straight bevels lie in quieter running at high speed and greater load-carrying capacity.

The angle between the shafts is generally a right angle, but may be greater or less than 90°, as shown in Fig. 28.25.

Bevel gearing is used extensively in the automotive industry for the differential gearing connecting the drive shaft to the back axle of motor vehicles.

Bevel-gear terms and definitions

The following are additions to those terms used for spur gears.

The pitch angle is the angle between the axis of the gear and the cone generating line.

The root-cone angle is the angle between the gear axis and the root generating line.

The face angle is the angle between the tips of the teeth and the axis of the gear.

The addendum angle is the angle between the top of the teeth and the pitch-cone generator.

The dedendum angle is the angle between the bottom of the teeth and the pitch-cone generator.

The outside diameter is the diameter measured over the tips of the teeth.

The following figures show the various stages in drawing bevel gears. The approximate construction for the profile of the teeth has been described in the section relating to spur gears.

Gear data: 15 teeth, 20° pitch-cone angle, 100 mm pitch-circle diameter, 20° pressure angle.

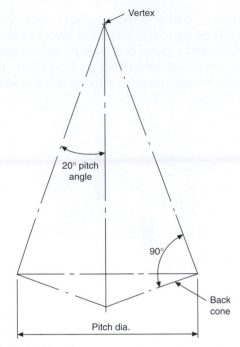

FIGURE 28.26   Stage 1.

Stage 1 Set out the cone as shown in Fig. 28.26.
Stage 2 Set out the addendum and dedendum. Project part of the auxiliary view to draw the teeth (Fig. 28.27).

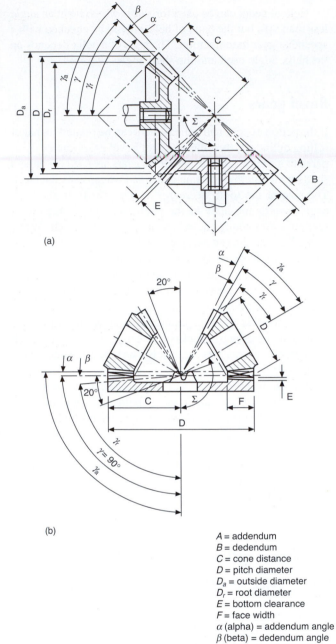

A = addendum
B = dedendum
C = cone distance
D = pitch diameter
D_a = outside diameter
D_r = root diameter
E = bottom clearance
F = face width
α (alpha) = addendum angle
β (beta) = dedendum angle
γ (gamma) = pitch angle
γ_a = back cone angle
γ_r = root angle
Σ (sigma) = shaft angle

FIGURE 28.24   Bevel-gear terms. (a) Bevel gears. (b) Crown wheel and pinions.

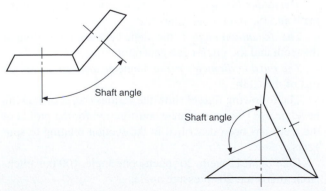

FIGURE 28.25   Bevel-gear cones.

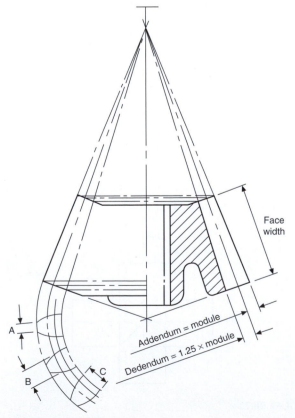

FIGURE 28.27   Stage 2.

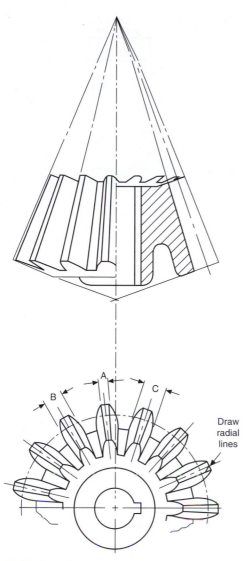

FIGURE 28.28 Stage 3.

Stage 3 Project widths A, B, and C on the outside, pitch, and root diameters, in plan view. Complete the front elevation (Fig. 28.28).

Worm gearing

Worm gearing is used to transmit power between two non-intersecting shafts whose axes lie at right angles to each other. The drive of a worm gear is basically a screw, or worm, which may have a single- or multi-start thread, and this engages with the wheel. A single-start worm in one revolution will rotate the worm wheel one tooth and space, i.e. one pitch. The velocity ratio is high; for example, a 40 tooth wheel with a single-start worm will have a velocity ratio of 40, and in mesh with a two-start thread the same wheel would give a velocity ratio of 20.

A worm-wheel with a single-start thread is shown in Fig. 28.29. The lead angle of a single-start worm is low, and the worm is relatively inefficient, but there is little tendency for the wheel to drive the worm. To transmit power, multi-start thread forms are used. High mechanical advantages are obtained by the use of worm-gear drives.

Worm-gear drives have many applications, for example indexing mechanisms on machine tools, torque converters, and slow-motion drives.

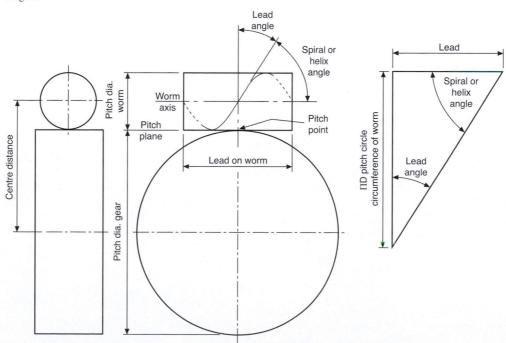

FIGURE 28.29

Figure 28.30 shows typical cross-sections through part of a worm and wheel. Note the contour of the wheel, which is designed to give greater contact with the worm.

Recommendations for the representation of many types of gear assembly in sectional and simplified form are given in BS 8888.

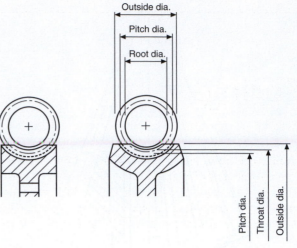

FIGURE 28.30 Worm-gearing terms applied to a worm and part of a worm-wheel.

Chapter 29

Springs

Mechanical springs may be defined as elastic bodies the primary function of which is to deform under a load and return to their original shape when the load is removed. In practice, the vast majority of springs are made of metal, and of these the greatest proportion are of plain-carbon steel.

Plain-carbon steels

These steels have a carbon-content ranging from about 0.5% to 1.1%. In general it may be taken that, the higher the carbon-content, the better the spring properties that may be obtained.

In the manufacture of flat springs and the heavier coil springs, it is usual to form the spring from annealed material and subsequently to heat-treat it. However, it is sometimes possible to manufacture flat springs from material which is already in the hardened and tempered condition, and this latter technique may give a lower production cost than the former.

For light coil springs, the material loosely known as piano wire is used; this is a spring wire which obtains its physical properties from cold-working, and not from heat-treatment. Springs made from this wire require only a low-temperature stress-relieving treatment after manufacture. Occasionally wire known as 'oil-tempered' is used – this is a wire which is hardened and tempered in the coil form, and again requires only a low-temperature stress relief after forming.

Plain-carbon steel springs are satisfactory in operation up to a temperature of about 180 °C. Above this temperature they are very liable to take a permanent set, and alternative materials should be used.

Alloy steels

Alloy steels are essentially plain-carbon steels to which small percentages of alloying elements such as chromium and vanadium have been added. The effect of these additional elements is to modify considerably the steels' properties and to make them more suitable for specific applications than are the plain-carbon steels. The two widely used alloy steels are

(a) chromium–vanadium steel – this steel has less tendency to set than the plain-carbon steels;

(b) silicon–manganese steel – a cheaper and rather more easily available material than chrome–vanadium steel, though the physical properties of both steels are almost equivalent.

Stainless steels

Where high resistance to corrosion is required, one of the stainless steels should be specified. These fall into two categories.

(a) *Martensitic*. These steels are mainly used for flat springs with sharp bends. They are formed in the soft condition and then heat-treated.

(b) *Austenitic*. These steels are cold-worked for the manufacture of coil springs and flat springs, with no severe forming operations.

Both materials are used in service up to about 235 °C.

High-nickel alloys

Alloys of this type find their greatest applications in high-temperature operation. The two most widely used alloys are

(a) Inconel – a nickel–chrome–iron alloy for use up to about 320 °C.

(b) Nimonic 90 – a nickel–chromium–cobalt alloy for service up to about 400 °C, or at reduced stresses to about 590 °C.

Both of these materials are highly resistant to corrosion.

Copper-base alloys

With their high copper-content, these materials have good electrical conductivity and resistance to corrosion. These properties make them very suitable for such purposes as switch-blades in electrical equipment.

(a) *Brass* – an alloy containing zinc, roughly 30%, and the remainder copper. A cold-worked material obtainable in both wire and strip form, and which is suitable only for lightly stressed springs.

(b) *Phosphor bronze* – the most widely used copper-base spring material, with applications the same as those of brass, though considerably higher stresses may be used.

(c) *Beryllium copper* – this alloy responds to a precipitation heat-treatment, and is suitable for springs which contain sharp bends. Working stresses may be higher than those used for phosphor bronze and brass.

Compression springs

Figure 29.1 shows two alternative types of compression springs for which drawing conventions are used. Note that the convention in each case is to draw the first and last two turns of the spring and to then link the space in between with a long dashed dotted narrow line. The simplified representation shows the coils of the springs drawn as single lines.

Note. If a rectangular section compression spring is required to be drawn then the appropriate shape will appear in view (e), view (d) will be modified with square corners and the ∅ symbol in view (f) replaced by □.

A schematic drawing of a helical spring is shown in Fig. 29.2. This type of illustration can be used as a working drawing in order to save draughting time, with the appropriate dimensions and details added.

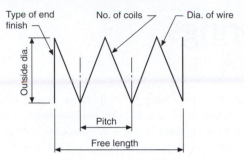

FIGURE 29.2   Schematic drawing of helical spring.

Figure 29.3 shows four of the most popular end formations used on compression springs. When possible, grinding should be avoided, as it considerably increases spring costs.

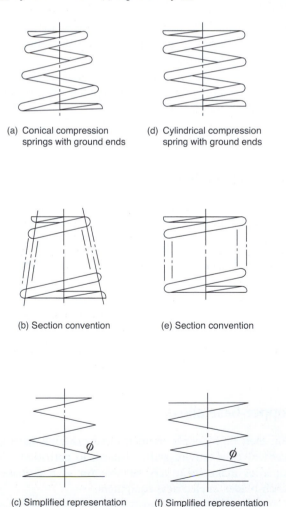

FIGURE 29.1   (a) Conical compression springs with ground ends, (b) section convention, (c) simplified representation, (d) cylindrical compression spring with ground ends, (e) section convention and (f) simplified representation.

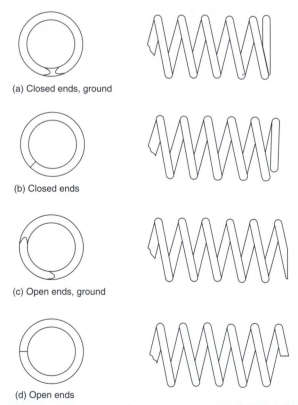

(a) Closed ends, ground

(b) Closed ends

(c) Open ends, ground

(d) Open ends

FIGURE 29.3   (a) Closed ends, ground, (b) closed ends, (c) open ends, ground and (d) open ends.

Figure 29.4 shows a selection of compression springs, including valve springs for diesel engines and injection pumps.

Flat springs

Figure 29.5 shows a selection of flat springs, circlips, and spring pressings. It will be apparent from the selection that it would be difficult, if not impossible, to devise a drawing

FIGURE 29.4

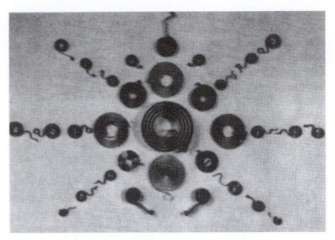

FIGURE 29.6

FIGURE 29.5

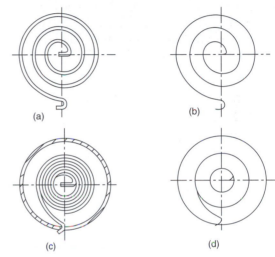

(a)   (b)

(c)   (d)

FIGURE 29.7

standard to cover this type of spring, and at present none exists.

Flat springs are usually made from high-carbon steel in the annealed condition, and are subsequently heat-treated; but the simpler types without bends can frequently be made more economically from material pre-hardened and tempered to the finished hardness required. Stainless steels are used for springs where considerable forming has to be done. For improved corrosion-resistance, 18/8 stainless steel is recommended; but, since its spring temper is obtained only by cold-rolling, severe bends are impossible. Similar considerations apply to phosphor bronze, titanium, and brass, which are hardened by cold-rolling. Beryllium copper, being thermally hardenable, is a useful material as it can be readily formed in the solution-annealed state.

Figure 29.6 shows a selection of flat spiral springs, frequently used for brush mechanisms, and also for clocks and motors. The spring consists of a strip of steel spirally wound and capable of storing energy in the form of torque.

The standard for spiral springs is illustrated in Fig. 29.7, parts (a) and (b) show how the spring is represented in conventional and simplified forms.

If the spring is close wound and fitted in a housing then the illustrations in (c) and (d) are applicable.

Torsion springs

Various forms of single and double torsion springs are illustrated in Fig. 29.8.

Figure 29.9 gives a schematic diagram for a torsion spring. This type of drawing, adequately dimensioned, can be used for detailing.

The drawing conventions for a cylindrical right-hand helical torsion spring are shown in Fig. 29.10. Part (a) shows the usual drawing convention, (b) how to show the spring in a section and (c) gives the simplified representation.

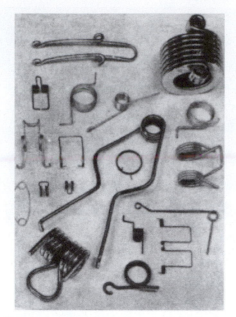

FIGURE 29.8

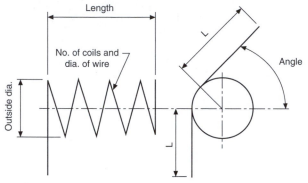

FIGURE 29.9

Although torsion springs are available in many different forms, this is the only type to be represented in engineering-drawing standards. Torsion springs may be wound from square-, rectangular-, or round-section bar. They are used to exert a pressure in a circular arc, for example in a spring hinge and in door locks. The ends of the wire in the spring may be straight, curved, or kinked.

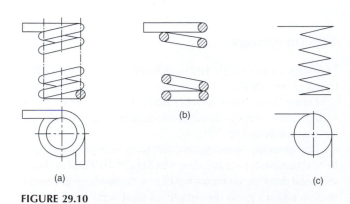

FIGURE 29.10

Leaf springs

The two standards applicable to leaf springs are shown in Fig. 29.11. These springs are essentially strips of flat metal formed in an elliptical arc and suitably tempered. They absorb and release energy, and are commonly found applied to suspension systems.

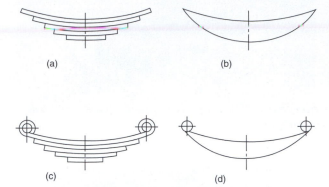

FIGURE 29.11 (a, b) Conventional and simplified representations for a semi-elliptic leaf spring (c, d) conventional and simplified representations for a semi-elliptic leaf spring with fixing eyes.

Helical extension springs

A helical extension spring is a spring which offers resistance to extension. Almost invariably they are made from circular-section wire, and a typical selection is illustrated in Fig. 29.12.

The conventional representations of tension springs are shown in Fig. 29.13 and a schematic drawing for detailing is shown in Fig. 29.14.

Coils of extension springs differ from those of compression springs in so far as they are wound so close together that a force is required to pull them apart. A variety of end loops are available for tension springs, and some of the types are illustrated in Fig. 29.15.

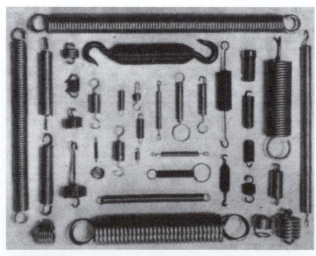

FIGURE 29.12

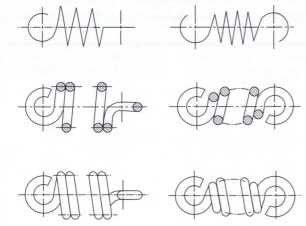

FIGURE 29.13 Conventional representation of tension springs.

Free length
Dia. of wire
Outside dia.
Pitch
No. of coils
Overall length

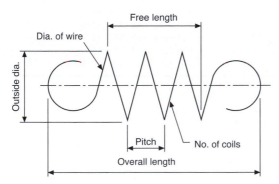

FIGURE 29.14 Schematic drawing of tension spring.

A common way of reducing the stress in the end loop is to reduce the moment acting on it by making the end loop smaller than the body of the spring, as shown in Fig. 29.16.

Disc springs

For bolted connections a very simple form of compression spring utilizes a hollow washer manufactured from spring steel, although other materials can be specified.

Table 29.1 shows a selection of Belleville washers manufactured to DIN 6796 from spring steel to DIN 17222.

If the disc has its top and bottom surfaces ground to approximately 95% of the appropriate thickness in the table above then bearing surfaces will be formed. These surfaces improve guidance where several discs are used together. Figure 29.17 shows the disc spring with flats.

A disc spring stack with six springs in single series is given in Fig. 29.18. In this arrangement six times the number of individual spring deflections are available. The force available in the assembly is equivalent to that of a single disc spring. In single series concave and convex surfaces are opposed alternatively.

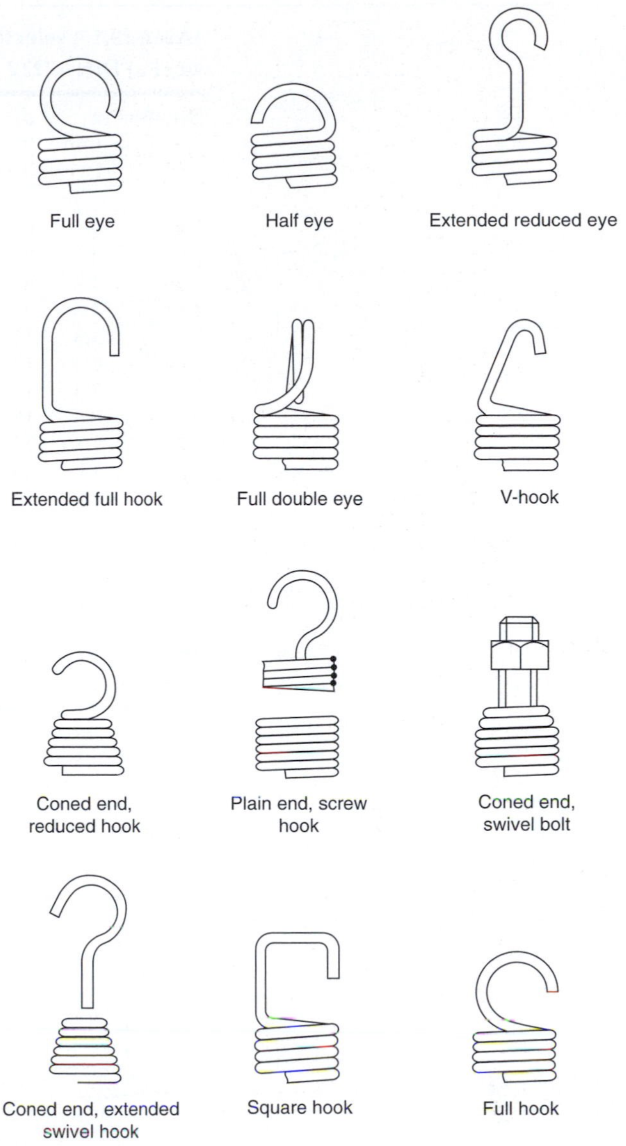

Full eye
Half eye
Extended reduced eye
Extended full hook
Full double eye
V-hook
Coned end, reduced hook
Plain end, screw hook
Coned end, swivel bolt
Coned end, extended swivel hook
Square hook
Full hook

FIGURE 29.15 Types of end loops.

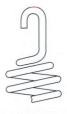

FIGURE 29.16

Figure 29.19 shows three disc springs assembled in parallel with the convex surface nesting into the concave surface. Here the deflection available is equivalent to that of a single spring and the force equal to three times that of an individual disc.

The methods of assembly illustrated in Figs. 29.18 and 29.19 can be combined to give many alternative selections of force and deflection characteristics. In the stack given in Fig. 29.20 there are four disc spring components assembled in

TABLE 29.1 A selection of Belleville washers manufactured to DIN 6796 from spring steel to DIN 17222

Notation	d_1 (mm) H 14	d_2 (mm) h 14	h (mm) max.[a]	h (mm) min.[b]	s[c] (mm)	Force[d]	Test force[e]	Weight (kg/1000 $\approx$)	Core diameter
2	2.2	5	0.6	0.5	0.4	628	700	0.05	2
2.5	2.7	6	0.72	0.61	0.5	946	1 100	0.09	2.5
3	3.2	7	0.85	0.72	0.6	1 320	1 500	0.14	3
3.5	3.7	8	1.06	0.92	0.8	2 410	2 700	0.25	3.5
4	4.3	9	1.3	1.12	1	3 770	4 000	0.38	4
5	5.3	11	1.55	1.35	1.2	5 480	6 550	0.69	5
6	6.4	14	2	1.7	1.5	8 590	9 250	1.43	6
7	7.4	17	2.3	2	1.75	11 300	13 600	2.53	7
8	8.4	18	2.6	2.24	2	14 900	17 000	3.13	8
10	10.5	23	3.2	2.8	2.5	22 100	27 100	6.45	10
12	13	29	3.95	3.43	3	34 100	39 500	12.4	12
14	15	35	4.65	4.04	3.5	46 000	54 000	21.6	14
16	17	39	5.25	4.58	4	59 700	75 000	30.4	16
18	19	42	5.8	5.08	4.5	74 400	90 500	38.9	18
20	21	45	6.4	5.6	5	93 200	117 000	48.8	20
22	23	49	7.05	6.15	5.5	113 700	145 000	63.5	22
24	25	56	7.75	6.77	6	131 000	169 000	92.9	24
27	28	60	8.35	7.3	6.5	154 000	221 000	113	27
30	31	70	9.2	8	7	172 000	269 000	170	30

[a] Max. size in delivered condition.
[b] Min. size after loading tests to DIN 6796.
[c] Permissible range of tolerance of s to DIN 1544 and DIN 1543 respectively for s > 6 mm.
[d] This force applies to the pressed flat condition and corresponds to twice the calculated value at a deflection $h_{mm} - s$.
[e] The test force applies for the loading tests to DIN 6796.

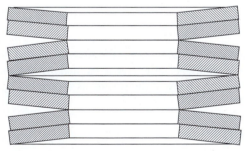

FIGURE 29.20

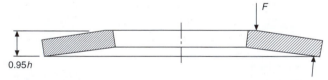

0.95h

FIGURE 29.17

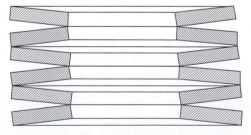

FIGURE 29.18

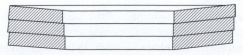

FIGURE 29.19

series and they each contain two disc springs assembled in parallel. This combination will give a force equal to twice that of an individual disc and a deflection of four times that of an individual disc.

Belleville washers are manufactured by Bauer Springs Ltd., of Eagle Road, North Moon Moat Industrial Estate, Redditch, Worcestershire, B98 9HF where full specifications are freely available.

Drawing conventions for these springs are given in Fig. 29.21, and show (a) the normal outside view, (b) the view in section and (c) the simplified representation. These conventions can be adapted to suit the disc combination selected by the designer.

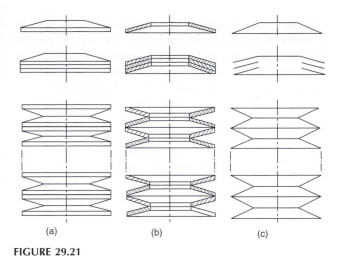

FIGURE 29.21

Spring specifications

A frequent cause of confusion between the spring supplier and the spring user is lack of precision in specifying the spring. This often results in high costs due to the manufacturer taking considerable trouble to meet dimensions which are not necessary for the correct functioning of the spring.

It is recommended that, while all relevant data regarding the design should be entered on the spring detail drawing, some indication should be given as to which of the particular design points must be adhered to for the satisfactory operation of the component; this is necessary to allow for variations in wire diameter and elastic modulus of the material. For example, a compression spring of a given diameter may, if the number of coils is specified, have a spring rate which varies between 15% of the calculated value. In view of this, it is desirable to leave at least one variable for adjustment by the manufacturer, and the common convenient variable for general use is the number of coils.

A method of spring specification which has worked well in practice is to insert a table of design data, such as that shown below, on the drawing. All design data are entered, and the items needed for the correct functioning of the spring are marked with an asterisk. With this method the manufacturer is permitted to vary any unmarked items, as only the asterisked data are checked by the spring user's inspector. The following are specifications typical for compression, tension and torsion springs.

Compression spring

Total turns	7
Active turns	5
Wire diameter	1 mm
*Free length	12.7 ± 0.4 mm
*Solid length	7 mm max.
*Outside coil diameter	7.6 mm max.
*Inside coil diameter	5 mm
Rate	7850 N/m
*Load at 9 mm	31 ± 4.5 N
Solid stress	881 N/mm^2
*Ends	Closed and ground
Wound	Right-hand or left-hand
*Material	S202

*Protective treatment	Cadmium-plate

Tension spring

Mean diameter	11.5 mm
*O.D. max.	13.5 mm
*Free length	54 ± 0.5 mm
Total coils on barrel	16$\frac{1}{2}$
Wire diameter	1.62 mm
*Loops	Full eye, in line with each other and central with barrel of spring
Initial tension	None
Rate	2697 N/m
*Load	53 ± 4.5 N
*At loaded length	73 mm
Stress at 53 N	438 N/mm^2
Wound	Right-hand or left-hand
*Material	BS 1408 B
*Protective treatment	Lanolin

Torsion spring

Total turns on barrel	4
Wire diameter	2.6 mm
*Wound	Left-hand close coils
Mean diameter	12.7 mm
*To work on	9.5 mm diameter bar
*Length of legs	28 mm
*Load applied at	25.4 mm from centre of spring
*Load	41 ± 2 N
*Deflection	20°
Stress at 41 N	595 N/mm^2
*Both legs straight and tangential to barrel	
*Material	BS 5216
*Protective treatment	Grease

Wire forms

Many components are assembled by the use of wire forms which are manufactured from piano-type wire. Figure 29.22 shows a selection, though the number of variations is limitless.

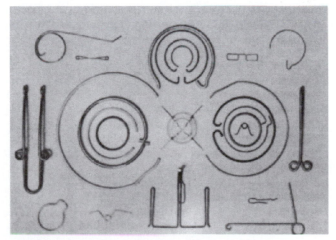

FIGURE 29.22

Corrosion prevention

Springs operating under severe corrosive conditions are frequently made from phosphor bronze and stainless steel, and occasionally from nickel alloys and titanium alloys. For less severe conditions, cadmium- or zinc-plating is recommended; alternatively, there are other electroplatings available, such as copper, nickel, and tin. Phosphate coatings are frequently specified. Organic coatings, originally confined to stove enamels, now include many plastics materials such as nylon and polythene, as well as many types of epoxy resins.

Fatigue conditions

Many springs, such as valve springs, oscillate rapidly through a considerable range of stress and are consequently operating under conditions requiring a long fatigue life. The suitability of a spring under such conditions must be considered at the detail design stage, and a satisfactory design and material must be specified. Special treatments such as shot-peening or liquid-honing may be desirable. In the process of shot-peening, the spring is subjected to bombardment by small particles of steel shot; this has the effect of workhardening the surface. Liquid honing consists of placing the spring in a jet of fine abrasive suspended in water. This has a similar effect to shot-peening, and the additional advantage that the abrasive stream removes or reduces stress raisers on the spring surface.

Chapter 30

Welding and welding symbols

In general, welding may be described as a process of uniting two pieces of metal or alloy by raising the temperature of the surfaces to be joined so that they become plastic or molten. This may be done with or without the application of pressure and with or without the use of added metal. This definition excludes the more recently developed method of *cold-welding*, in which pressure alone is used. *Cold-welding*, however, has a limited application, and is used principally for aluminium and its alloys, and not for steel.

There are numerous methods of welding, but they can be grouped broadly into two categories. *Forge welding* is the term covering a group of welding processes in which the parts to be joined are heated to a plastic condition in a forge or other furnace, and are welded together by applying pressure or impact, e.g., by rolling, pressing, or hammering. *Fusion welding* is the process where the surfaces to be joined are melted with or without the addition of filler metal. The term is generally reserved for those processes in which welding is achieved by fusion alone, without pressure.

Forge welding will be dealt with the first. *Pressure welding* is the welding of metal by means of mechanical pressure whilst the surfaces to be joined are maintained in a plastic state. The heating for this process is usually provided by the process of *resistance welding*, where the pieces of metal to be joined are pressed together and a heavy current is passed through them.

Projection welding is a resistance-welding process in which fusion is produced by the heat obtained from the resistance to flow of electric current through the work parts, which are held together under pressure by the electrodes providing the current. The resulting welds are localized at predetermined points by the design of the parts to be welded. The localization is usually accomplished by projections or intersections.

Spot welding is a resistance-welding process of joining two or more overlapping parts by local fusion of a small area or 'spot'. Two copper-alloy electrodes contact either side of the overlapped sheets, under known loads produced by springs or air pressure. *Stitch welding* is spot welding in which successive welds overlap. *Seam welding* is a resistance-welding process in which the electrodes are discs. Current is switched on and off regularly as the rims of the discs roll over the work, with the result that a series of spot welds is at such points. If a gas-tight weld is required, the disc speed and time cycle are adjusted to obtain a series of overlapping welds.

Flash-butt welding is a resistance-welding process which may be applied to rod, bar, tube, strip, or sheet to produce a butt joint. After the current is turned on, the two parts are brought together at a predetermined rate so that discontinuous arcing occurs between the two parts to be joined. This arcing produces a violent expulsion of small particles of metal (flashing), and a positive pressure in the weld area will exclude air and minimize oxidation. When sufficient heat has been developed by flashing, the parts are brought together under heavy pressure so that all fused and oxidized material is extruded from the weld.

Fusion-welding process can now be dealt with. The heat for *fusion welding* is provided by either gas or electricity. *Gas welding* is a process in which heat for welding is obtained from a gas or gases burning at a sufficiently high temperature produced by an admixture of oxygen. Examples of the gases used are acetylene (oxy-acetylene welding), hydrogen (oxy-hydrogen welding), and propane (oxy-propane welding). In *air–acetylene welding*, the oxygen is derived from the atmosphere by induction.

Electrical fusion welding is usually done by the process of 'arc welding'. *Metal-arc welding* is welding with a metal electrode, the melting of which provides the filler metal. *Carbon arc welding* is a process of arc welding with a carbon electrode (or electrodes), in which filler metal and sometimes flux may be used. *Submerged-arc welding* is a method in which a bare copper-plated steel electrode is used. The arc is entirely submerged under a separate loose flux powder which is continually fed into and over the groove which is machined where the edges to be welded are placed together. Some of the flux powder reacts with the molten metal: part fuses and forms a refining slag which solidifies on top of the weld deposit; the remainder of the powder covers the weld and slag, shielding them from atmospheric contamination and retarding the rate of cooling.

Argon-arc welding is a process where an arc is struck between an electrode (usually tungsten) and the work in an inert atmosphere provided by directing argon into the weld area through a sheath surrounding the electrode. *Heliarc welding* uses helium to provide the inert atmosphere, but this process is not used in the United Kingdom, because of the non-availability of helium. Several proprietary names are used for welding processes of this type, e.g. *Sigma (shielded inert-gas metal-arc) welding* uses a consumable electrode in an argon atmosphere. *Atomic-hydrogen arc welding* is a process where an alternating current arc is maintained between tungsten electrodes, and each electrode is surrounded by an annular stream of hydrogen. In passing through the arc, the molecular hydrogen is dissociated into its atomic state. The recombination of the hydrogen atoms results in a very great liberation of heat which is used for fusing together the metals to be joined. *Stud welding* is a process in which an arc is struck between the bottom of a stud and the base metal. When a pool of molten metal has formed, the arc is extinguished and the stud is driven into the pool to form a weld.

The application of welding symbols to working drawings

The following notes are meant as a guide to the method of applying the more commonly used welding symbols relating to the simpler types of welded joints on engineering drawings. Where complex joints involve multiple welds it is often easier to detail such constructions on separate drawing sheets.

Each type of weld is characterized by a symbol given in Table 30.1 Note that the symbol is representative of the shape of the weld, or the edge preparation, but does not indicate any particular welding process and does not specify either the number of runs to be deposited or whether or not a root gap or backing material is to be used. These details would be provided on a welding procedure schedule for the particular job.

It may be necessary to specify the shape of the weld surface on the drawing as flat, convex or concave and a supplementary symbol, shown in Table 30.2, is then added to the elementary symbol. An example of each type of weld surface application is given in Table 30.3.

A joint may also be made with one type of weld on a particular surface and another type of weld on the back and in this case elementary symbols representing each type of weld used are added together. The last example in Table 30.3 shows a single-V butt weld with a backing run where both surfaces are required to have a flat finish.

A welding symbol is applied to a drawing by using a reference line and an arrow line as shown in Fig. 30.1. The reference line should be drawn parallel to the bottom edge of the drawing sheet and the arrow line forms an angle with the reference line. The side of the joint nearer the arrowhead is known as the 'arrow side' and the remote side as the 'other side'.

The welding symbol should be positioned on the reference line as indicated in Table 30.4.

Sketch (a) shows the symbol for a single-V butt weld below the reference line because the external surface of the weld is on the arrow side of the joint.

Sketch (b) shows the same symbol above the reference line because the external surface of the weld is on the other side of the joint.

Sketch (c) shows the symbol applied to a double-V butt weld.

TABLE 30.1 Elementary weld symbols

Form of weld	Illustration	BS symbol
Butt weld between flanged plates (the flanges being melted down completely)		
Square butt weld		||
Single-V butt weld		V
Single-bevel butt weld		
Single-V butt weld with broad root face		Y
Single-bevel butt weld with broad root face		
Single-U butt weld		
Single-J butt weld		
Backing or sealing run		
Fillet weld		
Plug weld (circular or elongated hole, completely filled)		
Spot weld (resistance or arc welding) or projection weld	(a) Resistance / (b) Arc	○
Seam weld		

Sketch (d) shows fillet welds on a cruciform joint where the top weld is on the arrow side and the bottom weld is on the other side.

TABLE 30.2 Supplementary symbols

Shape of weld surface	BS symbol
flat (usually finished flush)	—
convex	⌒
concave	⌣

TABLE 30.3 Some examples of the application of supplementary symbols

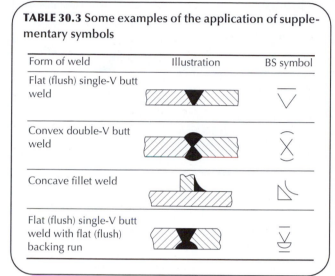

Form of weld	Illustration	BS symbol
Flat (flush) single-V butt weld		
Convex double-V butt weld		
Concave fillet weld		
Flat (flush) single-V butt weld with flat (flush) backing run		

TABLE 30.4 Significance of the arrow and the position of the weld symbol

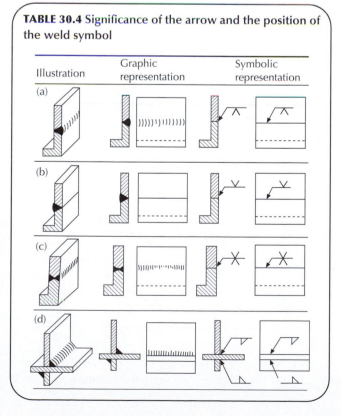

Illustration	Graphic representation	Symbolic representation
(a)		
(b)		
(c)		
(d)		

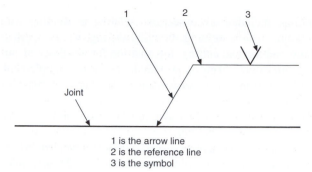

1 is the arrow line
2 is the reference line
3 is the symbol

FIGURE 30.1

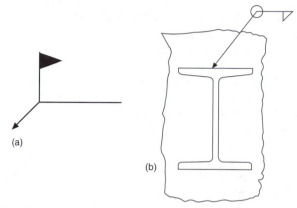

FIGURE 30.2 Indication of (a) site welds and (b) continuous welds.

The positioning of the symbol is the same for drawings in first or third angle projection.

Additional symbols can be added to the reference line as shown in Fig. 30.2. Welding can be done in the factory or on site when plant is erected. A site weld is indicated by a flag. A continuous weld all round a joint is shown by a circle at the intersection of the arrow and the reference line. Note that if a continuous weld is to be undertaken at site then both symbols should be added to the drawing.

The introductory notes relating to welding processes are of a general nature. There are many specialized methods listed in BS 499. Each process is given an individual identification number and group headings are as follows: (a) arc welding, (b) resistance welding, (c) gas welding, (d) solid phase welding; pressure welding, (e) other welding processes, (f) brazing, soldering and braze welding.

A welding procedure sheet will usually give details of the actual process to be used on a particular joint. On the drawing, a reference line with an arrow pointing towards the joint at one end, will have a 'fork' added at the other containing the selected number. In the example given below, the figure 23 indicates that projection welding is the chosen method.

Useful standards for the draughtsman are as follows: BS 499–1 gives a glossary for welding, brazing and thermal

cutting. Includes seven sections relating to welding with pressure, fusion welding, brazing, testing, weld imperfections, and thermal cutting. Information for welding and cutting procedure sheets is provided. BS 499–1 Supplement. Gives definitions for electrical and thermal characteristics of welding equipment.

European arc welding symbols in chart form are illustrated in BS 499–2C:1999.

Symbolic representation on drawings for welded, brazed and soldered joints are illustrated in BS EN 22553. Welded and allied processes, nomenclature of processes and reference numbers are given in BS EN ISO 4063:2000.

Dimensioning of welds

The dimensions of a weld may be added to a drawing in the following manner.

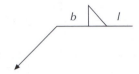

Dimensions relating to the cross-section of the weld are symbolized by b and are given on the left-hand side of the symbol. The cross-sectional dimension to be indicated for a fillet weld is the leg length. If the design throat thickness is to be indicated then the leg-length dimension is prefixed with the letter b and the design throat thickness with the letter a.

Longitudinal dimensions are symbolized by l and are given on the right-hand side of the symbol. If the weld is not continuous then distances between adjacent weld elements are indicated in parentheses. Unless dimensional indication is added to the contrary, a fillet weld is assumed to be continuous along the entire length of the weld. Leg-length dimensions of fillet welds of 3, 4, 5, 6, 8, 10, 12, 16, 18, 20, 22, and 25 mm are the preferred sizes.

Applications of dimensions to different types of fillet welds are given in Table 30.5 in order to indicate the scope of the British Standard, which should be consulted to fully appreciate this topic. Table 30.5(1) shows dimensions applied to continuous fillet welds, (2) shows dimensions applied to intermittent fillet welds, and (3) shows dimensions applied to staggered intermittent fillet welds.

TABLE 30.5 The dimensioning of welds

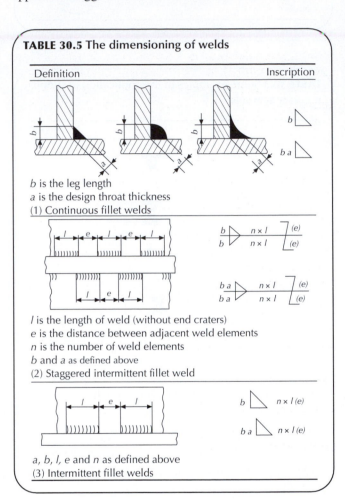

Chapter 31

Engineering diagrams

The following list gives details of Standards and publications, which will provide a major source of reference material for use in the preparation of various types of engineering diagram.

General engineering graphical symbols

Construction Drawing Series lists the following:

BS 1192	Collaborative production of architectural engineering and construction informative code of practice
BS EN ISO 6284	Indication of limit deviations
BS EN ISO 8560	Representations of modular sizes, lines and grids
BS EN ISO 9431	Spaces for drawing for text and title block on drawing
BS EN ISO 3766	Simplified representation of concrete reinforcement
BS EN ISO 7518	Simplified representation of demolition and rebuilding
BS EN ISO 11091	Landscape drawing practice

General engineering graphical symbol series

BS 1553: Specification for graphical symbols for general engineering.

Part 1 *Piping systems and plant*: This section deals with graphical symbols for use in the creation of flow and piping plant and heating and ventilation installations.

Part 2 *Graphical symbols for power generating plant*: Includes steam and internal combustion engines and turbines, also auxiliary plant.

Part 3 *Graphical symbols for compressing plant*: Includes applications to air operated switchgear.

Fluid power systems and components
BS 2917–1 specification for graphical symbols
Electrical power
IEC 60617–2: Graphical symbols for diagrams. Symbol elements, qualifying symbols, and other symbols having general application.

Note: In 2002 the IEC (International Electrotechnical Commission) launched an 'on-line' database format for the symbol library (available on subscription from the IEC website). Following this decision, in 2002 CENELEC (European Committee for Electrotechnical Standardization) decided to cease publication of EN 60617 in 'paper form', to withdraw the then existing standards and formally to adopt the IEC database without changes for use in Europe. Consequently the British Standard versions were withdrawn.

The database is the official source of IEC 60617, and it currently includes some 1750 symbols.

Engineering diagram drawing practice is covered by the following standards:

BS 61082–1 Preparations of documents used in electrotechnology. Rules (This standard replaces the now withdrawn standards BS 5070–1 and BS 5070–2).

BS 5070–3. Recommendations for mechanical/fluid flow diagrams.

Gives principles and presentation for mechanical, hydraulic, pneumatic, topographic, block, circuit, piping, interconnection, and supplementary diagrams.

BS 5070–4. Recommendations for logic diagrams.

Principles and presentation. Covers signal names, characteristics and logic circuit diagrams.

The diagrams which follow are representative of various branches of engineering and obviously every application will be different. The examples can only indicate the type of diagram one is likely to encounter. The standards listed will provide a valuable source of information relating to layout content and the appropriate symbols to be used (Table 31.1).

Engineered systems

All of the engineering specialities referred to at the start of this chapter need diagrams and circuits in order to plan and organize the necessary work. It is very difficult to standardize aspects of work of such a varied nature, however, the following general notes are applicable in most circumstances.

Block diagrams

Block symbols or outlines are used to indicate the main separate elements in an installation and how they are functionally linked together. The diagram needs to be simple, so that the basics of the operation it represents, can be appreciated quickly.

TABLE 31.1

Symbol	Description
—	Direct current
~	Alternating current
+	Positive polarity
–	Negative polarity
→	Propagation, energy flow, signal flow, one way
1 2 3 4 5	Terminal strip, example shown with terminal markings
	Junction of conductors
	Double junction of conductors
	Crossing of conductors with no electrical connection
	Data highway
	Primary cell or accumulator

Note– The longer line represents the positive pole, the short line the negative pole. The short line may be thickened for emphasis

Symbol	Description
	Inductor, coil, winding, choke
	Inductor with magnetic core
	Inductor with tappings, two shown
	Transformer with magnetic core
*	Machine, general symbol. The asterisk is replaced by a letter designation as follows: C synchronous converter G generator GS synchronous generator M motor MG machine capable of use as a generator or motor MS synchronous motor
	Socket (female), pole of a socket
	Plug (male), pole of plug
	Plug and jack, telephone type, two-pole *Note* – The longest pole on the plug represents the tip of the plug, and the shortest the sleeve

Symbol	Description
	Battery of accumulators or primary cells
	Fuse, general symbol Form 1
	Fuse with the supply side indicated Form 2
	Connecting link, closed
	Connecting link, open
	Circuit breaker
	Polarized capacitor, for example, electrolytic
	Variable capacitor
	Make contact normally open, also general symbol for a switch
	Break contact
	Change-over contact, break before make
	Contactor, normally open
	Contactor, normally closed
	Operating device (relay coil), general symbol
	Coil of an alternating current relay
- - - -	Mechanical coupling
	Wattmeter

Symbol	Description
	Conductor, group of conductors, line, cable, circuit, transmission path (for example, for microwaves)
/// -01-03	Three conductors
	Conductors in a cable, three conductors shown
	Earth or ground, general symbol
Y	Antenna
(A)	Ammeter
(~)	Oscilloscope
(V)	Voltmeter
(↑)	Galvanometer
	Earphone
	Loudspeaker
	Transducer head, general symbol
	Clock, general symbol
	Laser (optical maser, general symbol)
	Amplifier
	Semiconductor diode, general symbol
	Light emitting diode, general symbol
	Tunnel diode

TABLE 31.1 (*Continued*).

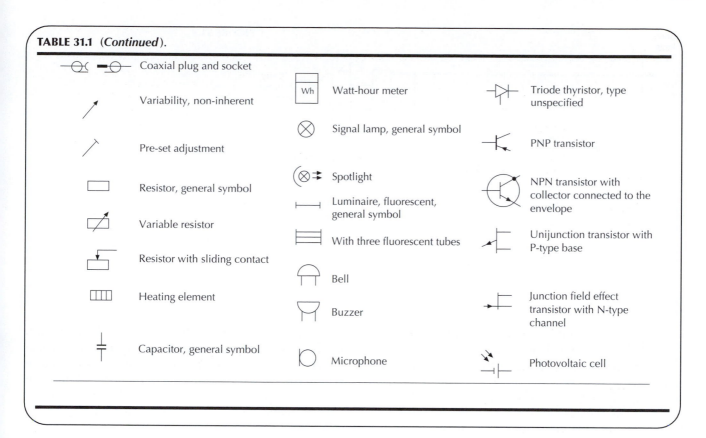

Coaxial plug and socket	
Variability, non-inherent	
Pre-set adjustment	
Resistor, general symbol	
Variable resistor	
Resistor with sliding contact	
Heating element	
Capacitor, general symbol	
Watt-hour meter	
Signal lamp, general symbol	
Spotlight	
Luminaire, fluorescent, general symbol	
With three fluorescent tubes	
Bell	
Buzzer	
Microphone	
Triode thyristor, type unspecified	
PNP transistor	
NPN transistor with collector connected to the envelope	
Unijunction transistor with P-type base	
Junction field effect transistor with N-type channel	
Photovoltaic cell	

The block symbols refer to single hardware items or self-contained units without necessarily indicating all of the exact connections.

A block diagram can be presented to show a sequence of events to the reader and be used for fault diagnosis.

Circuit diagrams

The term circuit suggests electrical components wired together but this need not be the only case. The circuit could show parts of a central heating system connected by water piping or units in an air conditioning system joined together by fabricated ductwork.

Theoretical circuit diagrams

Design staff will prepare theoretical circuit diagrams where all the necessary connections for the correct operation of the system are included. Different sections of industry freely use other terms, such as schematics, sequence diagrams and flow charts (see Fig. 31.1).

In all these diagrams, the component parts are arranged neatly and if possible horizontally or vertically. If several diagrams form a set, then the style of presentation should be consistent.

One of the conventions with this type of diagram is that components should be arranged so that the sequence of events can be read from left to right, or top to bottom, or perhaps a combination of both.

The diagram does not differentiate between the physical sizes of the separate components. The actual component shape may not be reflected in the standardized symbols and the arrangement on the diagram will not attempt to indicate the true layout of all the items.

Basic engineering practice follows where specifications will be produced for all parts of the system covering, for example, the components in detail, materials, manufacturing processes, relevant standards, inspection procedures, delivery dates and costs. The customer needs to know exactly what is being supplied, and details of financial arrangements. Contracts will be exchanged when supplier and client are satisfied. Obviously failure of any aspect of an agreement may involve either party in financial loss and litigation could follow. It is of course in nobody's interest that this should occur.

Construction diagrams

When the system is engineered, the actual position of each component part will dictate the arrangement of wiring, piping and general services, etc. The engineer will need to

FIGURE 31.1

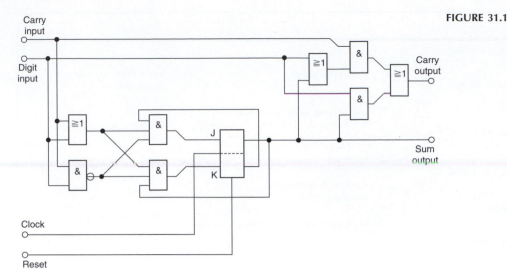

divide the work according to the scale of the contract and define which diagrams are necessary.

A production circuit diagram for an electrical control panel would show how the panel was built, with all the necessary line and neutral wiring connections in their exact places together with earth links. An exact construction record is essential for service requirements. Wiring must be sized. Standards for current capacity dictate the wire dimensions. Wires are often colour coded to facilitate tracing. Wires are run singly, in multicored cables, in looms and conduits, inside and outside, and in almost every conceivable ambient situation.

Clear, accurate and presentable layouts are essential in the production of engineering diagrams. Clarity depends on sufficient thought being given to spacing not only the symbols but associated notes and references.

Figure 31.2 shows a thyristor control system.

Part (a) outlines the basic blocks.

Part (b) provides added details to the four parts.

Part (c) gives the component connections for the zero voltage trigger with waveforms at various points.

Part (d) is an example of a supplementary diagram where the waveforms are related to a common datum.

Wiring diagrams for motor vehicles

The following diagrams are reproduced by kind permission of the Ford Motor Company Ltd. and show part of the wiring circuit for the Transit van. Service manuals need to be presented so that the technician can easily check each function for satisfactory operation. The manual is therefore written with each circuit shown completely and independently in one chapter or 'cell'. Other components which are connected to the circuit may not be shown unless they influence the circuit operation. For the benefit of the user, the diagram needs to be reasonably large to be used while work proceeds.

Figure 31.3 shows part of the circuitry for the headlamps. Each cell normally starts with the fuse, ignition switch, etc. that powers the circuit. Current flows from the power source at the top of the page towards earth at the bottom. Within the schematic diagram, all switches, sensors and relays are shown 'at rest' as if the ignition switch was in the Off position.

Page numbering system

The Ford Motor Company procedure is to divide the electrical system into individual sections. For example, the Engine Control Section is 29. The section is further broken down into cells, where cell 29.00 is the 2.OL2V engine, cell 29.10 is the 2.OLEFI engine and so on. All the engine information can be found in cells 29.00–29.50. Finally the pages in the manual are numbered using the cell number, a dash, and then a consecutive number. If there are two pages in cell 29.00, they will be numbered 29.00–00 and 29.00–01.

The headlamp circuit is continued at points 'A' & 'B' on to sheet number 32.00–02 down to the ground point G1001. The location of ground points is given on a separate diagram.

Connections to the dim light relay, component K20, at the centre left side of diagram 32.00–01 are made via a connector, numbered C122. Connectors are grouped together on separate sheets and 91.00–05 (Fig. 31.4) is typical. The dim light relay connector is shown at the right-centre side of this sheet. Looking at the connector, each wire is numbered to correspond with the circuit diagram, and recognizable by the colour coding.

Vehicle wiring is made up into looms and at the end of each of the bunches the connectors are fitted. Diagram 90.10–22 (Fig. 31.5) shows a pictorial view of wiring looms in the cab, and this is one of a series.

The circuits for signal and hazard indicators, also the wiper/washer control are given on diagrams 32.40–00 and 32.60–01 (Fig. 31.3). Note that the switch N9 serves three functions. On 32.00–01 its function is to control side/tail and

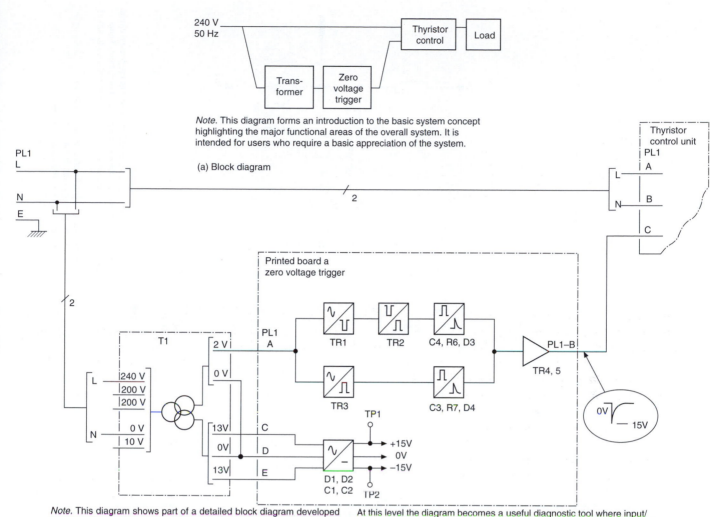

Note. This diagram forms an introduction to the basic system concept highlighting the major functional areas of the overall system. It is intended for users who require a basic appreciation of the system.

(a) Block diagram

Note. This diagram shows part of a detailed block diagram developed from the simple form of block diagram shown in (a). In this diagram functional information has been expanded and specific information in respect of input/output terminations has been added.

At this level the diagram becomes a useful diagnostic tool where input/output parameters may be monitored and hence faulty operation detected at unit, printed boards, etc. level. Maintenance at this level involves the replacement of the faulty unit or printed circuit board thus restoring normal working fairly rapidly.

(b) Detailed block diagram

FIGURE 31.2

dipped beam on either side of the vehicle. On 32.60–01, the four wipe functions are actuated. Note also that on 32.40–00 the multi-function switch is illustrated in full. However, four connections on this switch relate to headlamp functions so these details are given in full from points 8, 9, 10, 11 on 32.00–01. The location of connector C3 at position D1 on diagram 90.10–22 shows the connection to the direction indicators at the top left corner on 32.40–00.

There are many pages similar to these in a vehicle service manual. This type of simplified layout has many advantages during servicing and fault finding operations.

Heating, ventilation and air conditioning systems

Control systems are devised to suit each individual application. Generally, each part of the system will contain air of different types. With reference to Fig. 31.6 the room air (RA) is extracted by a fan, a proportion of the air is exhausted to atmosphere and the remainder is returned and mixed with a fresh supply of air. The mixed air will then be returned to the room via a supply air fan after its temperature has been corrected to suit the design requirements.

In most cases this involves a heating operation. However, if the outside air temperature is high, or if there are considerable heat gains within the controlled space, then a cooling operation may well be necessary. In addition, full air conditioning specifications require control of the relative humidity in the space.

Personal physical comfort conditions depend on adjusting air and surface temperatures, humidity and air movement. By balancing these four factors, the engineer can design a climate to suit any type of activity.

In Fig. 31.6 the air is heated by passage through a heat exchanger supplied with hot water. Hot water from a boiler

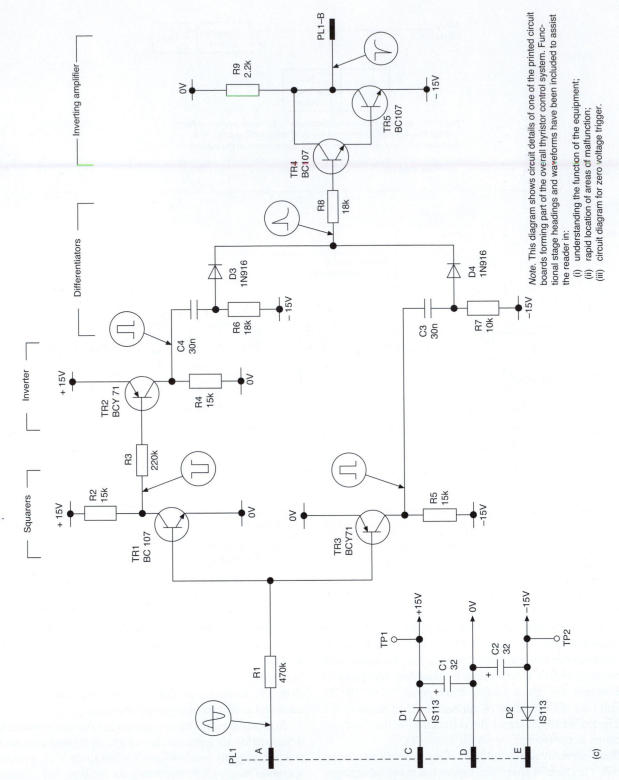

FIGURE 31.2 (*Continued*).

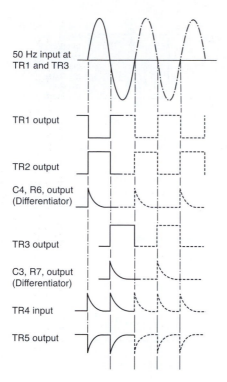

50 Hz input at
TR1 and TR3

TR1 output

TR2 output

C4, R6, output
(Differentiator)

TR3 output

C3, R7, output
(Differentiator)

TR4 input

TR5 output

Note. Although the waveforms shown in (c) are of assistance to the reader in establishing that the correct signals are present at various parts of the diagram, thus assisting preliminary fault location, it does not show the timing relationship that exists between waveforms. This diagram is a supplementary diagram that defines this time relationship.

(d) Supplementary diagram for waveform development for zero voltage trigger

FIGURE 31.2 (*Continued*).

operating at normal atmospheric pressure is low temperature hot water (LTHW). If the boiler operates at pressure, its output is high temperature hot water (HTHW). The heat exchanger could also be supplied with steam or operated by electricity.

The volume of outside air supplied will vary considerably with the occupancy density within the space and the activity. For example, theatres, public houses, conference rooms, areas with large solar heat gains, industrial premises with processing equipment, swimming pools and incubators, to name just a few, all require very special attention. Hence, various degrees of air purification and levels of sophistication exist. Three typical schemes now follow and Fig. 31.7 shows a ventilation system. Here a controller adjusts the position of a three-way valve so that more, or less, water passes through the heat exchanger in response to supply air temperature demands. The air supplies are controlled by electrically operated dampers fitted in the ducts. Note that the air into the space has its temperature measured by the sensor f1. An alternative position for the sensor could be in the outlet duct when it could take note of any temperature increases generated within the space, or it could be positioned within the space itself, shown by u2. Many choices need to be considered.

A scheme for partial air conditioning is illustrated in Fig. 31.8 where in addition to ventilation and heating, the humidity has been given a degree of control. For full air conditioning it is necessary to provide equipment to cool the air and typical plant has been added to the layout in Fig. 31.9.

All of the installed plant needs to be carefully sized to ensure that specifications for air quality are met. The engineer uses a psychrometric chart to determine the physical properties of the air to be handled.

Functions

Temperature control

The duct sensor f1 measures the temperature t_{zu}. The controller u0010 compares this value with the selected setpoint X_K on the controller u0010 or on the remote setpoint potentiometer u0020 and adjusts the heating coil valve s1 in accordance with the difference between the two.

Safety devices

When there is danger of frost, the frost protection thermostat f2 must switch off the fan, close the damper s2, open the heating coil valve s1 and, where appropriate, switch on the heating pump.

Controller output temperature

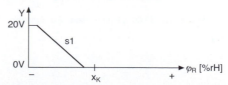

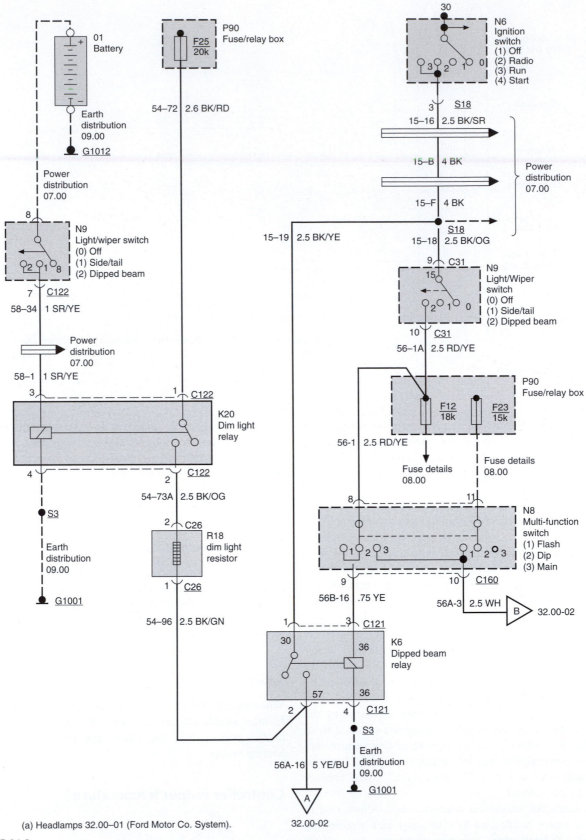

(a) Headlamps 32.00–01 (Ford Motor Co. System).

FIGURE 31.3

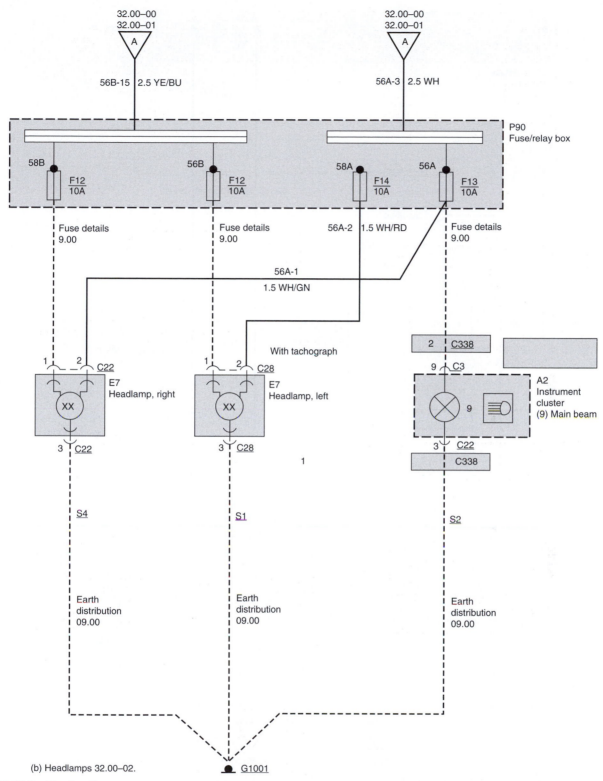

(b) Headlamps 32.00–02.

FIGURE 31.3 (*Continued*).

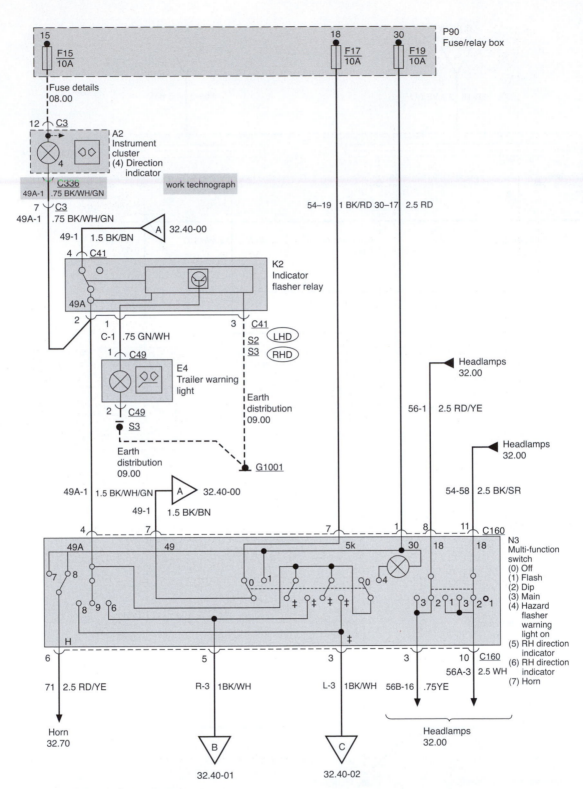

(c) Signal and hazard indicators 32.40–00.

FIGURE 31.3 (*Continued*).

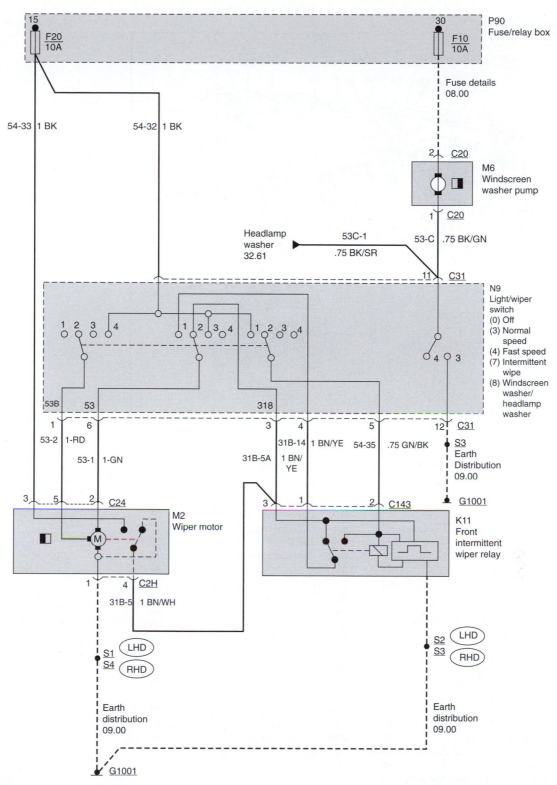

(d) Wiper/washer control 32.60–01.

FIGURE 31.3   (*Continued*).

Component location views

FIGURE 31.4 Component location views 90.10–22.

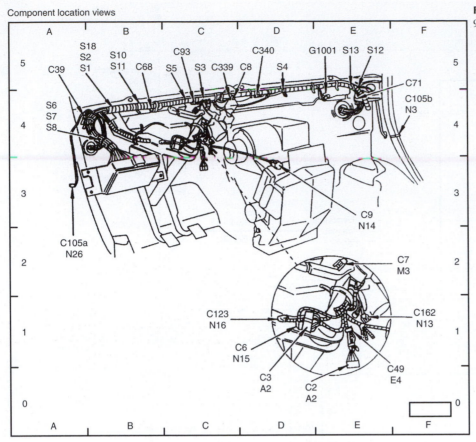

Behind dash (LH drive shown, RH drive similar)

A2 .... Instrument cluster D1	N13 .... Electrical headlamp alignment	
C2 .... E0	.... system switch E1	
C3 .... D1	N14 .... Heater blower switch D3	
C6 .... D1	N15 .... Stop light switch D1	
C7 .... E2	N16 .... Safety switch D1	
C8 .... C4	N26 .... Door ajar sensor, driver side A3	
C9 .... D3	S1 .... B4	
C39 .... A4	S2 .... B4	
C49 .... E1	S3 .... C4	
C68 .... B4	S4 .... D4	
C71 .... E4	S5 .... C4	
C93 .... C4	S6 .... B4	
C123 .... D1	S7 .... B4	
C162 .... E1	S8 .... B4	
C339 .... C4	S10 .... B4	
C340 .... D4	S11 .... B4	
C105a A3	S12 .... E4	
C105b F4	S13 .... E4	
E4 .... Trailer warning light E1	S18 .... B4	
G1001 .... E4		
M3 ...... Heater blower E2		
N3 .... Door ajar sensor, passenger side F4		

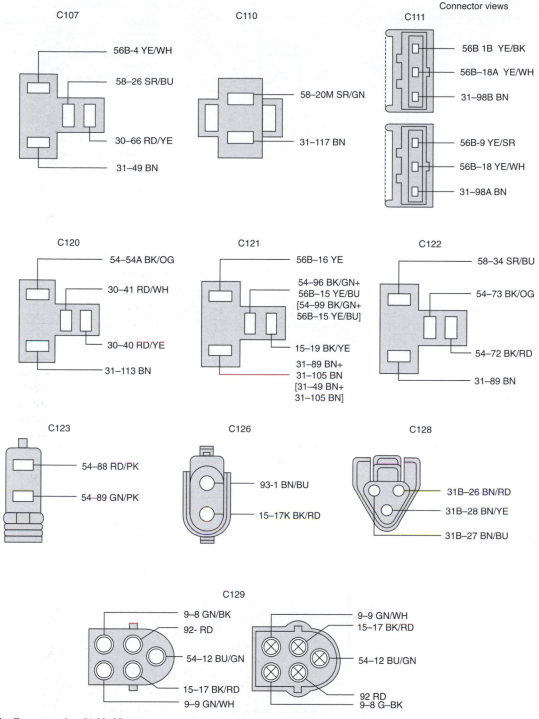

FIGURE 31.5 Connector view 91.00–05.

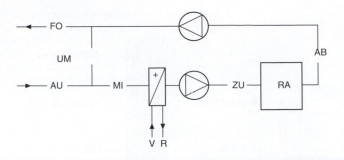

Air types: (DIN 1946).	
Air type	Abbreviation
Outside air	AU
Penetrated outside air	VAU
Exhaust air	FO
Aftertreated exhaust air	NFO
Extract air	AB
Room air	RA
Aftertreated extract air	NAB
Return air	UM
Mixed air	MI
Supply air	ZU
Pretreated supply air	VZU

Water:	
Flow	V
Return	R

FIGURE 31.6

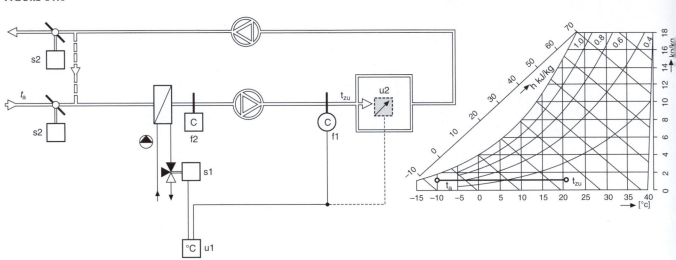

FIGURE 31.7 Ventilation system diagram. System designed to control the temperature of supply air into a space with heating from LTHW, HTHW or a steam heated coil. Variant with remote setpoint potentiometer.

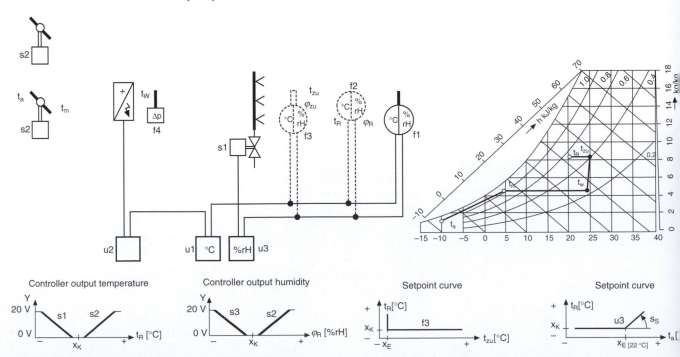

FIGURE 31.8 Partial air conditioning system. System designed to control the extract air from a room. The air into the space is heated with an electric heating coil and humidified with steam. Alternative: room sensor instead of extract air sensor. Variant: with low-limit supply air temperature control, and high-limit supply air humidity control.

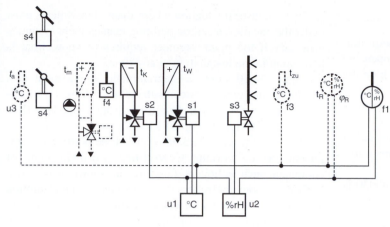

FIGURE 31.9 Air conditioning system diagram. System designed to control the extract air from a room. The air into the space is heated by LTHW, HTHW or a steam heated coil. Cooling and dehumidifying with CHW cooling coil. Humidifying with steam. Alternative: room sensor instead of extract air sensor. Variants: with low-limit supply air temperature control, and with summer compensation.

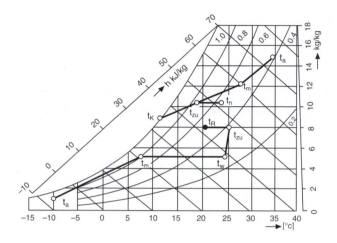

Functions

Temperature control

The duct sensor f1 or the room sensor f2 measures the temperature t_R. The controller u0010 compares this value with the selected setpoint X_K and adjusts the heating coil step controller (or power controller) u0020 in accordance with the difference between the two.

Humidity control

The duct sensor f1 or the room sensor f2 measures the humidity φ_R. The controller u0030 compares this value with the selected setpoint X_K and adjusts the humidfying valve s1 in accordance with the difference between the two.

Safety devices

If there is an air flow failure, the differential pressure switch f4 must cut off the control voltage of the electric heating coil.

With electric heating coils, it is advisable to incorporate a timer so that the fan will run on for approximately 5 min and dissipate any residual heat.

Functions

Temperature control

The duct sensor f1 or the room sensor f2 measures the temperature t_R. The controller u0010 compares this value with the selected setpoint X_K and adjusts the heating coil valve s1 or the cooling coil valve s2 in sequence in accordance with the difference between the two.

Humidity control

The duct sensor f1 or the room sensor f2 measures the humidity φ_R. The controller u0020 compares this value with the selected setpoint X_K and adjusts the humidfying value s3 or the cooling valve s2 in sequence in accordance with the difference between the two.

Safety devices

When there is danger of frost, the frost protection thermostat f4 must switch off the fan, close the damper s4, open the heating coil valve s1 and, where appropriate, switch on the heating pump.

Variant: with low-limit supply air temperature control.

The low-limit supply air temperature sensor f3 prevents the supply air temperature t_{zu} from dropping below the cut-in point X_E set on the controller u0010 (draught elimination).

Variant: with summer compensation.

The outside temperature compensation sensor u0030 is used to increase the room temperature t_R in summer. If the outside temperature t_a rises above the cut-in point X_E (22 °C), the setpoint X_K is increased continuously by the selected steepness S_S.

Engineer uses a psychrometric chart to determine the physical properties of the air to be handled.

Controller output temperature

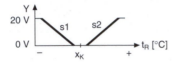

Controller output humidity

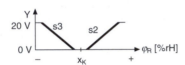

Setpoint curve

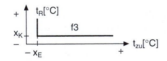

Setpoint curve

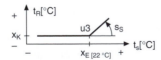

Building management

Figures 31.8 and 31.9 show possible schemes for partial and full air conditioning. In addition to the controls indicated there will be all the associated ductwork, filters, water, steam, and electrical services. An electrical control panel is usually necessary for the wiring of interconnected equipment, instrumentation, and to isolate plant for servicing. Although operations are generally fully automatic, emergency hand control facilities are often specified.

In an industrial situation where many departments exist under the one roof, a central building management system is necessary if the plant engineer requires to know what is happening in the installation at any time.

Figure 31.10 shows a diagrammatic arrangement of a building management system with software designed specifically for the installation. Control is from a standard IBM compatible PC(1), see Fig. 31.11.

The operator can check how any of the peripheral components (4) are working. Control units for major items of equipment such as boilers, fans, etc. are shown as item (3). Individual room controllers (5) control energy consumption as a function of room occupancy, the time of day and season. System controllers (2) coordinate process control tasks such as the overall management of energy.

The plant manager has instant access to data using a mouse operation and pull down menus. Individual schematic diagrams can be displayed. Recorded data over a period of time can be displayed or printed out (Fig. 31.12).

The psychrometric chart (Fig. 31.13)

In air conditioning technology, it is necessary to define thermodynamic processes and the properties of moist air.

This may be achieved by a good knowledge of physics, with theoretical calculations using complicated formulae and tables. The procedure can be time consuming.

By presenting the interrelated factors on a psychrometric chart, an immediate decision can be made regarding the feasibility of controlling an air conditioning system and the means required to carry this out.

For a given air sample nine different parameters are shown on the psychrometric chart.

A position on the chart can be established at the intersection of two ordinates for known conditions and the others obtained.

Since the properties and behaviour of moist air depend on barometric pressure, a psychrometric chart can only be drawn for a specific barometric pressure. Allowances may be made for changes in barometric pressure by using correction factors.

Note that the chart indicates a condition of 21 °C dry bulb temperature and 48% Relative humidity. These are typical values to provide comfort in an office.

Example: Find the missing values for the following case (see Fig. 33.13).

1. Dry bulb temperature $t_{sic} = 20\ °C$
2. Absolute humidity $x = g/kg$
3. Partial water vapour pressure $P_D = mbar$ or kPa
4. Saturation pressure $P_{sat} = mbar$ or kPa
5. Saturation temperature (dew point) $t_{sat} = °C$
6. Relative humidity $\varphi = 50\%$
7. Enthalpy $h = kJ/kg$
8. Wet bulb temperature $t_{hyg} = °Cs$
9. Density $p = kg/m^3$

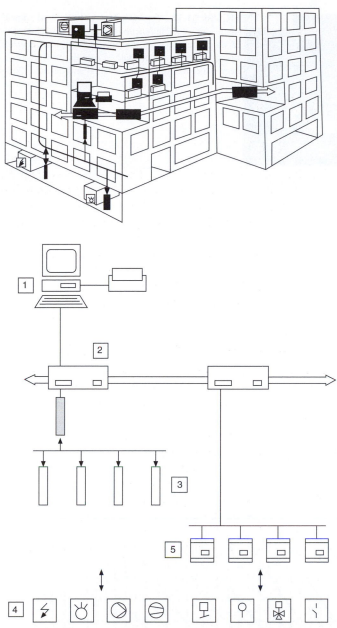

FIGURE 31.10

FIGURE 31.11

The point on the chart is defined by the temperature and the relative humidity given above.

Solution:

The point of intersection P between the 20 °C isotherm from the dry bulb temperature (1) and the line of 50% constant relative humidity (6) clearly defines the position of the required condition.

The absolute humidity (2) is found by drawing a horizontal line through the point P and extending it until it meets the ordinate on the right.

If this horizontal is extended to the left it will intersect the scale for partial water vapour pressure p_D (3).

To obtain the saturation pressure (4) the isotherm from P must be extended until it intersects the 100% relative humidity line. At this point the air is saturated, i.e. it cannot absorb any further moisture without a dense mist forming. An extension of the horizontal line through this point of intersection to the left intersects the partial pressure scale at point (4). The pressure of the saturated air can now be read.

Where the horizontal line from (P) intersects the saturation curve a similar condition occurs, whereby the air cannot absorb any additional moisture (5). The dew point or saturation temperature can now be read on the saturation curve (and also on the dry bulb temperature scale).

By following the isenthalp (line of constant enthalpy) which passes through the condition (P) we can determine the enthalpy at the points of intersection (7) with the enthalpy scale. If an adiabatic line is drawn through the point (P) towards the saturation curve, the two intersect at point (8) to give the wet bulb temperature. This is lower than the starting temperature because the absorption of moisture has caused sensible heat to be converted into latent heat.

The density is determined from the nearest broken lines of constant density (9).

The required values are:

2.	$\times$	= 7.65	g/kg
3.	P_D	= 11.8	mbar = 1.18 kPa
4.	P_{sat}	= 23.4	mbar = 2.34 kPa
5.	t_{sat}	= 9.6	°C
7.	H	= 39.8	kJ/kg
8.	t_{hyg}	= 13.8	°C
9.	P	= 1.16	kg/m^3

Refrigeration systems and energy-saving applications

In order to appreciate the engineering diagram examples relating to refrigeration practice, we have included an explanation of a typical cycle of operations.

Refrigeration through evaporation

When you pour liquid ether on to the back of your hand, after a few seconds you feel your hand turn ice-cold. The liquid evaporates very quickly, but in order to do so it requires (a)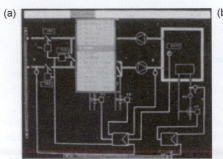

(b)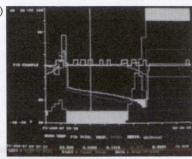

FIGURE 31.12 Typical displays. (a) Pull down menus for selecting functions relating to any schematic in the building layout. (b) Pictorial graph of control operations with shaded bands where limit values have been exceeded. (c) Data evaluation and display. Electricity consumption in various zones for one month is given in this example.

(c)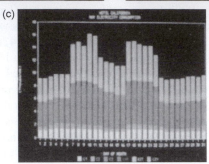

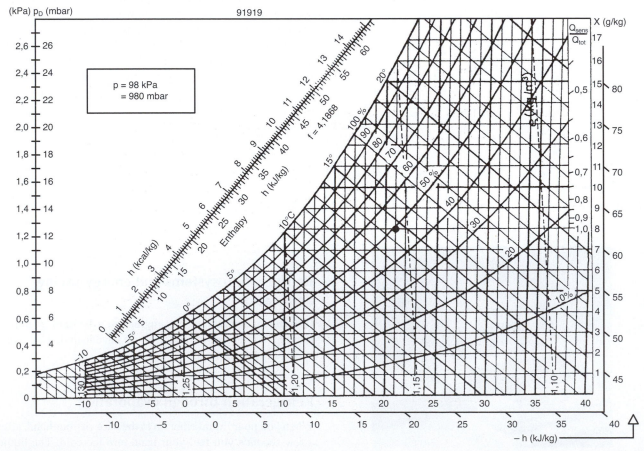

FIGURE 31.13

heat – so-called heat of evaporation. This heat is drawn in from all around the ether – including from your hand – and it is this which causes a sensation of cold.

If you could catch this evaporated ether and liquefy it again by compressing and cooling, the heat absorbed during evaporation would be released back into the surroundings.

This is precisely the principle on which the refrigeration cycle works. A special refrigeration agent, which is even more suitable for this purpose than ether, is evaporated close to the medium to be refrigerated. The heat necessary for this process is drawn in from all around, thereby cooling the air or water.

The most widely known refrigerating agent used to be ammonia but this has now been almost entirely superseded by halogen refrigerants, the best known of which are R12, R22 and R502.

The refrigerant cycle

The refrigerant circulates in a closed system. To produce this circulation, a very powerful pump is required – the compressor. This draws in the expanded refrigerant in vapour form and compresses it. On being compressed, the temperature of the vapour rises. It moves on to the condenser and is cooled by means of cold water. The cooling is so great that the condensation temperature is reached and

the condensation heat thus produced is given up to the water in the condenser.

The refrigerant is then pumped on in liquid form into a receiver and from there on to the evaporator. But just before this, it flows through an expansion valve. This valve reduces the pressure on the liquid so much that it evaporates, drawing in the required heat from its surroundings. This is precisely as intended for the air or water to be cooled and is led past the group of pipes in the evaporator. After leaving the evaporator, the refrigerant is once again drawn into the compressor.

To summarize: In one half of the cycle, the heat is removed by evaporation (i.e. cooled where cooling is required) and in the other half, heat has been released by condensing. Thus energy (heat) is moved from where it is not wanted to a (different) place where it is tolerated or, in fact, required.

When the theoretical cycle of operations is applied in practice it is necessary to include controls and safety devices. In a domestic system, the motor and compressor are manufactured in a sealed housing.

The heat extracted from the inside of the refrigerator, where the evaporator is positioned, passes to the condenser, generally at the back of the cabinet, where natural convection currents release the heat into the surroundings.

In a large industrial installation it may be economically viable to recover heat from a condenser and use it for another process (Figs. 31.14–31.16).

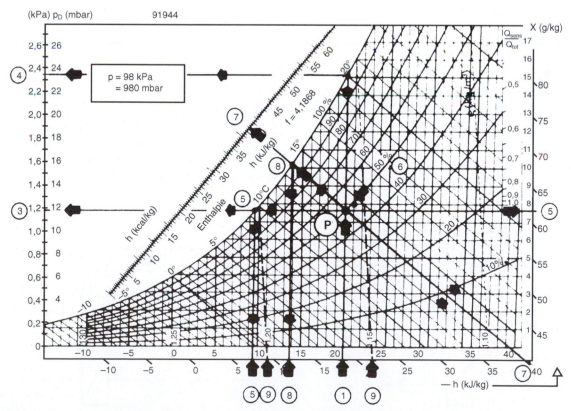

FIGURE 31.14

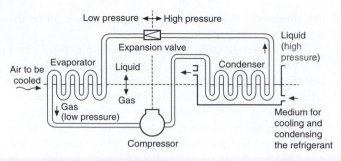
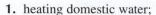

FIGURE 31.15 The refrigerant cycle.

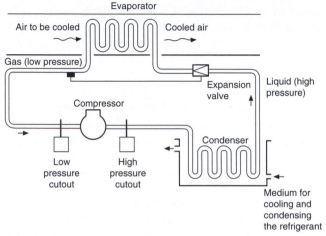

FIGURE 31.16 The safety devices in the refrigerant cycle.

Heat recovery control in refrigeration systems

The principle of the refrigeration cycle is shown in Fig. 31.17. Where the refrigerant passes through the following four phases:

1. evaporation (with heat absorption);
2. compression (with energy absorption);
3. condensation (with heat emission);
4. expansion.

In a refrigeration system, a second condenser can be incorporated in the cycle and the recovered heat used for:

1. heating domestic water;
2. supply air heating;
3. reheating in the case of supply air dehumidification.

Control options

The heat absorbed by the evaporator and that generated by the compressor are emitted in the condenser. To recover this heat an additional condenser can be connected in parallel, and alternative positions of a three-way valve give two options: 'Hot gas diversion' or 'Condensate control'.

Figure 31.18 shows the second condenser fed from a three-way diverting valve installed in the hot gas pipe to illustrate hot gas diversion.

Option 1: When the controller transmits a demand for heating (e.g. via the supply air temperature in the air conditioning system), the modulating diverting valve opens the flow to the heat recovery condenser and closes the supply to the other condenser. If not all the heat recovered is required, the remainder is dissipated via the second condenser.

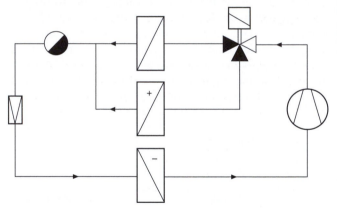

FIGURE 31.18

Figure 31.19 shows the connections for condensate control with the three-way value positioned in the liquid pipe.

Option 2: The entire refrigerant flow is normally directed through the valve (see Fig. 31.19). The three-way liquid valve opens the condensate pipe of the heat recovery

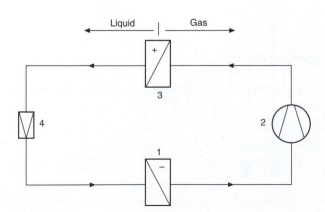

FIGURE 31.17

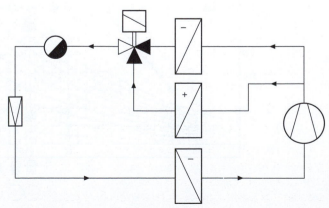

FIGURE 31.19

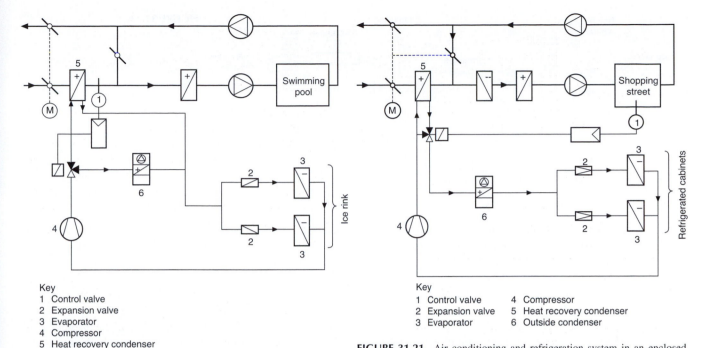

Key
1 Control valve
2 Expansion valve
3 Evaporator
4 Compressor
5 Heat recovery condenser
6 Outside condenser

FIGURE 31.20 Indoor swimming pool/ice rink.

Key
1 Control valve   4 Compressor
2 Expansion valve 5 Heat recovery condenser
3 Evaporator   6 Outside condenser

FIGURE 31.21 Air conditioning and refrigeration system in an enclosed shopping complex.

condenser and closes the pipe of the second condenser when the heat recovery condenser has to emit effective heat. This condenser then releases all condensate thus making its heat exchange surface free for condensation of fresh hot gas. At the same time, the second condenser is filled with liquid thus making its exchange surface inactive. When the demand for effective heat decreases, the process is reversed.

The following examples show how the two options have been applied in a swimming pool/ice rink and a covered shopping complex.

The system shown in Fig. 31.20 relates to an air conditioning system in an indoor swimming pool combined with a low-temperature refrigeration system for an ice rink. It is designed on the principle of the schematic in Fig. 31.18 (hot gas diversion) in order to provide a fast response time of the supply air control.

Indoor swimming pools normally require large quantities of heat and have high-relative humidity. If cold outside air is mixed with the humid return air, condensation results. To avoid this, the outside air is preheated by a heat recovery condenser (5) (Fig. 31.21).

Systems installed in enclosed shopping complexes have the following characteristics:

● large heat demand for air conditioning of shopping street;
● large heat output from refrigerated cabinets.

The two condensers in the system illustrated are connected in series. The heat recovery condenser (5) can be switched off completely in summer. In winter, spring and autumn it is used to pre-heat the outside in accordance with demand (Fig. 31.22).

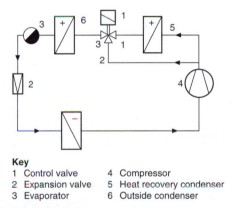

Key
1 Control valve   4 Compressor
2 Expansion valve 5 Heat recovery condenser
3 Evaporator   6 Outside condenser

FIGURE 31.22 Refrigeration cycle of the shopping complex (simplified).

The authors wish to express their thanks for the assistance given and permission to include examples of applications engineered by Staefa Control System Ltd., www.staefacontrol.com.

Pneumatic systems

Pneumatic systems require a supply of clean compressed air to motivate cylinders, tools, valve gear, instruments, delicate air controls and other equipment. Most factory and plant installations operate between 5.5 and 7 bar.

A typical compressor installation is shown in Fig. 31.23.

Compressors are sized according to the amount of free air delivered. Air flow is measured in cubic decimetres per second (dm^3/s) at standard atmospheric conditions of 1013 mbar and 20 °C as specified in ISO 554. The

FIGURE 31.23

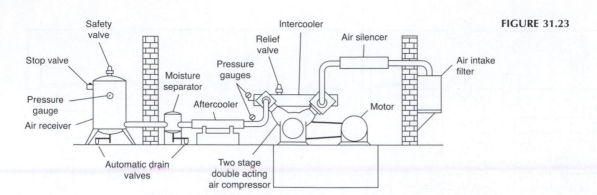

compressed air is stored in an air receiver and, for a system operating at pressures in the region of 7 bar gauge, the size of the receiver in litres should be approximately equal to 30 times the rated free air delivery of the compressor in dm^3/s. Thus, a compressor rated at $50\,dm^3/s$ free air delivery requires a receiver of approximately 1500 l capacity.

Compressed air in any normal supply mains contains contaminants which need to be either completely or partially removed depending on the ultimate use of the compressed air. Naturally, the cleaner the air has to be the greater the expense. The contaminants are:

a. water in liquid and vapour form;
b. oil which can exist in three forms, oil/water emulsions, minute droplets suspended in the air as an aerosol and oil vapours;
c. atmospheric dirt particles and solid particles formed by the heat of compression.

Having considered the types of contaminant present in an air system, one can decide upon the degree of cleanliness needed for any particular process and the means required to obtain this. These include after-coolers, receivers, air line filters, air dryers, coalescing filters, vapour adsorbers, and ultra high-efficiency dirt filters. Each application must be considered on its merits. Figure 31.24 shows a typical air line installation for a factory. Further cooling may occur in the distribution mains themselves; these should be installed with a pitch in the direction of the air flow, so that both gravity and air flow will carry water to drain legs at appropriate intervals. These legs should be fitted with automatic drain valves. Note that take-off points are connected to the top of the distribution mains to prevent water entering the take-off lines.

The quality of air required for plant use will now dictate which accessories are to be fitted at each take off point. These range from a selection of filters and pressure regulators; and if lubrication is required in air actuated components, then lubricant can be metered and atomized in the air line in the form of a fine fog to coat all operating parts with a thin protective film. The equipment is lubricated automatically through its operating cycle.

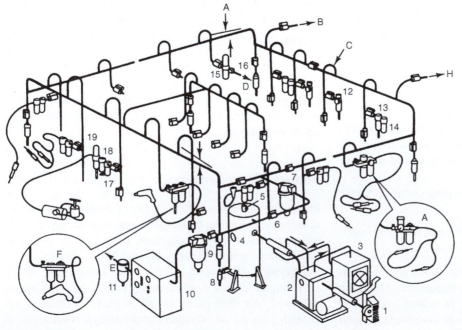

1. Air intake filter	17. Primary filter*
2. Air compressor	18. Precision
3. Air compressor	regulator*
water to air	19. 'Ultraire' filter
heat exchanger	A. Pitch with flow
4. Air receiver	B. To machine
5. Safety valve	shop
6. Isolating valve	C. Wide pattern
7. Main line air	return bends
filter	D. To gauging
8. Automatic air	equipment
receiver drain	E. Dry air to
9. Drip-leg drain	process control
10. Dryer	F. Olympian
11. 'Puraire' after-	'plug-in'
filter	Vitalizer unit
12. Filter-regulator	G. Olympian
13. Automatic	'plug-in' lubro-
drain filter	control unit
14. Lubricator	H. To future
15. 'Ultraire' filter*	extensions
16. Precision	
controller*	*Precision air sets

FIGURE 31.24

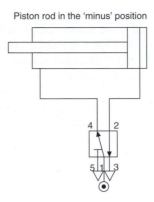

FIGURE 31.26

FIGURE 31.25 Circuit symbols can be drawn, stored in the computer database, and used repeatedly on other diagrams. Neat layout work manually often involves changes and a lot of repositioning. This is time consuming, but the computer handles alterations with speed and accuracy.

Regular maintenance will ensure trouble-free production facilities.

Industrial processes include: air agitation, air bearings, air conveying of foodstuffs and powders, air motors i.e. rotary, reciprocating and linear air cylinders, blow guns, cleaning and cooling nozzles, breathing masks and protective clothing, fluidics, food and drink processing, general machinery, instrumentation, pneumatic circuits and valves, and spray guns.

Circuit symbols can be drawn, stored in the computer database, and used repeatedly on other diagrams. Neat layout work manually often involves changes and a lot of repositioning. This is time consuming, but the computer handles alterations with speed and accuracy (Fig. 31.25).

Pneumatic circuit design

The first requirement in circuit design is a thorough understanding of the symbols to be used. The most important and frequently used symbols are the five-port and three-port valves. They are also the most frequently misunderstood symbols, and therefore we start by showing the build-up of a typical five-port valve. In Fig. 31.26 a double-acting cylinder is shown connected to a five-port valve. The square envelope represents the valve body and there are three ports on the bottom edge and two on the top edge. A compressed air supply is connected to the centre port 1. Air exhausts to atmosphere at ports 3 and 5. Air outlets to power the cylinder at ports 2 and 4. The lines within the envelope show the passages within the valve for the current valve state. The air supply 1 is connected to outlet 4 and outlet 2 is connected to exhaust 3. Exhaust 5 is sealed.

This means that the cylinder has air pressure pushing the piston to the instroked or 'minus' position. The other side of the piston is connected to exhaust (Table 31.2).

To make the cylinder move to the outstroked or 'plus' position the valve has to be operated to change to its new state. This is shown in Fig. 31.27. Note that the envelope and port connections are exactly the same and it is only the connection paths inside the valve that have changed.

The full symbol for a 5/2 valve (five ports, two positions) are these two diagrams drawn alongside each other. Only one half will have the ports connected.

Which half, will depend on whether the cylinder is to be drawn in the instroked or outstroked state. The method by which the valve is operated, i.e. push button, lever, foot pedal, etc. is shown against the diagram of the state that it produces.

Figure 31.28 shows a 5/2 push button operated valve with spring return. It is operating a double acting cylinder. In addition a pair of one-way flow regulators are included to control the speed of piston rod movement. The symbol for this type of flow regulator consists of a restrictor, an 'arrow' which indicates it is adjustable and a non-return valve in parallel, to cause restriction in one direction only. The conventional way to control the speed of a cylinder is to restrict the exhausting air. This allows full power to be developed on the driving side of the piston which can then work against the back pressure and any load presented to the piston rod.

Study Fig. 31.29 and imagine that when the push button is pressed the complete symbol moves sideways to the left, but leaves the pipe connections and port numbers behind so that they line up with the other half of the diagram. In this position the cylinder piston rod will move out to the 'plus' position. Imagine the spring pushing the symbol back again when the button is released. The numbers at the valve ends signify which output will be pressurised when the valve is operated at that end. If the button is pushed at end 12 then port 1 will be connected to port 2.

If the button is released, the spring at end 14 becomes dominant and port 1 will be connected to port 4.

A three port valve symbol works in a similar way. Two diagrams of the valve are drawn side by side. Figure 31.29 shows the full symbol for a 3/2 valve controlling a single acting cylinder. Port 1 is the normal inlet, port 2 the outlet and port 3 the exhaust. The valve end numbers 12 and 10 indicate that port 1 will be connected either to 2 or to 0 (nothing). Since there is only one pipe supplying a single-acting cylinder, speed control of the 'plus' motion has to be obtained by restricting the air into the cylinder. Speed of the 'minus' motion is effected conventionally by restricting the exhausting air. To provide independent adjustment two one-way flow regulators are used and these are connected in the line back-to-back.

Logic functions

Designers of pneumatic circuits are not usually consciously thinking in pure logic terms, but more likely designing intuitively from experience and knowledge of the result that is to be achieved. Any circuit can be analysed however, to show

TABLE 31.2 Selected symbols for fluid power systems (from BS 2917).

Description	Symbol	Description	Symbol	Description	Symbol
General symbols		Air-oil actuator (transforms pneumatic pressure into a substantially equal hydraulic pressure or vice versa)		spring loaded (opens if the inlet pressure is greater than the outlet pressure and the spring pressure)	
Basic symbols					
Restriction:		Directional control valves		pilot controlled (opens if the inlet pressure is higher than the outlet pressure but by pilot control it is possible to prevent:	
affected by viscosity		Flow paths:			
		one flow path			
unaffected by viscosity		two closed ports		closing of the valve	
Functional symbols		two flow paths			
hydraulic flow		two flow paths and one closed port		opening of the valve)	
pneumatic flow or exhaust to atmosphere		two flow paths with cross connection			
Energy conversion		one flow path in a by-pass position, two closed ports		with resistriction (allows free flow in one direction but restricted flow in the other)	
Pumps and compressors					
Fixed capacity hydraulic pump: with one direction of flow					
with two directions of flow		Directional control valve 2/2: with manual control			
Motors		controlled by pressure against a return spring		Shuttle valve (the inlet port connected to the higher pressure is automatically connected to the outlet port while the other inlet port is closed)	
Fixed capacity hydraulic motor: with one direction of flow					
Oscillating motor:		Direction control valve 5/2:		Pressure control valves	
hydraulic		controlled by pressure in both directions		Pressure control valve:	
Cylinders				one throttling orifice normally closed	or
Single acting:	Detailed	Simplified	NOTE. In the above designations the first figure indicates the number of ports (excluding pilot ports) and the second figure the number of distinct positions.	one throttling orifice normally open	or
returned by an unspecified force			Non-return valves, shuttle valve, rapid exhaust valve	two throttling orifices, normally closed	
Double acting:			Non-return valve:		
with single piston rod			free (opens if the inlet pressure is higher than the outlet pressure)	Sequence valve (when the inlet pressure overcomes the spring, the valve opens, permitting flow from the outlet port)	or
Cylinder with cushion:					
Single fixed					

TABLE 31.2 (*Continued*).

Description	Symbol	Description	Symbol	Description	Symbol
Flow control valves		Rotary connection		Over-centre device (prevents stopping in a dead centre position)	
Throttle valve: simplified symbol		one way		Pivoting devices: simple	
Example: braking valve		three way		with traversing lever	
Flow control valve (variations in inlet pressure do not affect the rate of flow): with fixed output	*Simplified	Reservoirs		with fixed fulcrum	
		Reservoir open to atmosphere:		Control methods	
Flow dividing valve (divided into a fixed ratio substantially independent of pressure variations)		with inlet pipe above fluid level		*Muscular control:* general symbol	
Energy transmission and conditioning		with inlet pipe below fluid level		by push-button	
Sources of energy		with a header line		by lever	
Pressure source		*Pressurized reservoir*		by pedal	
Electric motor		Accumulators		*Mechanical control:* by plunger or tracer	
		The fluid is maintained under pressure by a spring, weight or compressed gas		by spring	
Heat engine				by roller	
		Filters, water traps, lubricators and miscellaneous apparatus		by roller, operating in one direction only	
Flow line and connections		Filter or strainer		*Electrical control:* by solenoid (one winding)	
Flow line: working line, return line and feed line		Heat exchangers			
pilot control line		*Temperature controller (arrows indicate that heat may be either introduced or dissipated)*		by electric motor	
drain or bleed line				*Control by application or release of pressure*	
flexible pipe		*Cooler (arrows indicate the extraction of heat)*		*Direct acting control:*	
Pipeline junction		with representation of the flow lines of the coolant		by application of pressure	
Crossed pipelines (not connected)		*Heater (arrows indicate the introduction of heat)*		by release of pressure	
Air bleed		Control mechanisms		*Combined control:* by solenoid and pilot directional valve (pilot directional valve is actuated by the solenoid)	
Power take-off		Mechanical components			
plugged		*Rotating shaft:*		Measuring instruments	
with take-off line		in one direction		*Pressure measurement:* pressure gauge	
connected, with mechanically opened non-return valves		in either direction		Other apparatus	
uncoupled, with open end		*Detent (device for maintaining a given position)*		*Pressure electric switch*	
uncoupled, closed by free non-return valve		*Locking device (*symbol for unlocking control)*			

Note 1: The symbols for hydraulic and pneumatic equipment and accessories are functional and consist of one or more basic symbols and in general of one or more functional symbols. The symbols are neither to scale nor in general orientated in any particular direction.

Note 2: In circuit diagrams, hydraulic and pneumatic units are normally shown in the unoperated position.

Note 3: The symbols show connections, flow paths and the functions of the components, but do not include constructional details. The physical location of control elements on actual components is not illustrated.

Piston rod in the 'plus' position

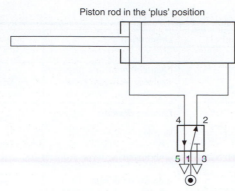

FIGURE 31.27

that it is made up of a combination of logic functions. The four most commonly used are illustrated in Figs. 31.30–31.33.

The AND function: The solenoid valve A (AND) the plunger operated valve B must both be operated before an output is given at port 2 of valve B (Fig. 31.30).

The OR function: For this a shuttle valve is required so that either of two push-button valves A (OR) B can provide a

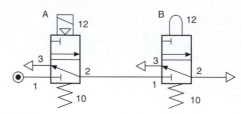

FIGURE 31.30

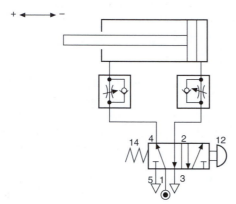

FIGURE 31.28

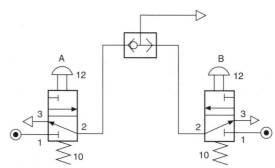

FIGURE 31.31

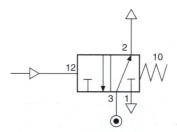

FIGURE 31.32

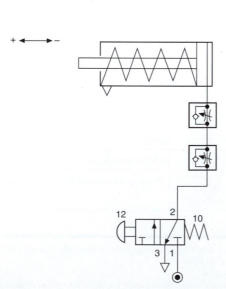

FIGURE 31.29

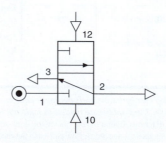

FIGURE 31.33

signal that is directed to the same destination. The shuttle valve contains a sealing element that is blown by the incoming signal to block off the path back through the other valve's exhaust port (Fig. 31.31).

The NOT function: This is simply a normally open valve. When it is operated by a pilot signal on port 12 it will NOT give an output. The outlet will be given when the valve resets to its normal state by removing the signal (Fig. 31.32).

The MEMORY function: When a double pressure operated three port valve is given a signal at port 12, an output is obtained at port 2. If the signal is now removed the output will remain, it has REMEMBERED its ON state even when the signal that caused it has gone.

If a signal is given to port 10 the valve will re-set and the output exhausted. If the signal is removed the new OFF state is REMEMBERED (Fig. 31.33).

The TIME DELAY: By using a flow regulator and a 3/2 pilot operated pressure switch, a signal can be slowed down to provide a time delay. Figure 31.34 shows that when a signal is fed through the flow regulator, it will slowly build up pressure in an air reservoir (R) and on the signal port 12 of the pressure switch. This will continue until the pressure is high enough to operate the pressure switch. Then, a strong unrestricted signal will be sent to operate a control valve or other device. The delay can be adjusted by changing the setting on the flow regulator. A reservoir, of approximately 100 cc in volume, would allow a delay range of between 2 and 30 s. Without the reservoir, the range will be reduced to approximately 3 s maximum. Note that the pressure switch is like a pilot operated 3/2 valve, but uses air pressure as a return spring. The pilot signal on port 12 overcomes this, as it is working on a larger area piston.

A semi-automatic circuit is shown in Fig. 31.35. When the push button is operated and released, the 3/2 valve will send a signal to operate the 5/2 double pilot valve. This will cause the cylinder to move to the 'plus' position. A cam on the piston rod will operate the roller plunger valve and this will give a signal to re-set the 5/2 valve. The piston rod will then automatically move to the 'minus' position and wait until a further operation of the push button is given.

Sequential circuits

In an automatic system where two or more movements are to occur in a specific order, a sequence is formed. A typical example is a special purpose automatic machine. This may be carrying out a manufacturing, or packaging operation where air cylinders are used to power the movements in a continuously repeating sequence.

Each movement in a sequence can be produced by a pneumatic cylinder. This will either be single acting, or double acting and the choice depends on whether there is

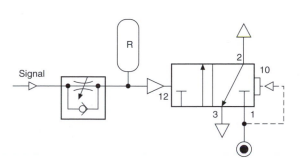

FIGURE 31.34

FIGURE 31.35

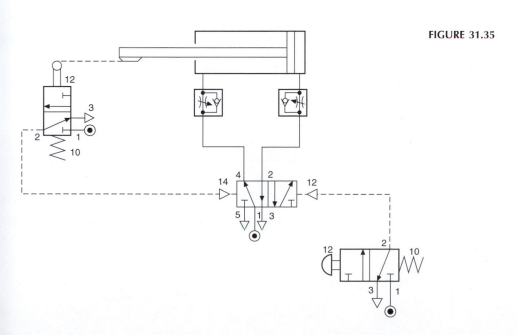

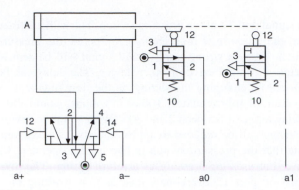

FIGURE 31.36

any return resistance or load requiring a powered return. Single acting cylinders are controlled by a 3/2 double pilot operated valve and double acting cylinders are controlled by a 5/2 double pilot operated valve.

For each cylinder used, a circuitry building block can be established (see Fig. 31.36). This illustrates a double acting cylinder building block for the cylinder labelled 'A'. Two command signals are required, one to move it 'plus' (a^+), the other to move it 'minus' (a^-). To prove that the movements have been completed, two feed-back signals are required. These are provided by the two roller operated 3/2 valves. One proving the 'plus' movement (a1), the other proving the 'minus' movement (a0).

Consider a two cylinder system where the cylinders are labelled A and B. The sequence required after selecting the RUN control is A + B + A − B −, it will then repeat continuously until the operator selects the END control. The circuit is constructed from two building blocks (see Fig. 31.37).

Note that flow regulators are included in the power lines to each end of the cylinders.

These provide adjustable speed control for each movement. To RUN and END the repeating cycle a 3/2 manually operated valve is included.

The two building blocks form a complete circuit by having their command and feedback lines connected together. The method of interconnection is achieved by application of this simple rule:

'The proof of position signal resulting from the completion of each movement is connected to initiate the next movement.'

The circuit can be traced as follows:

Start with the output given from the RUN/END valve when it is switched to RUN.
The a^+ command is given.
Cylinder A moves+.
The a1 proof of position signal results.
This becomes the b^+ command.
Cylinder B moves+.
The b1 proof of position signal results.
This becomes the a^- command.
Cylinder A moves−.
The a0 proof of position signal results.
This becomes the b^- command.
Cylinder B moves−.
The b0 proof of position signal results.
This becomes the supply to the RUN/END valve.

If the RUN/END valve is still switched to RUN a repeat cycle will be started.

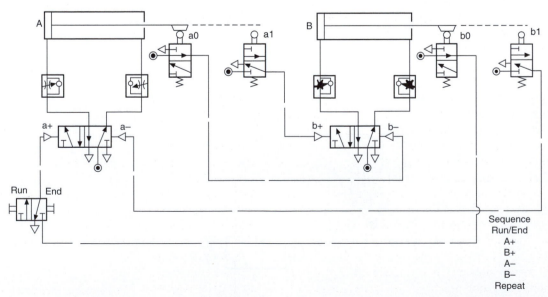

FIGURE 31.37

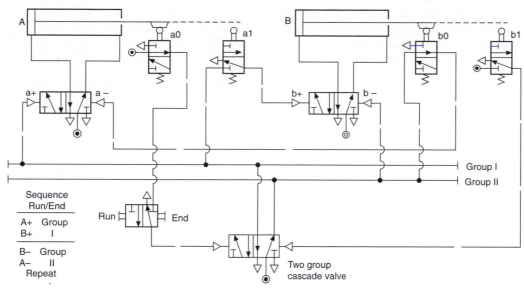

FIGURE 31.38

This simple daisy chain method of interconnection will work for any number of cylinders, provided the sequence allows their return movements to occur in the same order as their first movements. For this to be true, the first movement of a cylinder need not be plus nor is it necessary for the first half of the sequence to be in alphabetical order, e.g. the sequence B + A − D + C − B − A + D − C + conforms to these rules and can be solved with this simple daisy chain method.

If the cylinders do not return in exactly the same order as their first movements complications will arise. Take for example, the sequence A + B + B − A − and repeat. If we try to interconnect the equipment for this sequence in the same way as before, there will be two states where the 5/2 valves will have both a 'plus' and 'minus' command existing at the same time, therefore preventing operation. This condition is commonly known as opposed signals and can be cured in a variety of ways. For the most reliable and economical method we suggest the use of the Cascade system (see Fig. 31.38).

The cascade technique is to switch on and off the supply air to the critical trip valves in groups. The need for this will occur when a trip valve's mechanism is still held down, but the output signal has been used and requires removing. By switching off the group air that is supplying the valve, the output is also removed and achieves the desired result. After the valve's mechanism is naturally released in the sequence, the group supply is switched on again in time for its next operation. To determine the number of cascade groups for any sequence, the sequence must be split into groups starting at the beginning, so that no letter is contained more than once in any group. The group numbers are given roman numerals to avoid confusion with other numbering systems

that may exist on larger systems. The placing of the RUN/END valve should be in the line that selects group I. This determines that the first task of group I is to signal the first movement of the sequence. In addition, when the circuit is at rest, inadvertent operation of an uncovered trip valve will not risk an unwanted operation of a cylinder.

By studying Fig. 31.38, it can be seen that the sequence splits into two groups. These groups are supplied from a single, double pressure operated 5/2 valve, so that only one group can exist at any time. This is known as the cascade valve.

It can also be seen that neither of the 5/2 valves controlling the cylinders can have the + and − command lines as opposed signals, since their source is from different groups.

The circuit can be traced as follows:

To start, set RUN/END valve to RUN. This generates a command to select group I.

Group I gives a command a^+.

Cylinder A moves+.

Valve a1 is operated and generates a command b^+.

Cylinder B moves+.

Valve b1 is operated and generates a command to select group II.

Group II gives a command b^- (because group I has been switched off there is no opposing signal from a1).

Cylinder B moves −.

Valve b0 is operated and generates a command a^- (no opposed signal).

Cylinder A moves −.

Valve a0 is operated and generates a command to start the sequence again.

If at any time the RUN/END valve is switched to END, the current cycle will be completed, but the final signal will be blocked and no further operation will occur.

The rules for interconnection are as follows:

1. The first function in each group is signalled directly by that group supply.
2. The last trip valve to become operated in each group will be supplied with main air and cause the next group to be selected.
3. The remaining trip valves that become operated in each group are supplied with air from their respective groups and will initiate the next function.

Pneumatics and electronics

Systems of low complexity and those in use in hazardous areas, not compatible with electronics, will probably be designed as pure pneumatic systems.

A purely pneumatic system can be viewed as three main sections:

1. Generation and preparation of the compressed air source.
2. Power actuation of pneumatic cylinders through directional control valves.
3. Pneumatic signal processing or logic control.

Electronics can influence all of these sections, for example:

a. By electronic management control of compressors and controlled pressure regulation.
b. In Section 2 there are solenoid valves that provide proportional flow and pressure, together with air cylinders having electronic proportional feedback.
c. In Section 3, for many systems pneumatic logic has been replaced completely by electronic sequence or logic control.

Programmable sequence controllers (sequencers) and programmable logic controllers (PLCs) are commonly used devices and offer a wide range of features such as timing, counting, looping and logic functions. If a proposed scheme involves a sequence of events more complicated than that shown in Fig. 31.38, then electronic possibilities should be explored. In addition to sequence operations there may be the additional complications from long-counting operations, or a number of time delays, requiring a high degree of repeatable accuracy. Here the electronic controller will usually be the better choice. Inputs to the controller indicate the completion of the cylinder movement.

These are most conveniently achieved by using a magnetic cylinder fitted with reed switches. The reed switch consists of two spring like metal reeds within a sealed enclosure. When a magnet around the piston is within range,

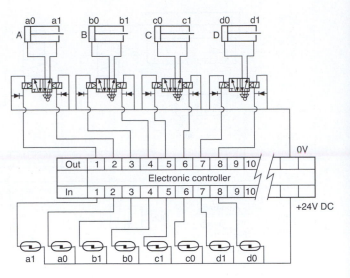

FIGURE 31.39

the reeds are magnetized, each having a N and S pole. As the free ends will be of the opposite polarity they snap together. For environments where strong magnetic fields exist mechanical limit switches may be used.

The scope of such a system will be appreciated from Fig. 31.39. Programming methods vary with the type of controller and for someone with no experience it is generally easier than they think. Sequencers are designed to be easy to program and are a good choice for machines where the actions are performed in a one-after-the-other interlock. Sequencers are able to jump from one part of the sequence to another, run sections of a sequence in a repeating loop, time, count and perform logic functions such as AND, OR, NOT, etc. It may also be possible to hold several sequences in a memory and select the desired one for a particular task. Sequencers will have a built in range of control buttons to provide facilities such as, run/end cycle, emergency stop, single cycle, auto cycle and manual over-ride.

It takes a little longer to program a PLC. This is produced by keying in a list of logic statements first determined by drawing a ladder diagram. A ladder diagram for a PLC is a logic circuit of the program as it relates to a machines function and sequence. The ladder diagram illustrated in Fig. 31.40 is derived from, and similar to the ladder electrical circuits used to design electro mechanical relay systems.

Pneumatic and electronic systems play an important part in production engineering and typical applications are the control of the main axes of variable pick and place arms and robotics.

The authors wish to express their thanks for the assistance given and permission to include examples of applications of pneumatic controls manufactured by Norgren Martonair Limited, www.norgren.com.

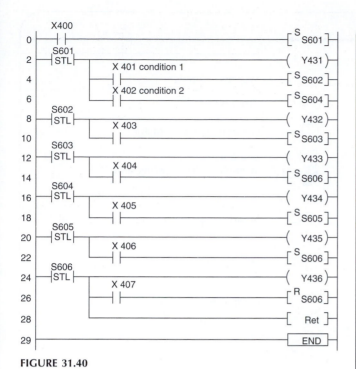

FIGURE 31.40

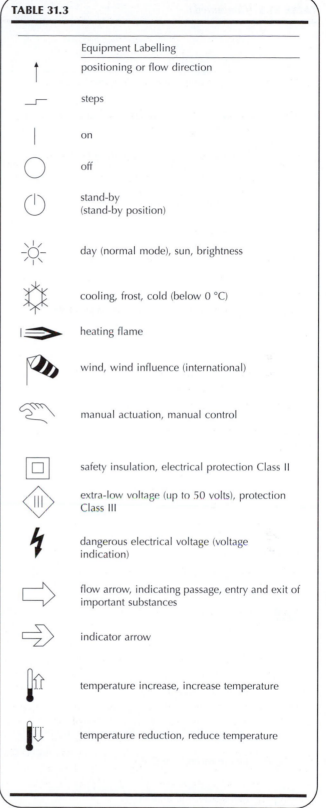

The BS 1533 series, parts 1, 2 and 3, specifies graphical symbols for use in general engineering. Within the European Community, many additional symbols are in common use and a selection of these are included here for reference purposes (Table 31.3).

TABLE 31.3

	Equipment Labelling
↑	positioning or flow direction
⌐_	steps
		on
○	off
⏻	stand-by (stand-by position)
☼	day (normal mode), sun, brightness
❄	cooling, frost, cold (below 0 °C)
⊢▷	heating flame
🚩	wind, wind influence (international)
✋	manual actuation, manual control
▣	safety insulation, electrical protection Class II
◇III	extra-low voltage (up to 50 volts), protection Class III
⚡	dangerous electrical voltage (voltage indication)
⇨	flow arrow, indicating passage, entry and exit of important substances
⇨	indicator arrow
🌡↑	temperature increase, increase temperature
🌡↓	temperature reduction, reduce temperature

TABLE 31.3 (*Continued*).

Systems symbols

heating energy, energy demand

heat exchanger, general with substance flows crossing

heat exchanger, general without substance flows crossing

tank, general (pressureless)

tank with convex bottom, general (for high pressure)

isolating valve (general) two way valve

three way valve

four way valve

1 — 3 flow symbols:
2

– variable

– constant

1 — 3 mixing
2

3 — 1 diverting
2

shower, nozzle

steam trap

filter

manual actuator

self operated actuator (or actuator in general)

electromotoric actuator

electrothermic actuator

hydraulic or pneumatic actuator

diaphragm actuator

Systems symbols

cam control

electromagnetic actuator

example:
1 — 3 magnetic valve showing flow
2

liquid pump, circulating pump, general

fan (general)

compressor (general)

compressor, 4 step

air damper

air filter (general)

heating coil

cooling coil

device or function unit, general

modulating controller (general)

keys, keyboard

sensor with on-off function (e.g. thermostat, hygrostat pressure switch etc.)

sensor with on-off function (e.g. thermostat, hygrostat pressure switch etc.) with immersion, duct or capillary pocket

immersion thermostat for temperature

Other references examples:
x   absolute humidity   h   enthalpy
p   pressure   aq   air quality (SCS)
Δp differential pressure   occupancy
V   flow, volume flow rate   etc.
v   velocity

TABLE 31.3 (*Continued*).

Symbols for electrical schematics	Symbols for electrical schematics
DC-current, also DC-voltage (general)	combined wires, general, any sequence on each side (wires should be coded)
alternative (use this symbol only where there is a risk of confusion on diagrams)	combined wires, general, as above but single line representation
AC-current, also frequency in general AC-voltage (frequency indicated where necessary – on the right of the symbol, e.g. ~50 Hz)	general symbol denoting a cable
suitability for use on either DC or AC supply	example: 2 core cable
	example: 2 core cable 'screened' (general)
positive polarity	coaxial line, screened
negative polarity	crossing of conductor symbols no electrical connection
definitions of electrical conductors	junction of conductors
– L Phase (formerly PH) – N Neutral (formerly N) – L$_1$ Phase 1 (formerly R) – L$_2$ Phase 2 (formerly S) – L$_3$ Phase 3 (formerly T) – PE Earth	
	general contact, in particular one that is not readily separable e.g. soldered joint
AC-current with m phases, frequency f and voltage U	readily separable contact e.g. terminal on controller base
Example: three-phase AC-current with neutral wire, 50 Hz, 380 V (220 V between phase wire and neutral wire)	terminals:
one wire or a group of wires	device terminals
flexible wires	control panel terminals:
line showing the number of wires e.g. 3 wires	
numbers of wires = n	– on connection diagram
example: 8 wires	– on circuit diagram
line showing the number of circuits e.g. 2 circuits	plug or plug pin
Combining wires for the sake of simplicity in wiring diagrams	socket outlet
	fuse general

(labels within diagrams: L1, L2, L3, N, PE; m ~ fU; 3N ~ 50 Hz 380 V; n; 8; U > 50 V; U < 50 V; 1 2 3; 16 17)

TABLE 31.3 (*Continued*).

Symbols for electrical schematics		Symbols for electrical schematics	
	fuse showing supply side		– anti-clockwise
	voltage fuse general over voltage discharge device surge arrestor		– both directions
	isolating point with plug-in connection		thermostat, hygrostat etc. e.g. p → pressure switch
	earth, general		Manually operated control, general
	safety conductor, safety earth		– this symbol is used when space is limited
	chassis, general		manual operation by pushing
	GND (ground, common chassis)		manual operation by pulling
	resistor, general		manual operation by turning
	inductor, inductive reactance		manual operation by toggle or lever
	capacitor, capacitive reactance		actuator general, e.g. for relay, contactor
	polarized (electrolytic) capacitor		electromechanical actuator, e.g. showing active winding
	motor, general		electromechanical actuator with two windings active in the same direction
	transformer with two separate windings		signal lamp general 'operation'
	as above (alternative representation)		signal lamp, flashing for fault
	battery of cells or accumulators (the long line represents the positive pole)		signal lamp 'fault', emergency lamp
	mechanical coupling:		
	– general symbol		buzzer
	– symbol used when space is limited		bell
	linear motion:		horn
	– to the right		siren
	– to the left		transducer, signal transducer, transmitter, general symbol
	– both directions		
	rotational motion:		
	– clockwise		

TABLE 31.3 (*Continued*).

Symbols for electrical schematics		Symbols for electrical schematics	
	rectifier, rectifying device general		Variability inherent linear variability under influence of a physical variable
	amplifier general symbol		inherent non-linear variability under influence of a physical variable
	oscillograph, general symbol		continuous variability by mechanical adjustment, general
	recording measuring device, recorder		adjustable in steps
	recording measuring device, printer		non-inherent non-linear variability
	remote operation, general		continuous variability by mechanical adjustment, linear
	adjuster		continuous variability by mechanical adjustment, non-linear
	communication (electronic)		pre-set mechanical adjustment, general symbol
	clock, general		
	synchronous clock		Example: temperature dependent resistor with negative temperature coefficient (thermistor)
	time clock		Wire colour abbreviations:
	dividing line (e.g. between two zones or to separate a space)		
	example: control panel		

Wire colour abbreviations:

bl	blue	ws	white
dbl	dark blue	sw	black
hbl	light blue	og	orange
rt	red	vl	violet
gb	yellow	gb/gn	yellow/green
gn	green	bn	brown
gr	grey		

(IEC)

semi-conductor rectifier diode

zener diode

E — C   PNP-transistor   E – emitter
                B   C – collector
                           B – base

E — C   NPN-transistor
                B   the collector is connected to the housing

optocoupler (SCS) combined symbol

International colour code

Colour reference for resistance value and its tolerance

black	0	0	–
brown	1	1	0
red	2	2	00
orange	3	3	000
yellow	4	4	0 000
green	5	5	00 000
blue	6	6	000 000
violet	7	7	0 000 000
grey	8	8	00 000 000
white	9	9	–

Tolerance class:
- without colour reference   ± 20%
* – silver   ± 10%
* – gold   ± 5%
  – red   ± 2%
  – brown   ± 1%
* As alternative colours the following are valid on the 4th ring:
  green instead of gold for   ± 5%
  white instead of silver for   ± 10%

e.g brown – green – red – gold
         1   5   00   = 1500 Ω ± 5%

Chapter 32

Bearings and applied technology

When surfaces rotate or slide, the rotational or sliding motion results in friction and heat. Energy is used, the surfaces wear, and this reduces component life and product efficiency. Friction may be reduced by lubrication which keeps the surfaces apart. At the same time, lubricants dissipate heat and maintain clean contact surfaces. Materials are carefully selected with appropriate mechanical and physical properties for bearings and their housings, to minimize the effects of friction, and particular care is taken with the accuracy of machining, surface finish and maintenance of all component parts associated with bearings.

In a plain bearing, the relative motion is by sliding in contrast with the rolling motion of ball and roller bearings.

Plain bearings

Plain bearings may be classified as follows:

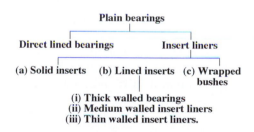

The bearing metal should have a low coefficient of sliding friction, be able to conduct heat generated away from the bearing surfaces, resist wear in use and be tough enough to withstand shock loading in service. In the event of breakdown due to lack of lubrication, it may be desirable when overheating occurs for the bearing material to run, preventing seizure and possible severe damage to associated mechanical parts.

Direct-lined housings

These housings are lined directly with bearing materials and the choice of material is limited by the practicality of keying or bonding the bearing material to the housing surface.

The dimensions of the housings, casting temperatures and bonding characteristics of the bearing materials will determine whether metallurgical bonding is possible without unacceptable distortion of the housing. Generally this technique is limited to ferrous housings with low-melting point whitemetal bearing surfaces. Light alloy and zinc base housings are difficult to line directly with whitemetal.

Insert liners

These are bearing elements which consist of a liner inserted into a previously machined housing and they can be divided into separate classes:

(a) Solid-insert liners.
(b) Lined inserts.
(c) Wrapped bushes.

Solid-insert liners: Manufactured wholly from suitable bearing materials such as aluminium alloy, copper alloy or whitemetal, these liners consist of machined bushes, half bearings and thrust washers.

The housings are machined to relatively close tolerances. An insert may be finished machined after assembly or a prefinished standard precision liner added as a final operation and this has the added advantage of spares replacement.

Typical applications of insert liners are to be found in diesel engine small bores, crank shaft main bearings, bushes for gearboxes, steering gear and vehicle suspensions.

Lined inserts: These consist of a backing material such as cast iron, steel or a coper alloy which has been lined with a suitable bearing surface of aluminium or copper alloy, or of whitemetal. This type can also be supplied as a solid insert, a split bush, half bearing or thrust washer.

Insert bearing half liners are manufactured as:

(a) Rigid or thick-walled bearings
(b) Medium-walled bearings
(c) Thin-walled bearings.

Thick-walled bearings: These are backing shells of cast iron, steel pressings and copper base alloys generally lined with whitemetal and copper alloys are used to produce bearings which are manufactured as pairs and used in turbines, large diesel engines and heavy plant machinery. Usually more economic than direct lined housings, these bearings may be provided with a finishing allowance for the bore and length which is adjusted during assembly.

Medium-walled insert liners: Normally a steel backing is used with a wide range of lining materials. Bearings are prefinished in bore and length and manufactured as interchangeable halves.

Thin-walled insert liners: These are high precision components with steel backing and whitemetal or copper and aluminium base alloy surfaces, and are suitable for universal application in large production products such as high-speed diesel engines and compressors.

Wrapped bushes: These are pressed from flat strip of rolled bronze, or steel lined with whitemetal, lead–bronze, copper–lead, or aluminium alloys. They are supplied as a standard range of prefinished bushes or with a bore allowance for finishing in situ by fine boring, reaming, broaching or burnishing. These are suitable for all bushing applications in which the tolerable wear will not exceed the thickness of the lining material.

Plain bearing lubrication

The requirements of a lubricant can be summarized as follows:

(1) To support the bearing when static and under all speed and load conditions.
(2) To have a low coefficient of friction.
(3) To be non-corrosive to the materials used in the bearings.
(4) To maintain viscosity over the operating range of temperature.
(5) Able to provide an effective bearing seal.
(6) Have the ability to adhere as a film to the bearing.
(7) Be able to conduct heat rapidly.

No single lubricant can satisfy all of these properties and the design of the equipment will determine which aspect needs priority before a choice from available types can be made.

Plain bearing materials

The application of the bearing, the bearing material and the lubricant used are all interdependent, but four basic requirements are necessary for the material:

(1) Strong enough to resist failure by fatigue or overheating.
(2) Good wear resistance and running properties.
(3) Good lubricant retention.
(4) High corrosion resistance which may arise due to temperature, the environment and lubricants used.

A wide range of materials consists of metallic, metallic backings with various bearing surfaces, reinforced synthetic resin, graphitic and sintered metallic. Various surface treatments are also available to improve wear resistance and reduce friction.

Whitemetals

These are a large range of either lead base or tin base alloys and are covered by British Standards. Antimony is used as a hardening agent since tin and lead are soft. Whitemetal is a low-melting point alloy which is compatible with virtually any type of mating surface. Bearing materials should not be subject to corrosion due to water or the products of oil oxidation and the resistance of tin base whitemetals is high but lead base alloys are susceptible to acidic corrosion from oil oxidation products. Whitemetals are nearly always lubricated under pressure. Loss of lubricant for a short period may cause the bearing to soften and 'wipe'. It loses its compressive strength at elevated temperatures.

Other bearing materials

Other materials for plain bearings include copper lead alloys, lead bronzes, tin bronzes (phosphor bronze), gunmetals and aluminium base alloys.

Before concluding this section it should be stated that metallic porous metal bearings are widely used which are manufactured by powder metallurgy where very fine metal powders are mixed and compressed in moulds to the correct form and sintered at high temperature in a reducing atmosphere. The product is in effect a metal sponge which can be impregnated with lubricating oil. The porosity depends on the initial compression and these products are designed for suitable applications where high volume is required. Self-lubricating materials are also available in tube and bar form for individual manufacture.

Figures 32.1–32.9 show a selection of different types of bearing from the range manufactured by The Glacier Metal Company Limited. It is generally the case that for

FIGURE 32.1 Selection of standard stock bushes manufactured in lead bronze.

FIGURE OF EIGHT
Mainly for grease.
Recommended where
lubrication is infrequent.

AXIAL
Mainly for
gravity oil feed.
Groove must be remote
from loaded area.

FIGURE 32.2 Types of groove which can be added for lubrication.

ELLIPTICAL
Mainly for grease–
facilitates initial
distribution of lubricant.

ANNULAR
For pressure oil feed.
Distributes oil
circumferentially
and axially.

**HELICAL—
LEFT OR RIGHT
HAND**
For grease or oil.
Used to distribute and
convey lubricant axially.

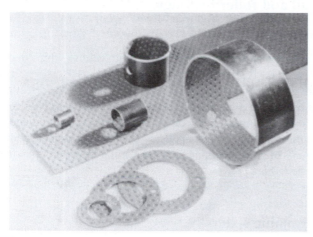

FIGURE 32.3 Prelubricated bearings have an acetal co-polymer lining. The indentations in the linings provide a series of grease reservoirs.

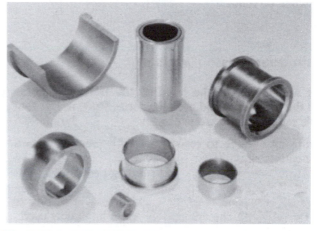

FIGURE 32.4 Fully machined components from self-lubricating materials produced by powder metallurgy.

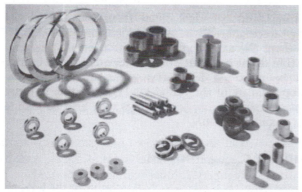

FIGURE 32.5 Components pressed to finished size by powder metallurgy techniques.

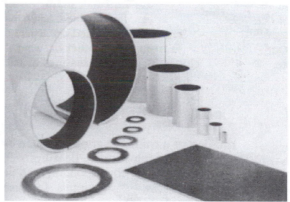

FIGURE 32.6 Dry bearings requiring no lubrication with a PTFE/lead bearing surface.

FIGURE 32.7 Standard wrapped bushes, steel backed lined with lead bronze.

FIGURE 32.8 Thick walled bearings are produced as bushes, half bearings and thrust washers in copper and aluminium base alloys, also tin and lead base whitemetals.

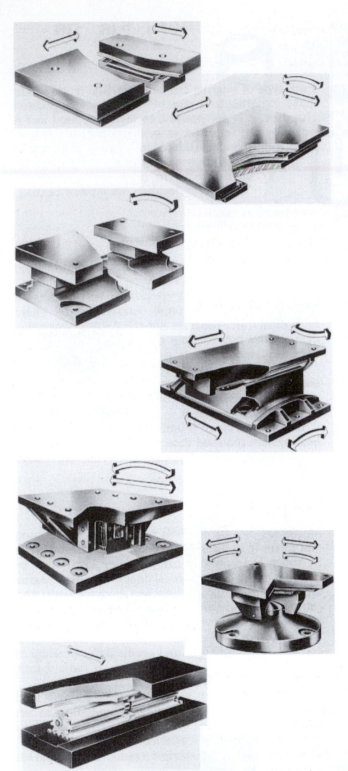

FIGURE 32.9 Structural plain bearings.

small quantities a design, for economic reasons, should incorporate a standard bearing as first choice if possible. Bearing manufacturers employ applications engineers to advise and ensure the correct application and use of their products.

Structural bearings

Plain bearings are also used in structural work and bridge construction to permit expansion and movement and a selection are shown below with an indication of possible motions. This type of bearing utilizes the low-friction properties of polytetrafluoroethylene (PTFE) and in the applications shown can withstand a maximum live loading up to 45 N/mm^2. Illustrations of bearings in Figs. 32.1–32.9 are reproduced by kind permission of the manufacturers, GGB Bearing Technologies – www.ggbearings.com.

Ball and roller bearings

Bearing selection

Each type of bearing has characteristic features which make it particularly suitable for certain applications. However, it is not possible to lay down hard and fast rules for the selection of bearing types since several factors must be considered and assessed relative to each other. The following recommendations will, for a given application, serve to indicate those details of greatest importance in deciding the type of bearing to be used.

Available space

In many instances at least one of the main dimensions of the bearing, usually the bore, is predetermined by the machine design. Deep-groove ball bearings are normally selected for small diameter shafts whereas cylindrical-roller bearings, spherical-roller bearings and deep-groove ball bearings can be considered for shafts of large diameter.

If radial space is limited then bearings with small sectional height must be selected, e.g. needle roller assemblies, certain series of deep groove-bearings and spherical-roller bearings.

Where axial space is limited and particularly narrow bearings are required then some series of deep-groove ball bearings and cylindrical-roller bearings can be used.

Bearing loads

Magnitude of load – This is normally the most important factor in determining the size of bearing. Generally, roller bearings can carry greater loads than ball bearings of the same external dimensions. Ball bearings are mostly used to carry light and medium loads, whilst roller bearings are often the only choice for heavy loads and large diameter shafts.

Direction of load – Cylindrical-roller bearings having one ring without flanges and needle-roller bearings can only carry radial loads. Other types of radial bearing can carry both radial and axial loads.

Thrust-ball bearings are only suitable for axial loads. Spherical-roller thrust bearings, in addition to very heavy axial loads, can also carry a certain amount of simultaneously acting radial load.

A combined load comprises a radial and an axial load acting simultaneously.

The most important feature affecting the ability of a bearing to carry an axial load is its angle of contact. The greater this angle the more suitable is the bearing for axial loading. Refer to maker's catalogue for individual values. Double and single row angular contact ball bearings are mainly used for combined loads.

Self-aligning ball bearings and cylindrical roller bearings can also be used to a limited extent. Duplex bearings and spherical roller thrust bearings should only be considered where axial loads predominate.

Where the axial component constitutes a large proportion of the combined load, a separate thrust bearing can be provided for carrying the axial component independently of the radial load. In addition to thrust bearings, suitable radial bearings may also be used to carry axial loads only.

Angular misalignment

Where a shaft can be misaligned relative to the housing, bearings capable of accommodating such misalignment are required, namely self-aligning ball bearings, spherical roller bearings, spherical roller thrust bearings or spherical plain bearings. Misalignments can, for example, be caused by shaft deflection under load, when the bearings are fitted in housings positioned on separate bases and large distances from one another or, when it is impossible to machine the housing seatings at one setting.

Limiting speeds

The speed of rotation of a rolling bearing is limited by the permissible operating temperature. Bearings with low frictional resistance and correspondingly little internal heat generation are most suitable for high-rotational speeds. For radial loads, the highest bearing speeds are obtainable with deep groove ball bearings or cylindrical roller bearings and for combined loads the highest bearing speeds are obtainable with angular contact ball bearings.

Precision

Rolling bearings with a high degree of precision are required for shafts where stringent demands are made on running accuracy, e.g. machine tool spindles and usually for shafts rotating at very high speeds.

Deep groove ball bearings, single row angular contact ball bearings, double row cylindrical roller bearings and angular contact thrust ball bearings are manufactured to high degrees of precision both as regards running accuracy and dimensions. When using high precision rolling bearings, shaft and housings must be machined with corresponding accuracy and be of rigid construction.

Rigidity

Elastic deformation in a loaded rolling bearing is very small and in most instances can be ignored. However the bearing rigidity is of importance in some cases, e.g. for machine tool spindles.

Due to the greater area of contact between the rolling elements and raceways, roller bearings, e.g. cylindrical roller bearings or taper roller bearings, deflect less under load than ball bearings. The rigidity of the bearings can be increased by suitable preloading.

Axial displacement

The normal bearing arrangement consists of a locating (fixed) bearing and a non-locating (free) bearing. The non-locating bearing can be displaced axially thus preventing cross location, e.g. by shaft expansion or contraction. Cylindrical roller bearings having one ring without flanges or needle roller bearings are particularly suitable for use as free bearings. Their internal design permits axial displacement of the inner and outer rings in both directions. The inner and outer rings can therefore be mounted with interference fits.

Mounting and dismounting

The rings of separable bearings (cylindrical roller bearings, needle roller bearings, taper roller bearings) are fitted separately. Thus, when an interference fit is required for both inner and outer rings or where there is a requirement for frequent mounting and dismounting, they are easier to install than non-separable bearings (deep groove ball bearings, angular contact ball bearings, self aligning ball bearings and spherical roller bearings).

It is easy to mount or dismount bearings with taper bores on tapered seatings or when using adapter withdrawal sleeves on cylindrical shaft seatings. Figure 32.10 gives a simplified guide showing the suitability of the more popular types of bearing for particular applications. The type of bearing indicated for each of the features should be considered as a first choice, but not necessarily the only choice. The bearings listed are described later.

A – Deep groove ball bearing.
B – Self aligning ball bearing.
C – Angular contact ball bearing.
D – Cylindrical roller bearing.
E – Needle roller bearing.
F – Spherical roller bearing.
G – Taper roller bearing.
H – Thrust ball bearing.

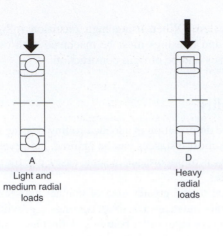

A
Light and
medium radial
loads

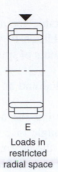

D
Heavy
radial
loads

E
Loads in
restricted
radial space

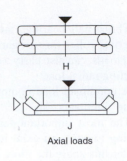

H

J

Axial loads

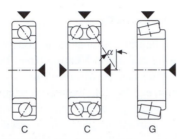

C   C   G

Radial and axial combined loads, α is the angle of contact

A   F   B

Radial and lighter axial combined loads

D   D   E

Axial displacement

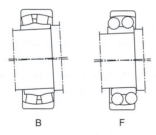

B   F

Angular misalignment

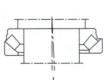

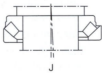

J

A   C   D   D

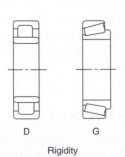

D   G

Rigidity

A   D   C

Limiting speeds

L

Precision

FIGURE 32.10

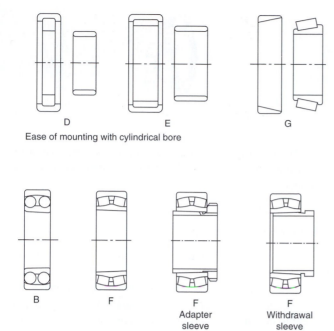

Ease of mounting with cylindrical bore

B   F   F
Adapter
sleeve
added

F
Withdrawal
sleeve
added

Ease of mounting with taper bore

FIGURE 32.10 (*Continued*)

J – Spherical roller thrust bearing.
K – Spherical plain bearing.
L – Double row angular contact thrust bearing.

Deep groove ball bearings (Fig. 32.11)

Deep groove ball bearings are available in both single row and double row designs. Single row ball bearings are the most popular of all rolling bearings. They are a simple design, non-separable, suitable for high-speed operation and require little attention in service. The deep grooves enable axial loads to be carried in either direction. Bearings are available with shields and seals and can be supplied with the correct quantity of lithium base grease and used in operating temperatures between $-30\,°C$ and $+110\,°C$. Special bearings operate over a wider range. Relubrication in service is not required. Shielded and sealed bearings are primarily intended for applications where the inner ring rotates. In cases where the outer ring rotates, there is a risk that lubricant will be lost and the manufacturer should be consulted.

Snap rings, fitted to bearings with snap ring grooves provide a simple means of location.

Deep groove ball bearings have very limited ability to accommodate errors of alignment.

FIGURE 32.11 Single and double row deep groove ball bearings.

Self-aligning ball bearings (Fig. 32.12)

Self-aligning ball bearings have two rows of balls and a common sphered raceway in the outer ring and this feature gives the bearing its self-aligning property which permits a minor angular displacement of the shaft relative to the housing. These bearings are particularly suitable for applications where misalignment can arise from errors in mounting or shaft deflection. A variety of designs are available with cylindrical and taper bores, with seals and adapter sleeves and extended inner rings.

FIGURE 32.12 Self-aligning ball bearings with cylindrical bore.

Angular contact ball bearings (Fig. 32.13)

In angular contact ball bearings the line of action of the load, at the contacts between balls and raceways, forms an angle with the bearings axis. The inner and outer rings are offset to each other and the bearings are particularly suitable for carrying combined radial and axial loads. The single row bearing is of non-separable design, suitable for high speeds and carries an axial load in one direction only. A bearing is usually arranged so that it can be adjusted against a second bearing.

A double row angular contact bearing has similar characteristics to two single bearings arranged back to back. Its width is less than two single bearings and can carry an axial load in either direction. These bearings are used for very accurate applications such as the shafts in process pumps.

FIGURE 32.13 Single and double row angular contact ball bearings.

Cylindrical roller bearings (Fig. 32.14)

In cylindrical roller bearings, the rollers are guided between integral flanges on one of the bearing rings. The flanged ring and rollers are held together by the cage to form an assembly which can be removed from the other ring. This separable feature of the bearing design facilitates mounting and

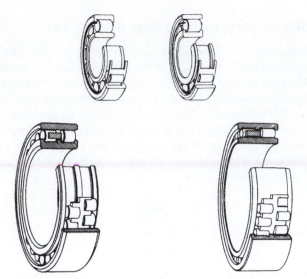

FIGURE 32.14   Single and double row cylindrical roller bearings.

dismounting, particularly where, because of loading conditions, interference fits for both rings are necessary. Single and double row bearings are available for heavy loads, high speeds and rigidity. Typical applications are for machine tools and heavy electric motors.

Needle roller bearings (Fig. 32.15)

The chief characteristic of needle roller bearings is that they incorporate cylindrical rollers with a small diameter/length ratio. Because of their low-sectional height these bearings are particularly suitable for applications where radial space is limited. Needle roller bearings have a high load carrying capacity in relation to their sectional height.

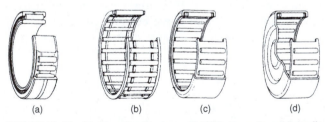

(a)   (b)   (c)   (d)

FIGURE 32.15   Needle roller bearings. (a) With inner ring, (b) needle roller cage assembly, (c) drawn cup needle roller bearings with open ends and (d) drawn cup needle roller bearings with closed end.

Spherical roller bearings (Fig. 32.16)

Spherical roller bearings have two rows of rollers which run on a common sphered raceway in the outer ring, the inner ring raceways each being inclined at an angle to the bearing axis. The bearings are self-aligning and permit minor angular displacements of the shaft relative to the housing which may occur in mounting or because of shaft deflection under

FIGURE 32.16   Spherical roller bearing.

load. Heavy duty types are available for severe operating conditions encountered in vibrating machinery such as soil compactors.

Taper roller bearings (Fig. 32.17)

In a taper roller bearing the line of action of the resultant load through the rollers forms an angle with the bearing axis. Taper roller bearings are therefore particularly suitable for carrying combined radial and axial loads. The bearings are of separable design, i.e. the outer ring (cup) and the inner ring with cage and roller assembly (cone) may be mounted separately.

Single row taper roller bearings can carry axial loads in one direction only. A radial load imposed on the bearing gives rise to an induced axial load which must be counteracted and the bearing is therefore generally adjusted against a second bearing.

Two and four row taper roller bearings are also made for applications such as rolling mills.

FIGURE 32.17

Thrust ball bearings (Fig. 32.18)

Thrust ball bearings are designed to accommodate axial loads. They are not suitable for radial loads. To prevent sliding at the ball to raceway contacts, caused by centrifugal forces and gyratory moments, thrust ball bearings must be subjected to a certain minimum axial load. The bearings are of separable design and the housing and shaft washers may be mounted independently.

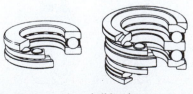

FIGURE 32.18   Single row thrust ball bearing.

Spherical roller thrust bearings (Fig. 32.19)

In spherical roller thrust bearings the line of action of the load at the contacts between the raceways and the rollers forms an angle with the bearing axis, and this makes them suitable for carrying a radial load. This radial load must not exceed 55% of the simultaneous acting axial load. The sphered raceway of the housing washer provides a self-aligning feature which permits, within certain limits, angular displacement of the shaft relative to the housing.

FIGURE 32.19 Spherical roller thrust bearing.

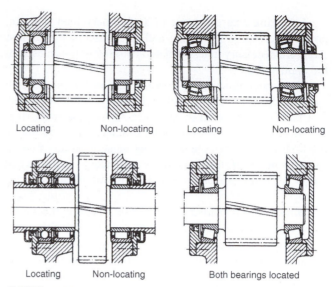

| Locating | Non-locating | Locating | Non-locating |

| Locating | Non-locating | Both bearings located |

FIGURE 32.20

Application of bearings

A rotating machine element, e.g. the shaft, generally requires two bearings to support and locate it radially and axially relative to the stationary part of the machine, e.g. the housing. Normally, only one of the bearings (the locating bearing) is used to fix the position of the shaft axially, whilst the other bearing (the non-locating bearing) is free to move axially.

Axial location of the shaft is necessary in both directions and the locating bearing must be axially secured on the shaft and in the housing to limit lateral movement. In addition to locating the shaft axially the locating bearing is also generally required to provide radial support and bearings which are able to carry combined loads are then necessary, e.g. deep groove ball bearings, spherical roller bearings and double row or paired single row angular contact ball bearings. A combined bearing arrangement, with radial and axial location provided by separate bearings can also be used, e.g. a cylindrical roller bearing mounted alongside a four-point contact ball bearing or a thrust bearing having radial freedom in the housing.

To avoid cross location of the bearings the non-locating bearing, which provides only radial support, must be capable of accommodating the axial displacements which arise from the differential thermal expansion of the shaft and housings. The axial displacements must be compensated for either within the bearing itself, or between the bearing and its seating on the shaft, or in the housing.

Typical examples of locating and non-locating bearings are shown on the applications in Fig. 32.20.

To prevent roll or creep it is important to maintain the correct fits between the bearings and seatings. Inadequate fits can result in damage to both the bearings and associated components. Normally, the only way to prevent movement at the bearing seatings is to provide a sufficient degree of interference for the bearing rings. Interference fits provide a further advantage in that relatively thin section bearing rings are properly supported around their circumference to give a correct load distribution and allow the load carrying

ability of the bearing to be fully utilized. However, where there is a requirement for easy mounting and dismounting of the bearing, or where a non-locating bearing must have freedom of movement axially on its seating, interference fits may not be possible.

Bearings with cylindrical bore

The most important factors to be considered when selecting bearing fits are as follows:

Conditions of rotation – The conditions of rotation refer to the direction of the load in relation to the bearing rings.

If the bearing ring rotates and the load is stationary, or if the ring is stationary and the load rotates so that all points on the raceway are loaded in the course of one revolution, the load on the ring is defined as a rotating load. Heavy oscillating loads such as apply to the outer rings of connecting rod bearings are generally considered as rotating loads.

If the bearing ring is stationary and the load is also stationary, or if the ring and load rotate at the same speed so that the load is always directed towards the same point on the raceway, the load on the ring is defined as a 'stationary load'.

Variable external loading, shock loading, vibrations and out of balance forces in high speed machines, giving rise to changes in the direction of the load which cannot be accurately established, are classified under the term 'direction of load indeterminate'.

A bearing ring subjected to a rotating load will creep on its seating if mounted with a clearance fit, and wear of the contacting surfaces will occur (fretting corrosion). To prevent this, an interference fit should be used. The degree of interference required is dictated by the operating conditions referred to below in the notes on internal clearance and temperature conditions.

A bearing ring subjected to a stationary load will not normally creep on its seating and an interference fit is not therefore necessary unless dictated by other requirements of the application.

When the direction of loading is indeterminate, and particularly where heavy loading is involved, it is desirable that both rings have an interference fit. For the inner ring the fit recommended for a rotating inner ring is normally used. However, when the outer ring must be axially free in its housing or if the loading is not heavy a somewhat looser fit than that recommended for rotating loads may be used.

Magnitude of the load – The load on a bearing inner ring causes it to expand resulting in an easing of the fit on the seating; under the influence of a rotating load, creep may then develop. The amount of interference between the ring and its seating must therefore be related to the magnitude of the load: the heavier the load the greater the interference required.

Internal clearance – When bearing rings are mounted with an interference fit, the bearing radial internal clearance is reduced because of the expansion of the inner ring and/or contraction of the outer ring. A certain minimum clearance should however remain. The initial clearance and permissible reduction depend on the type and size of bearing. The reduction in clearance due to the interference fit can be such that bearings with radial internal clearance greater than normal may be necessary.

Temperature conditions – In service, the bearing rings normally reach a higher temperature than the component parts to which they are fitted. This can result in an easing of the fit of the inner ring on its seating or alternatively the outer ring may expand and take up its clearance in the housing thereby limiting its axial freedom. Temperature differentials and the direction of heat flow must therefore be carefully considered in selecting fits.

Requirements regarding running accuracy – Where bearings are required to have a high degree of running accuracy, elastic deformation and vibration must be minimized and clearance fits avoided. Bearing seatings on shafts should be at least to tolerance IT5 and housing seatings to tolerance IT6. Accuracy of form (ovality and taper) is also very important and deviations from true form should be as small as possible.

Design and material of shaft and housing – The fit of the bearing ring on its seating must not lead to uneven distortion (out of round) of the bearing ring, which may for example be caused by surface irregularities of the seatings. Split housings are not suitable when outer rings are to have an interference fit and the limits of tolerance selected should not give a tighter fit than that obtained when tolerance groups H or J apply. To ensure adequate support for bearing rings mounted in thin walled housings, light alloy housings or on hollow shafts, heavier interference fits must be used than would normally be selected for thick walled steel or cast iron housings or solid shafts.

Ease of mounting and dismounting – Bearings having clearance fits are preferred for many applications to facilitate installation and removal. When operating conditions necessitate the use of interference fits and ease of mounting and dismounting is also essential, separate bearings or bearings having a tapered bore and an adapter or withdrawal sleeve can often provide a solution.

Displacement of a non-locating bearing – When a non-separable bearing is used at the non-locating position, it is necessary that under all conditions of operation one of the rings is free to move axially. This is ensured by using a clearance fit for that ring which carries a stationary load. Where for example, light alloy housings are used, it may sometimes be necessary to fit a hardened intermediate bush between the outer ring and the housing. If certain types of cylindrical roller bearings, or needle roller bearings are used at the non-locating position, then both inner and outer rings can be mounted with an interference fit.

Bearings with tapered bore

Bearings with a tapered bore are often used to facilitate mounting and dismounting and in some cases this type of bearing may be considered essential to the application. They can be mounted either directly on to a tapered shaft, or by means of an externally tapered sleeve on to a cylindrical shaft.

The axial displacement of a bearing on its tapered seating determines the fit of the inner ring and special instructions relating to the reduction of clearance of bearings with a tapered bore must be observed. The fit of the outer ring in the housing is the same as that for bearings having a cylindrical bore. Adapter and withdrawal sleeves allow greater shaft tolerances to be used (h9 or h10). Errors of form (ovality and taper) of the shaft seating must, however, still be closely controlled (tolerance IT5 or IT7).

Fits and tolerances

Tolerances for the bore and outside diameter of metric rolling bearings are internationally standardized. The desired fits are achieved by selecting suitable tolerances for the shaft and housing using the ISO tolerance system incorporated in data sheet BS 4500A and B.

For any particular bearing, the manufacturer's catalogue should be consulted with regard to recommended fits because these must be related to the actual size of the bearings supplied.

Axial location of bearings

Interference fits in general only provide sufficient resistance to axial movement of a bearing on its seating when no axial forces are to be transmitted and the only requirement is that lateral movement of the ring should be prevented. Positive axial location or locking is necessary in all other cases. To prevent axial movement in either direction of a locating bearing it must be located at both sides. When non-separable bearings are used as non-locating bearings, only one ring, that having the tighter fit, is axially located, the other ring must be free to move axially in relation to the shaft or housing.

Where the bearings are arranged so that axial location of the shaft is given by each bearing in one direction only it is sufficient for the rings to be located at one side only.

Methods of location (Fig. 32.21)

Bearings having interference fits are generally mounted against a shoulder on the shaft or in the housing. The inner ring is normally secured in place by means of a locknut and locking washer (a), or by an end plate attached by set screws to the shaft end (b). The outer ring is normally retained by the housing end cover (c), but a threaded ring screwed into the housing bore is sometimes used (d).

Instead of shaft or housing abutment shoulders, it is frequently convenient to use spacing sleeves or collars between the bearing rings (e), or a bearing ring and the adjacent component, e.g. a gear (f). On shafts, location can also be achieved using a split collar which seats in a groove in the shaft and is retained by either a solid outer ring which can be slid over it, or by the inner ring of the bearing itself.

Axial location of rolling bearings by means of snap rings can save space, assist rapid mounting and dismounting and simplify machining of shaft and housings. An abutment collar should be inserted between the snap ring and the bearing if heavy loads have to be carried, in order that the snap ring is not subjected to large bending moments across its section. If required, the axial clearance, which is generally present between the snap ring and the snap ring groove can be reduced by selecting an abutment collar of suitable width or by using shims. Deep groove ball bearings with a snap ring groove in the outer ring and fitted with a snap ring sometimes provide a simplified and compact housing arrangement.

Bearings with a tapered bore mounted directly on tapered shafts are usually retained by a locknut on the shaft (g), or the locknut may be screwed on to an externally threaded split ring inserted into a groove in the shaft (h). With adapter sleeve mounting, the locknut positions the bearing relative to the sleeve (j). When bearings with an adaptor sleeve are mounted on shafts without an abutment shoulder, the axial load which can be applied depends on the resulting friction between shaft and sleeve. When bearings with a tapered bore are mounted on withdrawal sleeves the inner ring of the bearing must be mounted against an abutment (k). A suitable abutment can be provided by a collar which can frequently serve as part of a labyrinth seal. The withdrawal sleeve must be secured in position either by means of a locknut or an end plate and set screws.

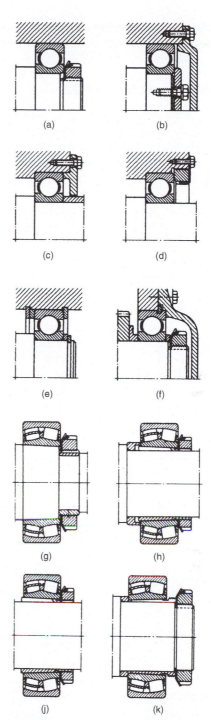

FIGURE 32.21   Bearing location methods.

Seals

Bearings must be protected by suitable seals against the entry of moisture and other contaminants and to prevent the loss of lubricant. The effectiveness of the sealing can have a decisive effect on the life of a bearing.

Many factors must be considered when deciding on the best sealing arrangements for a given bearing application, e.g. the type of lubricant (oil or grease), peripheral speed at the sealing surface, misalignment of the shaft, available space, friction of the seal and resultant temperature rise and cost. Two basic designs are normally used for rolling bearings.

Non-rubbing seals (Fig. 32.22)

Non-rubbing seals depend for their effectiveness on the sealing efficiency of narrow gaps, which may be arranged axially, radially or combined to form a labyrinth. This type of seal has negligible friction and wear and is not easily damaged. It is particularly suitable for high speeds and temperatures.

This simple gap type seal which is sufficient for machines in a dry, dust free atmosphere comprises a small radial gap formed between the shaft and housing (a). Its effectiveness can be improved by providing one or more grooves in the bore of the housing cover (b). The grease emerging through the gap fills the grooves and helps to prevent the entry of contaminants. With oil lubrication and horizontal shafts,

right or left hand helical grooves can be provided in the shaft or the seal bore (c). These serve to return any oil which may tend to leak from the housing. However, with this arrangement it is essential that the direction of rotation does not vary.

Single or multiple labyrinths give appreciably more effective sealing than gap seals but they are generally more expensive to manufacture. They are chiefly used with grease lubrication. Their effectiveness can be still further improved by providing a grease duct connecting with the labyrinth passage and periodically pumping in a quantity of water insoluble grease, e.g. a calcium soap base grease. In solid housings the tongues of the labyrinth seal are arranged axially (d), and in split housing, radially (e). The radial clearance between the shaft and the housing seal components is not affected by axial displacement of the shaft during running and can be made very small. If angular misalignment of the shaft relative to the housing has to be accommodated, labyrinths of the form shown at (f) are normally used.

An inexpensive and effective labyrinth seal can be made using pressed steel sealing washers (g). The effectiveness of this type of seal increases in direct proportion to the number of washers used. To increase the sealing efficiency of non-rubbing seals, the shaft can be fitted with rotating discs (h) and in case of oil lubrication, flinger rings (i) are often used. The oil flung from the ring is collected in a channel in the housing wall and returned to the sump through suitable ducts.

Rubbing seals (Fig. 32.23)

Rubbing seals rely for their effectiveness essentially on the elasticity of the material exerting and maintaining a certain pressure at the sealing surface. The choice of seal and the required quality of the sealing surface depend on the peripheral speed.

Felt washers (a) are mainly used with grease lubrication, e.g. in plummer blocks. They provide a simple seal suitable for peripheral speeds up to 4 m/s and temperatures of about 100 °C. The effectiveness of the seal is considerably improved if the felt washer is supplemented by a simple labyrinth ring (b). The felt washers or strips should be soaked in oil at about 80 °C before assembly.

Where greater demands are made on the effectiveness of the rubbing seal, particularly for oil lubricated bearings, lip seals are often used in preference to felt seals. A wide range of proprietary lip type seals is available in the form of ready to instal units comprising a seal of synthetic rubber or plastics material normally enclosed in a sheet metal casing. They are suitable for higher peripheral speeds than felt washers. As a general guide at peripheral speeds of over 4 m/s the sealing surface should be ground, and above 8 m/s hardened or hard chrome-plated and fine ground or polished if possible. If the main requirement is to prevent leakage of lubricant from the bearing then the lip should face inwards (c). If the main purpose is to prevent the entry of dirt, then the lip should face outwards (d).

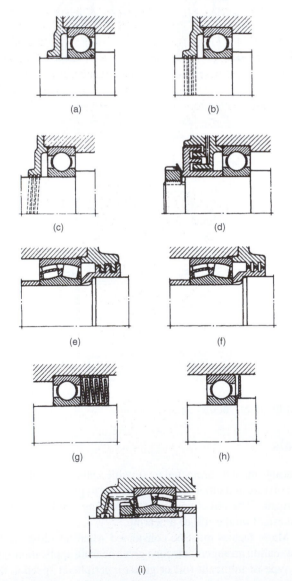

FIGURE 32.22

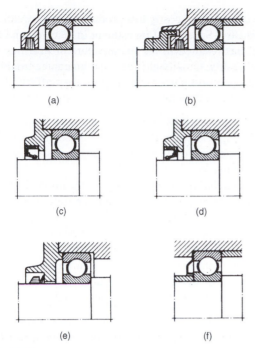

FIGURE 32.23

The V-ring seal (e) can be used for grease or oil lubricated bearing arrangements. It comprises a rubber ring with a hinged rubber lip which is pressed axially against the sealing surface. It is easy to fit, can accommodate fairly large angular misalignments of the shaft relative to the housing at slow speeds, and in certain circumstances is suitable for high speeds. The effectiveness of the seal owes much to the fact that dirt and liquids tend to be flung off by the rotating seal. The V-ring seal is normally fitted on the inside rotating seal. The V-ring seal is therefore normally fitted on the outside of the housing when grease lubrication is used and on the inside with oil lubrication.

Spring steel sealing washers provide a cheap and space saving seal, especially for grease lubricated deep groove ball bearings. They can either be clamped against the outer ring (f) or against the inner ring and are designed so that the sealing face is constrained to press against the face of the other bearing ring.

Combined seals

In difficult operating conditions and where severe demands are placed on sealing, e.g. large amounts of dirt or water, rubbing and non-rubbing seals are often combined. In such cases the non-rubbing seals (labyrinths, flinger rings, etc.) are arranged to supplement the rubber seals and protect them from wear.

Sealed and shielded bearings

Simple space saving arrangements can be achieved by using bearings incorporating seals or shields at one or both sides

which are supplied lubricated with the correct quantity of grease. Relubrication is not normally required and they are primarily intended for applications where sealing is otherwise inadequate or where it cannot be provided for reasons of space.

Lubrication

Grease lubrication is generally used where ball and roller bearings operate at normal speeds, temperature and loading conditions. Grease has certain advantages by comparison with oil: it is more easily retained in the bearing housing and assists in sealing against the entry of dirt and moisture.

In general the free space in the bearing and housing should only be partly filled with grease (30–50%). Overfilling causes rapid temperature rise particularly if speeds are high. Manufacturers supply details regarding suitable weights of grease for particular bearings.

Bearings can be supplied which are sealed after prepacking with the correct type and quantity of grease. Where relubrication is more frequent, provision must be made by fitting grease nipples to the housing. Grease will then be applied by a grease gun and a lubrication duct should feed the grease adjacent to the outer ring raceway or between the rolling elements. Examples are shown in Fig. 32.24.

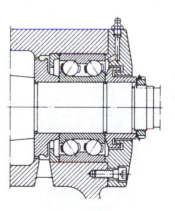

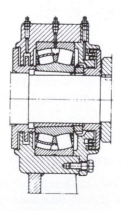

FIGURE 32.24

Oil lubrication

Oil lubrication is generally used where high speeds or operating temperatures prohibit the use of grease, when it is necessary to transfer frictional heat or other applied heat away from the bearing, or when the adjacent machine parts, e.g. gears, are oil lubricated.

Oil bath lubrication is only suitable for slow speeds. The oil is picked up by rotating bearing elements and after circulating through the bearing drains back to the oil bath. When the bearing is stationary, the oil should be at a level slightly below the centre of the lowest ball or roller. An application is shown in Fig. 32.25. At high speeds it is important that sufficient oil reaches the bearing to dissipate the heat generated by friction and oil jets provide an effective method (Fig. 32.26).

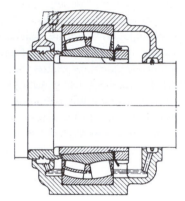

FIGURE 32.25

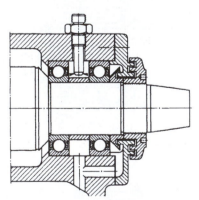

FIGURE 32.26

The illustrations in this section (Figs. 32.10–32.26) are reproduced by kind permission of SKF (UK) Limited – www.skf.co.uk.

Trouble-free bearing operation

When bearings fail, they can bring equipment to an unscheduled halt. Every hour of down time due to premature bearings failure can result in costly lost production in a capital intensive industry. Substantial investment in research and development has resulted in the manufacture of bearings of the highest quality. Quality alone cannot guarantee trouble-free bearing operation since other factors may affect life span including the following:

1. *Operating environment*: Machinery must be kept in peak operating condition. Bearings should be properly aligned and protected from extreme temperatures, moisture and contaminants.
2. *Proper installation*: Knowledge of the proper installation techniques and tools is required to ensure that the bearings are not damaged.
3. *Proper maintenance*: Following lubrication and maintenance schedules using recommended materials and time intervals is essential. A familiarity with operating procedures, basic trouble shooting, condition monitoring and vibration analysis is also desirable.

However, bearing manufacturers do have a full line of products and services to make installation and maintenance easy to perform and should be consulted. This will certainly contribute to long bearing life and ensure cost-effective operation.

General convention and simplified representation

Both types are illustrated in Fig. 32.27.

Simplified representations for both types are the same. The simplification shown here with crossed diagonal lines was the practice used by industry in the past.

Current practice introduces a free-standing upright cross referred to in ISO 8826-1.

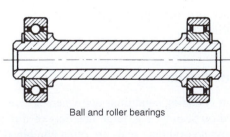

Ball and roller bearings

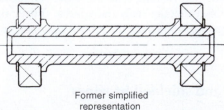

Former simplified representation

FIGURE 32.27

Engineering adhesives

The use of adhesives is now a well-established practice in manufacturing. New materials and production processes have considerably increased the options available to the engineering designer. Adhesive bonding is a proved cost-effective manufacturing method and can be used with confidence. A basic principle is that joints should be designed with this method of production in mind when the product is in the early stages of development.

The following are some advantages of using adhesives:

(a) Stress concentrations present in bolted, riveted or spot welded joints are avoided.

(b) The distribution of stresses achieved by adhesive bonding permits a reduction in weight and cost, especially relevant with fragile materials and lightweight structures. Joint strength and fatigue properties are improved.

(c) Production costs are reduced due to the elimination of drilled holes and other machining operations. Labour costs are reduced with automated assembly work.

(d) Structures are generally stiffer despite weight reduction since the bonding covers the whole area of the joint. Rivets, screws and spotwelds pin the surfaces together only at localized points. Loading may be increased before buckling occurs.

(e) Gap filling properties. Certain adhesives are gap filling, and this makes possible the continuous joining of materials where the gap along the joint is of irregular width.

(f) Delicate or brittle materials such as metal foils or ceramics are readily bonded.

(g) High-strength bonds can be formed at room temperature with minimal pressure by using cold-setting adhesives.

(h) The film formed by the adhesive resists corrosion, can form a leak-proof seal and insulate dissimilar metals against electrochemical action.

Designing for adhesives

For the best possible performance, joints should be specifically designed for adhesive bonding. Follow this principle and much better joints will be achieved than if bonding is adopted as a substitute for welding in a joint designed for that purpose. Bond stresses, materials, type of adhesive, surface preparations, method of application and production requirements can then all be considered in relation to each other at the outset. The designer should consider especially the effect of shear, tension, cleavage, and peel stresses upon the joint. Bonded joints perform best under conditions of tension (pure), compression or shear loading; less well under cleavage; and relatively poorly under peel loading. The loading conditions are shown in Fig. 33.1.

Designing a joint to take pure tensile or compressive stresses is normally impracticable with sheet materials, so all joints in sheet materials should be designed so that the main loading is in shear. Joints between massive parts perform well in tension or compression loading, provided this is uniform – a side load may set up excessive cleavage stresses in a tension-loaded bond (Fig. 33.1(d)). Cleavage loading will concentrate stress at one side of the joint. Bond area may have to be increased to withstand this load so the joint will not prove so economical in terms of material and/or adhesives as joints designed for shear and tension stresses. Peel strength is usually the weakest property of a joint. A wide joint will be necessary to withstand peel stresses, plus the use of an adhesive with high-peel strength.

For an adhesive to be used, a joint must allow the easy application of the adhesive, must allow for the adhesive to cure fully, and must be designed to give uniform stress. Even in a simple face-to-face joint it must be possible to apply adhesive to one surface and for it to remain there until the two parts are brought together and after that until curing takes place.

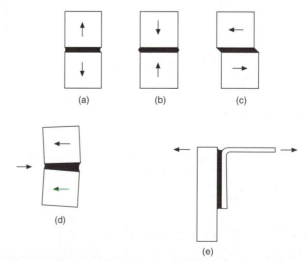

FIGURE 33.1 Loading conditions. (a) Tension, (b) Compression, (c) Shear, (d) Cleavage, and (e) Peel.

These requirements highlight the need for a choice of thin, thick or thixotropic adhesives. Design details which may also be significant include removal of sharp edges and substitution of a bevel or radius.

The bond line

The gap between the parts, and therefore the thickness of the adhesive film, has an important bearing on the characteristics of the joint. In terms of simple strength, a thick bond line will generally be a weakening feature, since the mechanical strength of the unsupported resin film is likely to be less than that of the substrates.

A thick bond line can however confer advantages. The adhesive is generally more flexible than the adherents or substrates. This is particularly so in most engineering applications where metals or other rigid materials can be bonded. Because of this, a thick bond line can offer a capacity to absorb some impact energy, thus increasing the strength of the bond under this type of loading.

Consideration of bond line thickness leads immediately to the question of environmental resistance.

Adhesive bonds will always be susceptible to environmental attack and it is essential that any such attack should not reduce the strength of the bond to an unacceptable level. The most important factor here is the correct choice of adhesive, but design of the joint can make a significant

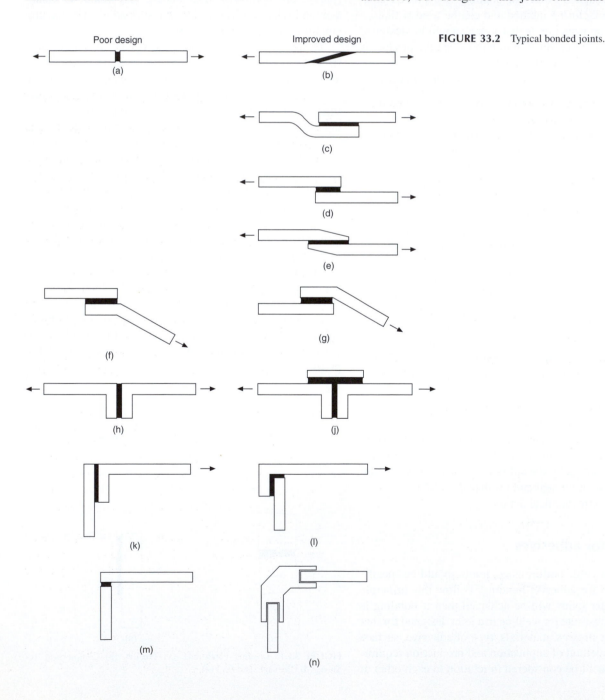

FIGURE 33.2   Typical bonded joints.

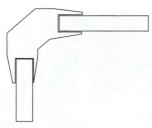

FIGURE 33.3 Where slotted joints are used, tapering removes the high-stress concentrations caused by abrupt changes in section. Example gives a possible modification to Fig. 33.2(n).

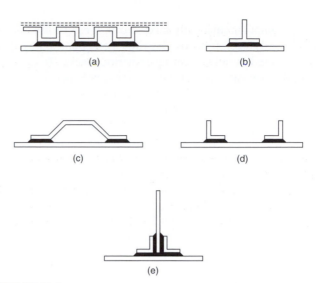

FIGURE 33.5

difference. Thus a thick bond line offers a ready path for access by moisture or other solvents which might be able to diffuse through the cured adhesive.

Typical bonded joints

Figure 33.2 shows a range of bonded joints and possible modifications which can be made to reduce or eliminate the effect of cleavage and peel stresses.

The following notes should be regarded as of a general nature:

(a) Avoid butt joints if bond area is small.
(b) Scarfed joint provides increased bonding area.
(c) Simple lap joint with in-line forces.
(d) Alternative lap joint with offset loading.
(e) Tapered lap joint.
(f) Bracket bonded to a fixed surface where peel is likely.
(g) Repositioned bracket strengthens joint.
(h) and (i) Cleavage loading eliminated by the addition of a component in shear.
(i) and (k) Simple improvement for safety.
(j) and (m) Increase in bond area reinforces corner joint.

Quite obviously practical considerations involve a study of the forces applicable and acceptable appearance of the finished assembly (Fig. 33.3).

Figure 33.4 shows two tubular applications.

In (a) a cylindrical plug is used to join two tubes in a structure. An example of a tapered tubular joint is given

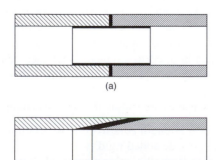

FIGURE 33.4

in (b). The taper ensures that the adhesive is not pushed out of the assembly.

The joint permits a long bond line and does not impede fluid flow.

A selection of bonded stiffeners is shown in Fig. 33.5. These can be used to reduce vibration and deflection of thin sheet materials. When the flanges on the stiffened sections deflect with the sheet, little difficulty from the peel results due to the area of the bond. Corrugated backings can provide complete flatness over the entire area. If a corrugated insert is sandwiched between two flat sheets (the second sheet is indicated by dotted lines) as indicated in example (a) then a structure of lightweight and high strength can be manufactured from adhesive bonding. There are many aircraft applications. Standard strip, angles, tee sections, and formed channels are used in structural engineering.

The types of adhesive which cover the vast majority of engineering assembly applications come from the following categories.

1. *Epoxies*: Two components are mixed in equal proportions. The adhesive and the hardener begin to cure immediately and have a usable 'pot life'. After this time the bond becomes less effective, often used for DIY repairs. Industry uses an alternative type of epoxy which incorporates rubber of low-molecular weight and is called a toughened adhesive. It has greater resistance to impact forces and peel.

    This is a single component epoxy which is hardened by heat curing while the parts being bonded are clamped.

    Used to bond composite materials, tubular frames and in the manufacture of components for double glazing assemblies.

2. *Acrylic adhesives*: Four basic types:
    (a) Toughened acrylics: These are two-part systems where a hardener and an adhesive are applied to the two surfaces being joined and the assembly of the

joint automatically mixes them. It can be used on oily steel. It bonds glass into metal frames. It is also used in railway carriage interior panels.

(b) Cyanoacrylate adhesives: These polymerize (solidify) by a chemical reaction which is usually initiated by atmospheric moisture, present as traces of water on the surfaces to be joined. Successful bonding depends upon ambient humidity, the type of material being bonded, the choice of adhesive, and the nature of the surface.

(c) 'Instant adhesives' and 'Superglues' are in this range of products.

(d) Anaerobic adhesives: These automatically harden in the absence of air and are used mainly in rigid metallic joints. Many applications follow. These products are manufactured normally as single component materials.

(e) UV curing acrylics: These are single component adhesives where cure is effected rapidly by placing the assembly under an ultraviolet lamp.

These adhesives are applied in the manufacture of printed circuit boards for surface sealing.

3. *Hot melt adhesives*: These are available in rod, sheet and powder forms. A convenient method of assembling small components which are lightly loaded. A heating gun raises the temperature of the rod and the adhesive is applied to one component. On cooling, the adhesive solidifies and the two surfaces are bonded together. These adhesives are also used in packaging equipment.

4. *Solvent based contact adhesives*: Here the adhesive is applied in a solvent solution to the two surfaces. The solvent evaporates leaving a tacky film and the surfaces are brought together. Applications include laminated sheet fixings in furniture manufacture.

A considerable range of options is available to the designer in the choice of suitable types of adhesive.

Precision measuring and dispensing is necessary so that the required volume, in the defined position, is applied at a given time and with consistently repeatable accuracy on a production line.

In the interests of satisfactory selection and operation, it is recommended that the manufacturer should be consulted to ensure that all technical considerations have been included in a proposed scheme.

Engineering applications

The following examples show varied uses of engineering adhesives in industry.

Locking screw threads: The liquid is applied to the cleaned thread of a bolt or stud. When the nut is tightened the liquid fills the gaps between mating threads and hardens to form a tough plastic joint which is shock, vibration, corrosion, and leak proof. The joint will remain in place until

FIGURE 33.6   Thread locking.

it needs to be undone again using normal hand tools (Fig. 33.6).

Threadsealing pipe fittings: The sealant is applied to the clean thread and screwed together as normal. The sealant will not creep or shrink and gives a constant and reliable seal. There is no need to wrench tight and the fitting can be positioned as required (Fig. 33.7).

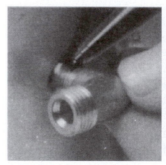

FIGURE 33.7   Thread sealing.
(a) Hydraulic sealant for fine threads in pneumatic and hydraulic systems – particularly those subject to vibration.
(b) Pipe sealant used to seal coarse threads of pipes and fittings up to 75 mm outside diameter.

Retaining: Traditional retaining methods using screws, splines, pins, keys, and press fits, etc., do not necessarily seal joints and eliminate the possibility of corrosion. Local stress concentrations may cause cracking. Retaining adhesives can be used to augment these methods. Often, a redesign will give a replacement with substantial cost savings.

These adhesives are supplied in various strengths:

(a) High-shear strength adhesives in association with press fits can provide added rigidity.

(b) Maximum strength retainers are used on parts which generally do not need to be taken apart.

(c) Medium strength adhesives suit parts which need frequent disassembly (Fig. 33.8).

FIGURE 33.8 Retaining.

Sealing with anaerobic gaskets: Gaskets are fitted between flanges to provide an effective seal against fluids and gases. It is cheaper to use a gasket than manufacture two perfectly flat mating surfaces with close flatness and surface finish tolerances.

Gaskets can be preformed from materials such as compressed asbestos, paper, fibre or cork. Alternatively, they can be formed where they are required with a liquid.

The principles of liquid gasketing are fundamentally different to preformed gaskets in that they allow metal-to-metal contact. There are several forms of liquid gasket such as anaerobic, non-setting solvent based and moisture curing.

The anaerobic principle: Anaerobic gaskets are available in a range of viscosities from thick liquids to non-slump pastes. Each can be applied directly from the original container, or by various application methods such as simple rollers, screen printing and computerized robotics. On assembly, the anaerobic gasket spreads between the flanges and is forced into surface irregularities to provide total con-

tact between the two faces. The product then polymerizes at ambient temperature into a tough thermoset plastic.

The strength of joints from anaerobics can be tailored to suit a specific application. Effective cure requires the absence of air and the presence of metal. At room temperature it takes just a few minutes.

Note: Anaerobic gaskets are thermosetting plastics; the temperature range in service can be from −50 °C up to 200 °C at the joint line. They seal against petroleum based fuels and lubricating oils, water/glycol mixtures and many other industrial chemicals. For compatibility of specific chemical environments the designer would be advised to consult the manufacturers.

Although anaerobic gaskets permit metal-to-metal contact, electrical continuity cannot be assumed.

Figure 33.9 shows the application of an anaerobic gasket to the backplate of a large diesel engine.

The flow of adhesive to the work surface is regulated by manual control of the air supply to a pneumatic cartridge gun.

It often happens during maintenance work that damaged or scored surfaces are found and an adhesive gasket can save the need and cost of re-machining.

Engineering adhesives for sealing flat faces have the following characteristics and applications

(a) They will seal on horizontal, vertical and overhead flanges and accommodate surface irregularities of up to 0.5 mm.

(b) Low-strength products are available for close fitting surfaces which will be frequently dismantled.

(c) In the illustrations overleaf (Fig. 33.10) many of the components are manufactured in aluminium alloys. The structural integrity of an assembly can be enhanced by the use of high-shear strength adhesives.

Engineering adhesives for retaining cylindrical assemblies have the following characteristics and applications:

FIGURE 33.9

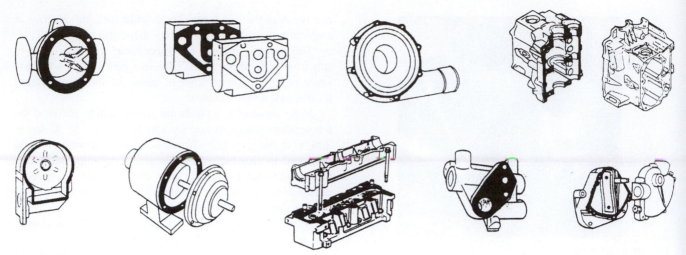

FIGURE 33.10

(a) The retention of shafts and rotors of electric motors,
     gears, pulleys, sleeves, bushes, and oil seals in
     housings.
(b) The ability to withstand fatigue and augment torsional
     strength.
(c) Suitable for parts that need easy disassembly, such as
     bearings on shafts and in housings, bushes and journals
     in soft metals.
(d) An oil-tolerant adhesive is available that gives high-
     strength retention of parts 'as received', i.e. no cleaning
     is needed before assembly. Oil impregnated bushes are
     retained with this grade. They are manufactured by the
     sintering process.
(e) An adhesive can be recommended for continuous
     working temperatures up to 175 °C. It combines the
     ability to fill gaps of up to 0.15 mm in diameter with
     high-shear strength and good solvent resistance.

Instant adhesives

As the name suggests, they work in seconds and are ideal for
bonding close fitting parts made from a variety of materials.
They offer advantages over other forms of joining, such as
plastic welding, two-part or heat-cured adhesives and mechanical fasteners. The benefits include faster assembly
times, better appearance, less outlay for capital equipment
and these adhesives can also be used to repair metal, plastic,
rubber or ceramic components which might otherwise be
scrapped.

Instant adhesives are available for the following applications:

(a) General purpose adhesive for plated metals, composite
     materials, wood, cork, foam, leather, paper – all surfaces
     which were once considered 'difficult' – can now be
     bonded quickly and permanently.
(b) A special rubber and plastics adhesive ensures fast-
     fixturing of elastomers and rubbers, especially EPDM

rubber. Bonds polyethylene, polypropylene and
polyolefin plastics.
(c) A gel-type adhesive can be used for fabrics, paper,
     phenolic, PVC, neoprene and nitrile rubber and bond
     them in 5 s; ceramic, leather, and balsa wood in 10 s;
     mild steel in 20 s; ABS and pine in 30 s. The gel form
     prevents absorption by porous materials and enables it to
     be applied to overhead and vertical surfaces without
     running or dripping.
(d) A low-odour, low-bloom adhesive has been developed
     where application vapours have been removed with no
     possibility of contamination. A cosmetically perfect
     appearance can be obtained. The absence of fumes
     during application means that it can be safely used close
     to delicate electrical and electronic assemblies,
     alongside optics and in unventilated spaces.
(e) A black rubber toughened instant adhesive gives
     superior resistance to peel and shock loads. Tests show
     bonds on grit blasted mild steel can expect a peel
     strength of 4 N/mm at full cure.

All adhesives can be applied direct from bottle, tube or
standard automatic application equipment on to surfaces
which require very little pretreatment.

In most cases, just one drop of instant adhesive is enough
to form an extremely strong, virtually unbreakable bond.
There is no adhesive mixing, and cure takes place in seconds
to give a joint with maximum surface-to-surface contact
(Fig. 33.11).

Four typical production applications are shown in
Fig. 33.12.

The illustration in Fig. 33.12(c) shows an operator using
a semi-automatic dispenser. The bonding product is contained in a bottle pack and dispensing regulated by an electronic timer controlled pinch valve mounted on the unit. The
dispense function can be initiated in a variety of ways, including a footswitch. The point of application is controlled
by hand.

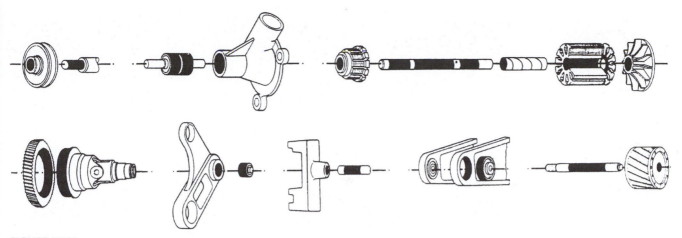

FIGURE 33.11

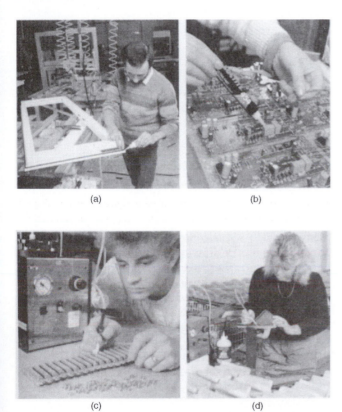

(a)　　　　　(b)

(c)　　　　　(d)

FIGURE 33.12 (a) Bonding rubber to uPVC double glazing units, (b) Bonding toroid to the PCB from a temperature control unit, (c) Bonding brass to PVC on a connector and (d) Bonding foam rubber to moulded polyurethane grouting tool.

Structural applications

Structural adhesives are ideal for bonding large areas of sheet materials. They can produce a much better finished appearance to an assembly than, say, rivets, or spot welding or screws. The local stress introduced at each fixing point will be eliminated. Furthermore, adhesives prevent the corrosion problems normally associated with joining dissimilar materials. This is a cost-effective method of providing high-strength joints (Figs. 33.13 and 33.14).

EN ISO 15785 Technical drawings – Symbolic presentation and indication of adhesive, fold and pressed joints. This

Standard includes examples of graphical symbols, indication of joints in drawings, basic conventions for symbolic presentation, and indication of joints. Also included are designation examples and the dimensioning of graphical symbols.

The authors wish to express their thanks for the assistance given and permission to include examples showing the application of adhesives in this chapter by Loctite UK, www.loctite.co.uk.

FIGURE 33.13 A structural adhesive used to bond a stiffener to an aluminium car bonnet. To line up the two parts a purpose made fixture is designed.

FIGURE 33.14 Mild steel stiffeners are bonded to up and over garage door. Result: rigidity, unblemished exterior surfaces.

Chapter 34

Related standards

Standards are essential tools for industry and commerce, influencing every aspect of the industrial process. They provide the basic ingredients for competitive and cost-effective production. Standards define criteria for materials, products and procedures in precise, authoritative and publicly available documents. They embrace product and performance specifications, codes of practice, management systems, methods of testing, measurement, analysis and sampling, guides and glossaries.

Thus they facilitate design and manufacture:

- establish safety criteria;
- promote quality with economy;
- assist communication and trade; and
- inspire confidence in manufacturer and user.

The role of standards in national economic life is expanding:

- they are increasingly referred to in contracts;
- called up in national and community legislation;
- used as a basis for quality management;
- required for product certification;
- required in public purchasing; and are used as a marketing tool.

The British Standards Institution

Established in 1901, The BSI was the world's first national standards body. Many similar organizations worldwide now belong to the International Organization for Standardization (ISO) and the International Electrotechnical Commission (IEC). BSI represents the views of British Industry on these bodies, working towards harmonizing world standards.

BSI has published approximately 20 000 standards; each year around 2000 new and revised standards are issued to encompass new materials, processes and technologies, and to keep the technical content of existing standards curent. BSI also provides services to its members and undertakes commercial activities, which help underwrite its core standards role.

The BSI Catalogue is published each year. BSI subscribing membership is designed to make keeping in touch with developments in world standardization easy and cost-effective. Membership benefits include:

- discounts on products and services;
- free standards catalogue, BSI magazines and use of the library;

- access to PLUS (see below) licensed electronic products;
- library loans, credit facilities, loans, Members, Days and AGM voting rights.

BSI Knowledge Centre

Contains over half a million standards as well as other reference material including documents on code of practice and technical requirements. Electronic access to many more documents are also available. These include, British Standards, National Standards, and European and International adoptions, ISO IEC, CEN, and CENELEC Standards, as well as standards published by other National standard bodies in Europe and worldwide, such as DIN, ANSI, JIS, GOST and many more.

Technical information group

For over 30 years BSI has run a technical help to exporters service and now covers more subjects and more countries than ever before. Technical barriers to trade (standards, regulations, certification and language) affect products in worldwide markets. BSI can support market research activities in a cost-effective and timely way. For more information log on to www.bsi-global.com/export.

Foreign standards and translations

BSI holds over 100 000 international and foreign standards and regulations in their original form, as well as expert translations that are regularly reviewed to ensure they are current. New translations from and into most languages can be arranged on request.

PLUS – private list updating service

PLUS monitors and automatically updates your standards collection. Exclusive to BSI subscribing members, PLUS not only saves time and effort, but can be an essential part of your quality control system.

Perinorm

Perinorm is the world's leading bibliographic database of standards. Available either on CD-ROM or online at www. perinorm.com, Perinorm contains approximately 520 000 records, including technical regulations from France and Germany, together with American, Australian, and Japanese standards (international version).

DISC

DISC – delivering information solutions to customers through international standardization – is the specialist IT arm of BSI. It enables UK businesses to exert influence when international standards are being formulated and responds directly by helping customers to use standards. It offers guidance, codes of practice, seminars and training workshops.

British Standards Online and CD-ROM

British Standards Publishing Ltd., supplies British Standards worldwide in hard copy, on CD-ROM and the Internet via British Standards Online. The exclusive, authoritative and most current source of British Standards delivers information on more than 38 000 BSI publications to your desktop. For more information on delivery options for British Standards contact:

UK customers: British Standards Publishing Sales Ltd.; e-mail: bsonline@techindex.co.uk.

Customers outside the UK: Information Handling Services; e-mail: info@ihs.com.

BSI quality assurance is the largest independent certification body in the UK and provides a comprehensive service including certification, registration, assessment, and inspection activities.

BSI testing at Hemel Hempstead offer services which are completely confidential and cover test work in many areas including product safety, medical equipment, motor vehicle safety equipment, and calibration work. Tests are undertaken to national or international standards.

Further information on BSI services can be obtained from www.BSI.global.com.

Complete sets of British Standards are maintained for reference purposes at many Public, Borough and County Libraries in the UK. Copies are also available in University and College of Technology Libraries.

The Standards-making process

The BSI Standards function is to draw up voluntary standards in a balanced and transparent manner, to reach agreement among all the many interests concerned, and to promote their adoption. Technical committees whose members are nominated by manufacturers, trade and research associations, professional bodies, central and, local government, academic bodies, user, and consumer groups draft standards.

BSI arranges the secretariats and takes care to ensure that its committees are representative of the interests involved.

Members and Chairmen of committees are funded by their own organizations.

Proposals for new and revised standards come from many sources but the largest proportion is from industry. Each proposal is carefully examined against its contribution to national needs, existing work programmes, the availability of internal and external resources, the availability of an initial draft and the required timescale to publish the standard. If the work is accepted it is allocated to a relevant existing technical committee or a new committee is constituted.

Informed criticism and constructive comment during the committee stage are particularly important for maximum impact on the structure and content of the future standard.

The draft standards are made available for public comment and the committee considers any proposals made at this stage. The standard is adopted when the necessary consensus for its application has been reached.

Strategy, policy, work programmes and resource requirements are formulated and managed by Councils and Policy Committees covering all sectors of industry and commerce.

International Organization for Standardization (ISO)

What ISO offers

ISO is made up of national standards institutes from countries large and small, industrialized and developing, in all regions of the world. ISO develops voluntary technical standards, which add value to all types of business operations. They contribute to making the development, manufacturing and supply of products and services more efficient, safer and cleaner. They make trade between countries easier and fairer. ISO standards also serve to safeguard consumers, and users in general, of products and services – as well as making their lives simpler.

ISO's name

Because the name of the International Organization for Standardization would have different abbreviations in different languages (ISO in English, OIN in French), it was decided to use a word derived from the Greek ISOS, meaning, and 'equal'. Therefore, the short form of the Organization's name is always ISO.

How it started

International standardization began in the electrotechnical field: the International Electrotechnical Commission (IEC) was established in 1906. Pioneering work in other fields was carried out by the International Federation of the National Standardizing Associations (ISA), which was set up in 1926. The emphasis within ISA was laid heavily on mechanical engineering. ISA's activities came to an end in 1942.

In 1946, delegates from 25 countries met in London and decided to create a new international organization, of which the object would be 'to facilitate the international coordination and unification of industrial standards'. The new organization, ISO, officially began operating on 23 February 1947. ISO currently has some 140-member organizations on the basis of one member per country. ISO is a non-governmental organization and its members are not, therefore, national governments, but are the standards institutes in their respective countries.

Every participating member has the right to take part in the development of any standard which it judges to be important to its country's economy. No matter what the size or strength of that economy, each participating member in ISO has one vote. ISO's activities are thus carried out in a democratic framework where each country is on an equal footing to influence the direction of ISO's work at the strategic level, as well as the technical content of its individual standards. ISO standards are voluntary. ISO does not enforce their implementation. A certain percentage of ISO standards – mainly those concerned with health, safety or the environment – has been adopted in some countries as part of their regulatory framework, or is referred to in legislation for which it serves as the technical basis. However, such adoptions are sovereign decisions by the regulatory authorities or governments of the countries concerned. ISO itself does not regulate or legislate.

ISO standards are market-driven. They are developed by international consensus among experts drawn from the industrial, technical or business sectors, which have expressed the need for a particular standard. These may be joined by experts from government, regulatory authorities, testing bodies, academia, consumer groups or other organizations with relevant knowledge, or which have expressed a direct interest in the standard under development. Although ISO standards are voluntary, the fact that they are developed in response to market demand, and are based on consensus among the interested parties, ensures widespread use of the standards.

ISO standards are technical agreements, which provide the framework for compatible technology worldwide. Developing technical consensus on this international scale is a major operation. This technical work is co-ordinated from ISO Central Secretariat in Geneva, which also publishes the standards.

Quantity and quality

Since 1947, ISO has published some 13 000 International Standards. ISO's work programme ranges from standards for traditional activities, such as agriculture and construction, through mechanical engineering to the newest information technology developments, such as the digital coding of audio-visual signals for multimedia applications.

Standardization of screw threads helps to keep chairs, children's bicycles and aircraft together and solves the repair

and maintenance problems caused by a lack of standardization that were once a major headache for manufacturers and product users. Standards establishing an international consensus on terminology make technology transfer easier and can represent an important stage in the advancement of new technologies.

Without the standardized dimensions of freight containers, international trade would be slower and more expensive. Without the standardization of telephone and banking cards, life would be more complicated. A lack of standardization may even affect the quality of life itself: for the disabled, for example, when they are barred access to consumer products, public transport and buildings because the dimensions of wheelchairs and entrances are not standardized. Standardized symbols provide danger warnings and information across linguistic frontiers. Consensus on grades of various materials gives a common reference for suppliers and clients in business dealings.

Agreement on a sufficient number of variations of a product to meet most current applications allows economies of scale with cost benefits for both producers and consumers. An example is the standardization of paper sizes. Standardization of performance or safety requirements of diverse equipment makes sure that users' needs are met while allowing individual manufacturers the freedom to design their own solution on how to meet those needs. Consumers then have a choice of products, which nevertheless meet basic requirements, and they benefit from the effects of competition among manufacturers.

Standardized protocols allow computers from different vendors to 'talk' to each other. Standardized documents speed up the transit of goods, or identify sensitive or dangerous cargoes that may be handled by people speaking different languages. Standardization of connections and interfaces of all types ensures the compatibility of equipment of diverse origins and the interoperability of different technologies.

Agreement on test methods allows meaningful comparisons of products, or plays an important part in controlling pollution – whether by noise, vibration or emissions. Safety standards for machinery protect people at work, at play, at sea . . . and at the dentist's. Without the international agreement contained in ISO standards on quantities and units, shopping and trade would be haphazard, science would be – well, unscientific – and technological development would be handicapped.

Tens of thousands of businesses in more than 150 countries are implementing ISO 9000, which provides a framework for quality management and quality assurance throughout the processes of producing and delivering products and services for the customer.

Conformity assessment

It is not the role of ISO to verify that ISO standards are being implemented by users in conformity with the requirements of the standards. Conformity assessment – as this verification process is known – is a matter for suppliers and their clients in the private sector, and of regulatory bodies when ISO standards have been incorporated into public legislation. In addition, there exist many testing laboratories and auditing bodies, which offer independent (also known as 'third party') conformity assessment services to verify that products, services or systems measure up to ISO standards. Such organizations may perform these services under a mandate to a regulatory authority, or as a commercial activity of which the aim is to create confidence between suppliers and their clients.

However, ISO develops ISO/IEC guides and standards to be used by organizations which carry out conformity assessment activities. The voluntary criteria contained in these guides represent an international consensus on what constitutes best practice. Their use contributes to the consistency and coherence of conformity assessment worldwide and so facilitates trade across borders.

Certification

When a product, service, or system has been assessed by a competent authority as conforming to the requirements of a relevant standard, a certificate may be issued as proof. For example, many thousands of ISO 9000 certificates have been issued to businesses around the world attesting to the fact that a quality management system operated by the company concerned conforms to one of the ISO 9000 standards. Likewise, more and more companies now seek certification of their environmental management systems to the ISO 14001 standard. ISO itself does not carry out certification to its management system standards and it does not issue either ISO 9000 or ISO 14000 certificates.

To sum up, ISO standards are market-driven. They are developed on the basis of international consensus among experts from the sector, which has expressed a requirement for a particular standard. Since ISO standards are voluntary, they are used to the extent that people find them useful. In cases like ISO 9000 – which is the most visible current example, but not the only one – that can mean very useful indeed!

The ISO catalogue

The ISO catalogue is published annually. The catalogue for example, contains a list of all currently valid ISO standards and other publications issued up to 31 December of the previous year.

The standards are presented by subject according to the International Classification for Standards (ICS).

Lists in numerical order and in technical committee order are also given. In addition, there is an alphabetical index and a list of standards withdrawn. Requests for information concerning the work of ISO should be addressed to the ISO Central Secretariat or to any of the National Member Bodies listed below.

ISO Central Secretariat
1, rue de Varembe
Case postale 56
CH-1211 Geneve 20
Switzerland
E-mail central@iso.ch
Web www.iso.ch

ISO/IEC Information Centre
E-mail mbinfo@iso.ch
www.standsinfo.net

Sales department
e-mail sales@iso.ch

ISO membership
The following bodies, constitute the total membership of the International Standards Organization. The letters in parentheses after the name of the country signify that country's national standards. For example, DIN plugs used on hi-fi equipment are manufactured to German standards. Plant designed to ANSI standards will be in accordance with American practice (see ISO webpage for most up-to-date information at www.iso.org/iso/about/iso-members.htm).

Afghanistan (ANSA)
Afghanistan National Standardization Authority
Kabul-Jalababad Road
AF-Kabul, P.O. Box 5172
Central Post Office
AF-KABUL
e-mail: ansa.ceo@gmail.com

Albania (DPS)
General Directorate of Standardization
"Mine Peza" Str. Nr 143/3
P.O. Box 98
e-mail: info@dps.gov.al

Algeria (IANOR)
Institut Algerien de Normalisation
5 et 7 rue Abou Hamou Moussa (ex-rue Daguerre)
B P. 104 R.P
ALGER
e-mail: dg@ianor.org

Angola (IANORQ)
Instituto Angolano de Normalização e Qualidade
Ministério da Industria
Rua Cerqueira Lukoki N°25, 7° andar
AO-LUANDA
C.P. 3709
e-mail: ianorq@netangola.com

Antigua and Barbuda (ABBS)
Antigua and Barbuda Bureau of Standards
Redcliffe St. & Corn Alley
P.O. Box 1550
AG-St. John's
e-mail: abbs@antigua.gov.ag

Argentina (IRAM)
Instituto Argentino de Normalizacion
Peru 552/556 AR-C 1068 AAB BUENOS AIRES
e-mail: iram-iso@iram.org

Armenia (SARM)
National Institute of Standards
Ministry of Economy
Komitas Avenue 49/2, AM-AM-0051
e-mail: sarm@sarm.am

Australia (SA)
Standards Australia
Level 10
The Exchange Centre
20 Bridge Street
Sydney, NSW 2000
GPO Box 476
SYDNEY NSW 2000
e-mail: intsect@standards.org.au

Austria (ON)
Austrian Standards Institute
Osterreichsches Normungsinstut
Heinestrasse 38
A-1021 WIEN
e-mail: iso@on-norm.at

Bahrain (BSMD)
Bahrain Standards & Metrology Directorate
Ministry of Industry and Commerce
P.O. Box 5479, Manama
Ministry of Industry & Commerce
BH-Bahrain
e-mail: bsmd@commerce.gov.bh

Bangladesh (BSTI)
Bangladesh Standards and Testing Institution
116/A, Teigeon Industrial Area
DHAKA-1208
e-mail: bsti@bangla.net

Barbados (BNSI)
Barbados National Standards Institution
Flodden Culloden Road
ST. MICHAEL
e-mail: office@bnsi.com.bb

Belarus (BELST)
State Committee for
Standardization, Metrology and
93 Starovilensky Trakt
MINSK 220053
e-mail: belst@anitex.by

Belgium (NBN)
Bureau de Normalisation
Av. de Ia Brabanconne 29
BE-1000 BRUXELLES
e-mail: vanvaerenbergh@nbn.be

Benin (CEBENOR)
Centre Béninois de Normalisation et de Gestion de la
　Qualité
Immeuble Trinité

Quartier Avleketecondji, 1ère Von à droite
Après le Carrefour Caboma, 02 BP 1101
BJ-Cotonou
e-mail: cebenorbenin@yahoo.fr

Bhutan (SQCA)
Standards and Quality Control Authority
Ministry of Works and Human Settlement
Royal Government of Bhutan
BT-Thimphu
e-mail: stdunit@sqcd.gov.bt

Bolivia (IBNORCA)
Instituto Boliviano de Normalización y Calidad
Av. Busch N° 1196 (entre calles Guatemala y Haití)
Zona Miraflores
Casilla postal N° 5034
BO-La Paz
e-mail: info@ibnorca.org

Bosnia and Herzegovina (BAS)
Institute for Standardization of Bosnia and Herzegovina
UI. Vojvode Radomira Putnika 34
BA-71123 Istocno, SARAJEVO
e-mail: stand@bas.gov.ba

Botswana (BOBS)
Botswana Bureau of Standards
Main Airport Road
Plot No. 55745, Block 8
Private Bag BO 48
BW-GABORONE
e-mail: infoc@hq.bobstandards.bw

Brazil (ABNT)
Associacao Brasileira de Normas
Av. 13 de Maio, No 13, 28 andar
20003-900 – RIO DE JANEIRO-RJ
e-mail: abnt@abnt.org.br

Brunei Darussalam (CPRU)
Construction Planning and Research Unit
Ministry of Development
BN-Brunei Darussalam
e-mail: modcpru@brunet.bn

Bulgaria (BDS)
Bulgarian Institute for Standardization
"Izgrev" Complex, 165 Str., Nr 3A
BG-1797 SOFIA
e-mail: standards@bds-bg.org

Burkina Faso (FASONORM)
Direction de la Normalisation et de la Promotion de la
　Qualité
Avenue de l'UEMOA
Immeuble ONAC
01 BP : 389
BF-Ouagadougou 01
e-mail: fasonorm@onac.bf

Burundi (BBN)
Bureau burundais de Normalisation et Contrôle de la
　Qualité
500 Boulevard de la Tanzanie

B.P. 3535
BI-Bujumbura
e-mail: bbnorme@yahoo.fr
Canada (SCC)
Standards Council of Canada
270 Albert Street, Suite 200
OTTAWA, ONTARIO K1P 6N7
e-mail: nfo@scc.ca
Cambodia (ISC)
Department of Industrial Standards of Cambodia
Ministry of Industry, Mines and Energy
#45 Norodom Blvd.
KH-Phnom Penh
e-mail: discinfo@camnet.com.kh
Cameroon (CDNQ)
Division de la Normalisation et de la Qualité
Ministère de l'Industrie, des Mines
et du Développement technologique
B.P. 5674
CM-Yaoundé, Centre 00237
e-mail: bootoangon@yahoo.fr
Chile (INN)
Instituto Nacional de Normalizacion
Matias Cousino 64 – 6c piso
Casilla 995 – Correo Central
SANTIAGO
e-mail: normas@inn.cl
China (SAC)
Standardization Administration of China
No. 9 Madian East Road
BEIJING 100088
e-mail: sac@sac.gov.cn
Colombia (ICONTEC)
Instituto Colombiano de Normas
Tecnicas y Certificacion
Carrera 37 52-95, Edificio ICONTEC
P.O. Box 14237
BOGOTA
e-mail: cliente@icontec.org
Congo, The Democratic Republic of the (OCC)
Office Congolais de Contrôle
98, Avenue du Port
B.P. 8806
CD-Kinshasa/Gombe
e-mail: delegation.generale_occ@yahoo.fr
Congo, The Republic of the (ACONOR)
Association Congolaise de Normalisation
c/o Chambre consulaire de Pointe-Noire
3, Boulevard Général De Gaulle
B.P. 665
CG-Pointe-Noire
e-mail: aconorcongo@yahoo.fr
Costa Rica (INTECO)
Instituto de Normas Tecnicas de
Costa Rica Barrio Gonzalez Flores
P.O. Box 10004-1000

SAN JOSE
e-mail: cerodiquez@inteco.or.cr
Côte-d'Ivoire (CODINORM)
Côte d'Ivoire Normalisation
Angle Rue du Commerce/Bd Botreau Roussel
5e Etage, Immeuble le Général
Abidjan Plateau
CI-Abidjan 01
e-mail: codinorm@powernet.ci
Croatia (HZN)
Croatian Standards Institute
Ulica grada Vukovara 78
P.P. 67
10000 ZAGREB
e-mail: hzn@hzn.hr
Cuba (NC)
Oficina Nacional de Normalización (NC)
Calle E No. 261 entre 11 y 13
VEDADO, LA HABANA 10400
e-mail: nc@ncnorma.cu
Cyprus (CYS)
Cyprus Organization for Standardization
Limassol Avenue and Kosta Anaxagora 30
P.O. Box 16197
CY-Nicosia 20
e-mail: cystandards@cys.org.cy
Czech Republic (UNMZ)
Czech Office of Standards, Metrology and Testing
Biskupsky dvur 5
110 02 PRAHA 1
e-mail: extrel@cni.cz
Denmark (DS)
Dansk Standard (DS)
Kollegieve 6
DK-2920 CHARLOTTENLUND
e-mail: dansk.standard@ds.dk
Dominica (DBOS)
Dominica Bureau of Standards
9 Great Marlborough Street
P.O. Box 1015
DM-Roseau
e-mail: info@dominicastandards.org
Dominican Republic (DIGENOR)
Dirección General de Normas y Sistemas de Calidad
Edificio de Oficinas Gubernamentales Juan Pablo Duarte
Piso 11
Ave. Mexico esquina Leopoldo Navarro
DO-Santo Domingo, D.N.
e-mail: digenor@digenor.gob.do
Ecuador (INEN)
Instituto Ecuatoriano de
Normalisacion
Calle Baquerizo Moreno No. 454 y
Almagro Edificio INEN
P.O. Box 17-01-3999
QUITO

e-mail: furresta@inen.gov.ec

El Salvador (CONACYT)
Consejo Nacional de Ciencia y Tecnología
Colonia Médica
Ave. Dr. E. Alvarez y Pasaje
Dr. Guillermo Rodríguez Pacas no. 51
SV-San Salvador
E-mail: evanegas@conacyt.gob.sv

Egypt (EOS)
Egyptian Organizaton for Standardization and
Quality (EOS)
16 Tadreeb EL-Modarrebeen St.
EI-Ameriya CAIRO
e-mail: moi@idsc.net.eg

Eritrea (ESI)
Eritrean Standards Institution
P.O. Box 245
171-1 House No. 9
ER-Asmara
e-mail: eristand@tse.com.er

Estonia (EVS)
Eesti Standardikeskus
10, Aru Street
EE-10317 Tallinn
e-mail: info@evs.ee

Ethiopia (QSAE)
Quality and Standards Authority of Ethiopia
P.O. Box 2310
ADDIS ABABA
e-mail: qsae@ethionet.et

Fiji (FTSQCO)
Fiji Trade Standards and Quality Control Office
Ministry of Industry, Tourism, Trade and
  Communication Naibati House P O Box 2118
FJ-Suva
e-mail: seema.sharma@govnet.gov.fj

Finland (SFS)
Finnish Standards Associaton SFS
P.O. Box 116
FI-00241 HELSINKI
e-mail: sfs@sfs.fi

France (AFNOR)
Association Francaise de
Normalisation
F-93571 PARIS LA DEFENSE
CEDEX
e-mail: uari@org.afnor

Gabon (ANTT)
Agence de Normalisation et de Transfert de Technologies
Ministère du Commerce et du Développement
  industruel, Chargé du NEPAD
Quartier Louis Libreville
BP 561
GA-Libreville
e-mail: s_emane@yahoo.fr

Georgia (GEOSTM)
Georgian National Agency for Standards, Technical
  Regulations and Metrology
67 Chargali Street
GE-Tbilisi 0141
e-mail: gnim_metrology@yahoo.com

Germany (DIN)
DIN Deutsches Institut fur Normung
Burggrafenstrasse 6
D-10787 BERLIN
e-mail: directorate.international@din.de

Ghana (GSB)
Ghana Standards Board
P.O. Box M 245
ACCRA
e-mail: gsbdir@ghanastandards.org

Greece (ELOT)
Hellenic Organization for Standardization
313, Acharnon Street
Gr-111 45 ATHENS
e-mail: info@elot.gr

Guatemala (COGUANOR)
Comisión Guatemalteca de Normas
Octava Avenidad 10-43, Zona 1
GT-Guatemala C.A. 01001
e-mail: mbeteta@mineco.gob.gt

Guinea (IGNM)
Institut Guinéen de Normalisation et de Métrologie
Ministère de l'Industrie, du Commerce, du Tourisme et
  de l'Artisanat
Quartier Almamya
KA 003
GN-Conakry
e-mail: inm89@yahoo.fr

Guyana (GNBS)
Guyana National Bureau of Standards
Flat 15, Sophia Exhibition Complex
Sophia
GY-Georgetown
e-mail: gnbs@networksgy.com

Hungary (MSZT)
Magyar Szabvanyugyi Testulet
Ulloi ut 25, Pf. 24
H-1450 BUDAPEST 9
e-mail: isoline@mszt.hu

Honduras (COHCIT)
Consejo Hondureño de Ciencia, Tecnología e Innovación
  (COHCIT)
Centro Civico Gubernamental
Boulevar Fuerzas Armadas
Edificio COHCIT-CAD
HN-Tegucigalpa M.D.C.
e-mail: ohn@cohcit.gob.hn

Hong Kong, China (ITCHKSAR)
Innovation and Technology Commission

Quality Services Division
Product Standards Information Bureau
36/F Immigration Tower
HK-Hong Kong
e-mail: psib@itc.gov.hk

Iceland (STRI)
Icelandic Standards
Skulatun 2
IS-105 REYKJAVIK
e-mail: stadlar@stadlar.is

India (BIS)
Bureau of Indian Standards
Manak Bhavan
9 Bahadur Shah Zafar Marg
NEW DELHI 110002
e-mail: ird@bis.org.in

Indonesia (BSN)
Badan Standardisasi Nasional
(National Standardization Agency, Indonesia)
Manggala Wanabakti Blok 4, 4th
Floor JL. Jenderal Gatot Subroto, Senayan
JAKARTA 10270
e-mail: bsn@bsn.go.id

Iran Islamic Republic of (ISIRI)
Institute of Standards and Industrial
Research of Iran
PD. Box 14155-6139, TEHRAN
e-mail: standard@isiri.or.ir

Ireland (NSAI)
National Standards Authority of Ireland
Swift Square, Santry
DUBLIN-9
e-mail: nsai@nsai.ie

Israel (SII)
Standards Institution of Israel
42 Chaim Levanon Street
TEL AVIV 69977
e-mail: sio/iec@sii.org.il

Italy (UNI)
Ente Nazionale Italiano di
Unificazione
Via Sannio, 2
20137 MILANO
e-mail: direzione@uni.com

Jamaica (BSJ)
Bureau of Standards Jamaica
6 Winchester Road
P.O. Box 113
KINGSTON 10
e-mail: info@bsj.org.jm

Japan (JISC)
Japanese Industrial Standards Committee
1-3-1, Kasumigaseki, Chiyoda-ku
TOKYO 100-8901
e-mail: isojisc@meti.gov.jp

Jordan (JISM)
Jordan Institution for Standards and Metrology (JISM)
Dabouq area
50 Khair Al-Din Al-Ma'ani St.
P.O. Box 941287
JO-Amman-11194
e-mail: jism@jism.gov.jo

Kazakhstan (KAZMEMST)
Committee for Technical Regulation and Metrology
Orynbor. 11 Street
010000 ASTANA
e-mail: int_rel@memst.kz

Kenya (KEBS)
Kenya Bureau of Standards
Popo Road
P.O. Box 54974-00200
NAIROBI
e-mail: info@kebs.org

Korea, Democratic People's Republic of (CSK)
Committee for Standardization of the
Democratic People's Republic of Korea
Inhung-Dong No. 1
PYONGYANO
e-mail: pdk0301@163.com

Korea, Republic of (KATS)[*]
Korean Agency for Technology and Standards
96 Gyoyukwon Gil, Gwacheon-si
KYUNGGI-DO 427-723
e-mail: standard@ats.go.kr

Kuwait (KOWSMD)
Public Authority for Industry
Standards and Industrial Services
Affairs (KOWSMD)
Standards & Metrology Department
Post Box 4690 Safat
KW-1 3047 KUWAIT
e-mail: aziz1994@yahoo.com

Kyrgyzstan (KYRGYZST)
National Institute for Standards and Metrology of the
  Kyrgyz Republic
197, Panfilov street
KG-720040 Bishkek
e-mail: nism@nism.gov.kg

Lao People's Democratic Rep. (DISM)
Department of Intellectual Property, Standardization and
  Metrology
National Authority for Science and Technology (NAST)
  Nahaidiou Rd P.O. Box 2279
LA-VIENTIANE
e-mail: sisomphet@nast.gov.la

Latvia (LVS)
Latvian Standard
157, Kr. Valdemara Street
LV-Riga 1013
e-mail: lvs@lvs.lv

Lebanon (LIBNOR)
Lebanese Standards Institution
Sin El Fil
P.O. Box 55120
LB-Beirut
e-mail: libnor@libnor.org

Lesotho (LSQAS)
Standards and Quality Assurance Department
Ministry of Trade and Industry, Cooperatives and
  Marketing
P.O. Box 747
LS-Maseru 100
e-mail: lessqa@leo.co.ls

Libyan Arab Jamahiriya (LNCSM)
Libyan National Centre for
Standardization and Metrology
Tripoli El Fornaj
P.O. Box 5178
TRIPOLI
e-mail: info@Incsm.org.ly

Lithuania (LST)
Lithuanian Standards Board
T. Kosciuskos g.30
LT-01100 Vilnius
e-mail: lstboard@lsd.lt

Luxembourg (ILNAS)
Institut Luxembougeois de la Normalisation
34-40 av. de la Porte-Neuve
Boite Postale 10
LU-2227 LUXEMBOURG
e-mail: normalisation@ilnas.etat.lu

Macau, China (CPTTM)
Macau Productivity and Technology Transfer Center
Rua de Xangai, 175
Edificio Associaçao Comercial de Macau
6°andar
MO-Macau
e-mail: cpttm@cpttm.org.mo

Madagascar (BNM)
Bureau de Normes de Madagascar
B.P. 1316
MG-Antananarivo 101
e-mail: bnm@moov.mg

Malawi (MBS)
Malawi Bureau of Standards
Moirs Road:
P.O. Box 946
MW-Blantyre
e-mail: mbs@mbsmw.org

Malaysia (DSM)
Department of Standards Malaysia
Ministry of Science, Technology and Innovation
Block 2300, Jalan Usahawan
MY-63000 Cyberjaya
SELANGOR
e-mail: central@standardsmalaysia.gov.my

Malta (MSA)
Malta Standards Authority
Second Floor, Evans Building
Merchants Street
VALLETTA VLT 03
e-mail: standards@msa.org.mt

Mauritius (MSB)
Mauritius Standards Bureau
Villa Road, MOKA
e-mail: msb@intnet.mu

Mexico (DGN)
Direccion General de Normas
Calle Puente de Tecamachalco No. 6
Lomas de Tecamachalco
Seccion Fuentes
Naucalpan de Juarez 53950
MEXICO
e-mail: iso-mex@economia.gob.mx

Moldova, Republic of (INSM)
National Institute of Standardization and Metrology of
  the Republic of Moldova
28, E. Coca str.
MD-Chisinau MD-2064
e-mail: insm@standard.md

Mongolia (MASM)
Mongolian Agency for
Standardization and Metrology
P.O. Box 48
ULAANBAATAR 211051
e-mail: masm@mongol.net

Montenegro (ISME)
Institute for Standardization of Montenegro
91, Bul. Sv. Petra Cetinjskog
ME-81000 Podgorica
e-mail: isme@cg.yu

Morocco (SNIMA)
Service de Normalisation Industriel
Marocaine (SNIMA)
Ministere de lindustrie, du Commerce,
des Nouvelles Technologies
Angle Avenue Kamal Zebdi et Rue Dadi
Secteur 21 Hay Riad 10100 RABAT
e-mail: snima@mcinet.gov.ma

Mozambique (INNOQ)
Instituto Nacional de Normalização e Qualidade
Av. 25 de Setembro n° 1179, 2° andar
P.O. Box: 2983
MZ-Maputo
e-mail: innoq@emilmoz.com

Myanmar (MSTRD)
Myanma Scientific and Technological Research
  Department
Ministry of Science and Technology
No. 6 KabaAye Pagoda Road
Yankin P.O.
MM-YANGON

e-mail: most7@myanmar.com.mm

Namibia (NSI)
Namibian Standards Institution
Old Sanlam Building, First Floor, Suite 115
11-17 Dr Frans Indongo Street
P.O. Box 26364
NA-Windhoek
e-mail: kaakunga@nsi.com.na

Nepal (NBSM)
Nepal Bureau of Standards and Metrology
P.O. Box 985
Balaju
NP-Kathmandu
e-mail: nbsm@nbsm.gov.np

Netherlands (NEN)
Nederlands Normalisatie-Instituut
P.O. Box 5059
NL-2600 GB DELFT
e-mail: info@nen.nl

New Zealand (SNZ)
Standards New Zealand
Radio New Zealand House
155 The Terrace
Private Bag 2439
WELLINGTON 6020
e-mail: isoadmin@standards.co.nz

Nigeria (SON)
Standards Organisation of Nigeria
No. 52 Lome Crescent
Wuse Zone 7
NG-Abuja, WUSE 2349
e-mail: info@sononline-ng.org

Norway (SN)
Standards Norway
Strandveien 18
P.O. Box 242
No-1326 YSAKER
e-mail: info@standard.no

Oman (DGSM)
Directorate General for Specifications and
  Measurements
Ministry of Commerce and Industry
P.O. Box 550-Postal code No. 113
OM-Muscat
e-mail: dgsm123@omantel.net.om

Pakistan (PSQCA)
Pakistan Standards and Quality Control Authority
Block 77, Pakistan Secretariat
KARACHI-74400
e-mail: sqcadg@super.net.pk

Palestine (PSI)
Palestine Standards Institution
P.O. Box 2258
Hay El-Tal
Al-Quds St.
PS-Ramallah

e-mail: info@psi.gov.ps

Panama (COPANIT)
Comision Panamena de Normas
Industriales y Tecnicas
Edificio Plaza Edison, Tercer Piso
Avenida Ricardo J. Alfaro y Calle E1 Paical
Apartado 0815-0119, PANAMA 4
e-mail: dgnti@mici.gob.pa

Philippines (BPS)
Bureau of Product Standards
Department of Trade and Industry
361 Sen. Gil J. Puyat Avenue
Makati City
METRO MANILA 1200
e-mail: bps@dti.gov.ph

Poland (PKN)
Polish Committee for Standardization
ul. Swietokrzyska 4
PL-00-50 WARSZAWA
e-mail: pl.isonb@pkn.pl

Portugal (IPO)
Institute Portugues da Oualidade
Rua Antonio Giao, 2
P-2829-513 CAPARICA
e-mail: ipg@mail.ipq.pt

Romania (ASRO)
Associatia de Standardizare din Romania
Str. Mendelieev 21-25
RO-010362 BUCURESTI 1
e-mail: international@asro.ro

Russian Federation (GOST R)
Federation for Standardization and Metrology
Leninsky Prospekt 9
MOSKVA V-49, GSP-1, 119991
e-mail: iso@gost.ru

Rwanda (RBS)
Rwanda Bureau of Standards
P.O. Box 7099
RW-Kigali
e-mail: orn@rwanda1.com

Saint Lucia (SLBS)
Saint Lucia Bureau of Standards
Bisee Industrial Estate
CP 5412
LC-Castries
e-mail: slbs@candw.lc

Saint Vincent and the Grenadines (SVGBS)
St. Vincent and the Grenadines Bureau of Standards
Campden Park Industrial Estate
P.O. Box 1506
VC-Kingstown
e-mail: svgbs@vincysurf.com

Saudi Arabia (SASO)
Saudi Arabian Standards Organization
Imam Saud Bin Abdul Aziz Bin
Mohammed Road (West End)

P.O. Box 3437
RIYADH 11471
e-mail: saso@saso.org.sa
Senegal (ASN)
Association Sénégalaise de Normalisation
Ministère des Mines, de l'Industrie et des PME
122 bis, Avenue André Peytavin
Boîte postale 4037
SN-DAKAR
e-mail: asnor@orange.sn
Serbia (ISS)
Institute for Standardization of Serbia
Stevana Brakusa 2
Post. fah 2105
RS-11030 Belgrade
e-mail: iss-international@iss.rs
Seychelles (SBS)
Seychelles Bureau of Standards
Providence Industrial Estate
P.O. Box 953
Victoria
SC-Mahé
e-mail: sbsorg@seychelles.net
Singapore (SPRING SG)
Singapore Productivity and Standards Board
2 Bukit Merah Central
SG-Singapore 159835
e-mail: stn@spring.gov.sg
Slovakia (SUTN)
Slovak Stanadards Institute
P.O. Box 246
Karleveska 63
84000 Bratislava 4
e-mail: int@sutn.gov.sk
Slovenia (SIST)
Slovenian Institute for Standardization
Smartinska c. 152
SI-1000 LJUBLJANA
e-mail: sist@sist.si
South Africa (SABS)
South African Bureau of Standards
1 Dr. Lategan Rd., Groenkloof
Private Bag X191
PRETORIA 0001
e-mail: info@sabs.co.za
Spain (AENOR)
Asociacion Espanola de
Normalizacion y Certificacido
Genova, 6
E-28004 MADRID
e-mail: aenor@aenor.es
Sri Lanka (SISI)
Sri Lanka Standards Institution
17 Victoria Place
Off Elvitigala Mawatha
COLOMBO 08

e-mail: dg@sisi.slt.lk
Sudan (SSMO)
Sudanese Standards and Metrology Organization
P.O. Box 13573
SD-Khartoum
e-mail: rahbamohamed563@hotmail.com
Suriname (SSB)
Suriname Standards Bureau
Leysweg no. 10, Uitvlugt
SR-Paramaribo
e-mail: dirssb@gmail.com
Swaziland (SWASA)
Swaziland Standards Authority
2nd Floor Pension Fund Building
Mhlambanyatsi Road
P.O. Box 300, Eveni
SZ-MBABANE
e-mail: drmkhonta@swasa.co.sz
Sweden (SIS)
Swedish Standards Institute
Sankt Paulsgatan 6
S-11880 STOCKHOLM
e-mail: info@sis.se
Switzerland (SNV)
Swiss Association for Standardization
Burglistrasse 29
8400 WINTERTHUR
e-mail: info@snv.ch
Syrian Arab Republic (SASMO)
Syrian Arab Organization for
Standardization and Metrology
P.O. Box 11836
DAMASCUS
e-mail: sasmo@net.sy
Tajikistan (TJKSTN)
Agency of Standardization, Metrology, Certification and
  Trade Inspection
Oulitsa N. Karaboeva 42/2
TJ-734018 DUCHANBE
e-mail: info@standard.tj
Tanzania, United Republic of (TBS)
Tanzania Bureau of Standards
Ubuno Area
Juntion of Morogoro Road/Sam Nujema Road
DAR ES SALAAM
e-mail: info@tbstz.org
Thailand (TISI)
Thai Industrial Standards Institute
Ministry of Industry
Rama VI Street
BANGKOK 10400
e-mail: intrelat@tisi.go.th
The former Yugoslav Republic of Macedonia (ISRM)
Standardization Institute of the Republic of Macedonia
Vasil Glavinov bb, block 10-mezanin
1000 SKOPJE

e-mail: isrm@isrm.gov.mk

Togo (CSN)
Conseil Supérieur de Normalisation
CASEF 4e étage, Porte 456
B.P. 831
TG-Lomé
e-mail: togonormes@yahoo.fr

Trinidad and Tobago (TTBS)
Trinidad and Tobago Bureau of Standards
1-2 Century Drive
Trincity Industrial Estate
P.O. Box 467 PORT OF SPAIN
TUNAPUNA
e-mail: ttbs@ttbs.org.net

Tunisia (INNORPI)
Institut National de la Normalisatien et
de la Propriete Industrielle
B.P. 57
1003 TUNIS
e-mail: inorpi@planet.tn

Turkey (TSE)
Turk Standardlari Enstitusu
Necatibey Cad. 112
Bakanliklar
TR-06100 ANKARA
e-mail: usm@tse.org.tr

Turkmenistan (MSST)
The Major State Service "Turkmenstandartlary"
1995 str., Building-12
TM-744000 Ashgabat
e-mail: ggigns@online.tm

Ukraine (DSSU)
State Committee of Ukraine on Technical Regulation and
    Consumer Policy
(Derzhspozhivstandard of Ukraine)
174, Gorkiy Street, GSP 03680
UA-Kyiv-150
e-mail: dstu@dssu.gov.ua

United Arab Emirates (ESMA)
Emirates Authority for Standardization and
    Metrology
P.O. Box 2166, ABU DHABI
e-mail: esma@esma.ae

United Kingdom (BSI)
British Standards Institution
389 Chiswick High Road
LONDON W4 4AL
e-mail: standards.international@bsigroup.com

Uruguay (UNIT)
Institute Uruguayo de Normas Tecnicas
Pza. Independencia 812, Piso 2
MONTEVIDEO
e-mail: unit-iso@unit.org.uy

USA (ANSI)
American National Standards Institute
1819 L Street, NW
WASHINGTON, DC 20036
e-mail: info@ansi.org

Uzbekistan (UZSTANDARD)
Agency for Standardization, Metrology and Certification
    of Uzbekistan
333 "a", Farobiy Street
100049 TASHKENT
e-mail: uzst@standart.uz

Venezuela (FONDONORMA)
Fondo Pare a Normalizacion y
Certificacion de Ia Calidad
Avenida Andres Bello, Edf. Torre Fondo Comun
Pisos 11 y 12, Apartado Postal 51116
CARACAS 1050-A
e-mail: info@fondonorma.org.ve

Vietnam (TCVN)
Directorate for Standards and Quality
8 Hoang Quoc Viet Road
HANOI
e-mail: htqt@tcvn.gov.vn

Yemen (YSMO)
Yemen Standardization, Metrology and Quality Control
    Organization
AL-Zoberi Street
Industrial Complex
P.O. Box 15261
YE-Sana'a
e-mail: YSMQCO@Y.NET.YE

Zambia (ZABS)
Zambia Bureau of Standards
Lechwe House
Freedomway-South End
P.O. Box 50259, ZA 15101, Ridgeway
ZM-LUSAKA
e-mail: zabs@zamnet.zm

Zimbabwe (SAZ)
Standards Association of Zimbabwe
P.O. Box 2259
HARARE
e-mail: info@saz.org.zw

Chapter 35

Production drawings

The following three typical drawings are included as examples of draughtsmanship, layout, dimensioning, and tolerancing.

Figures 35.1 and 35.2 show a pulley and a shaft, and illustrate some aspects of general dimensioning and tolerancing.

Figure 35.3 shows a partly dimensioned elevation and plan view of a proposed gear-box cover, with a wide application of theoretically exact boxed dimensioning and the associated positional tolerances. To emphasize this style of dimensioning, other dimensions relating to the form of the cover have been omitted.

Completed drawings are usually presented on company standard sheets. Sizes, areas and designations are given at the start of Chapter 5. The layout of a typical sheet

contains information required for identification, administration and interpretation and a company could design several types of standard sheets, particularly appropriate to their products.

If drawings are produced manually, then the drawing sheets may be supplied with printed borders and blocks containing relevant notes and headings. With computer aided design (CAD) layouts, similar details are stored in the database.

Figures 35.4–35.7, clearly show that documentation can occupy much of the sheet area. The space provided for basic and supplementary information will be decided by the company; its actual position on the sheet may vary with the sheet size.

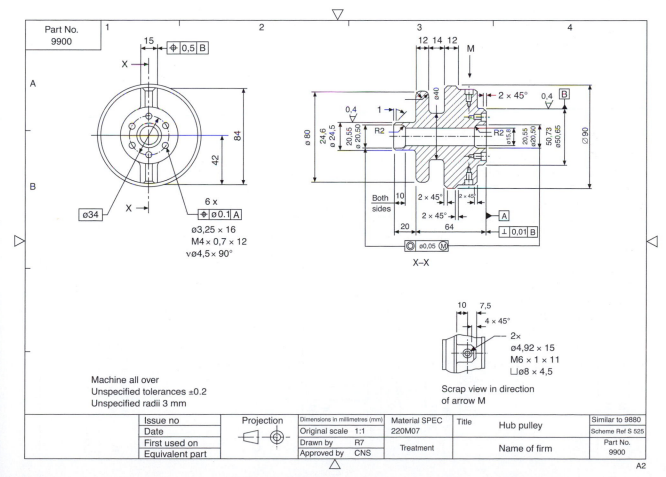

FIGURE 35.1

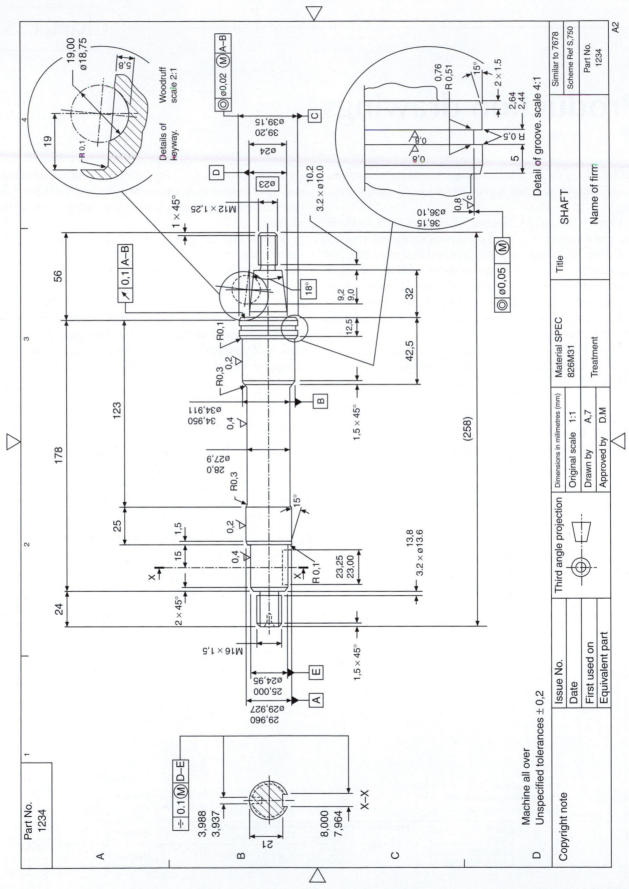

FIGURE 35.2

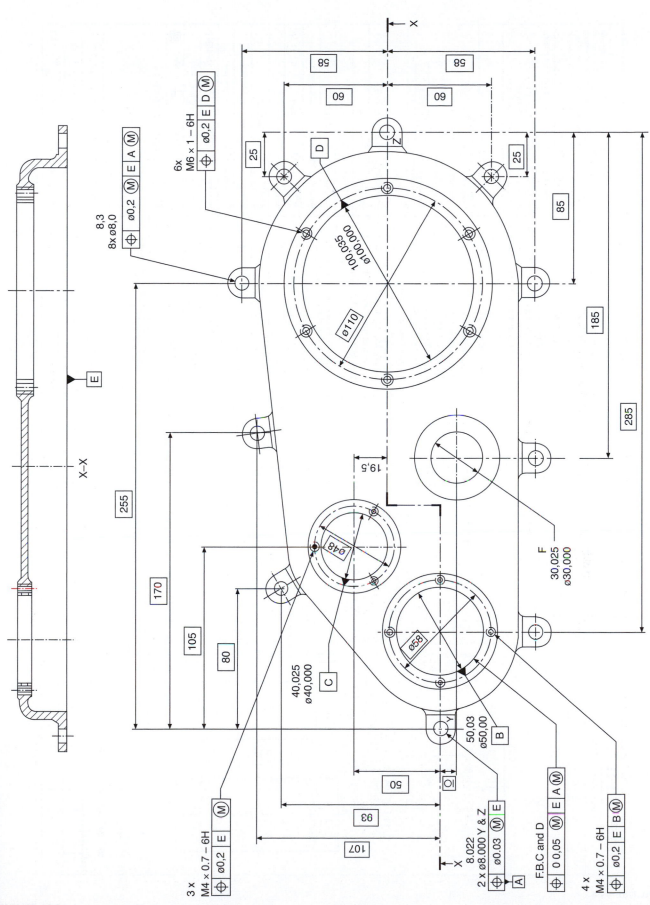

FIGURE 35.3

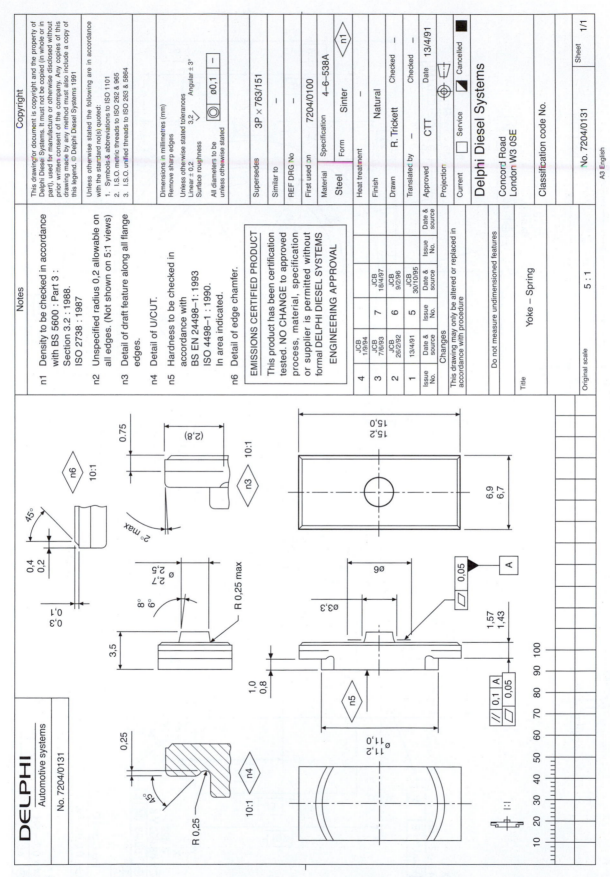

FIGURE 35.4

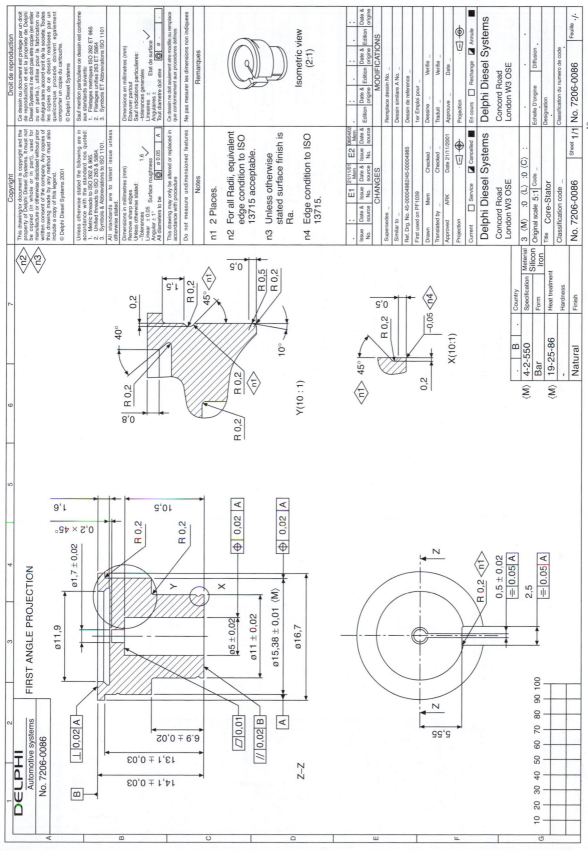

FIGURE 35.5

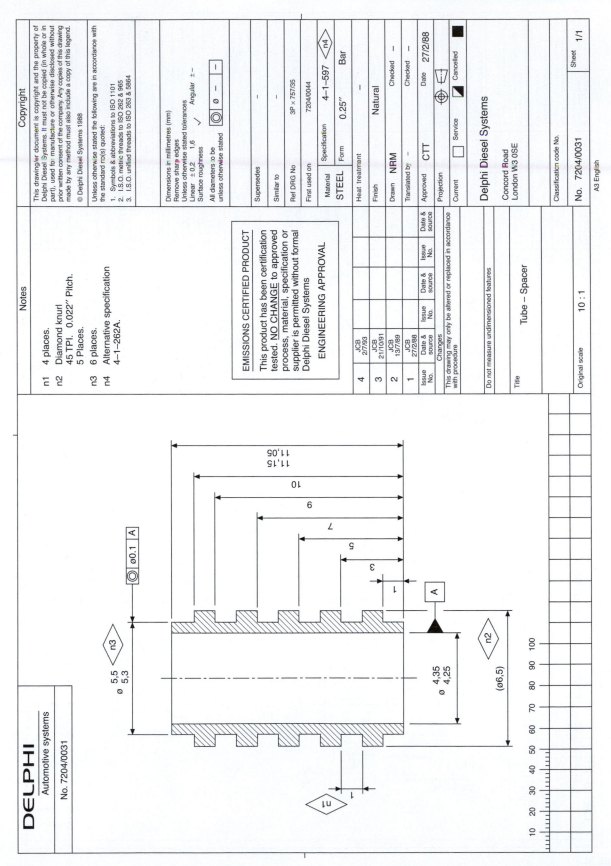

FIGURE 35.6

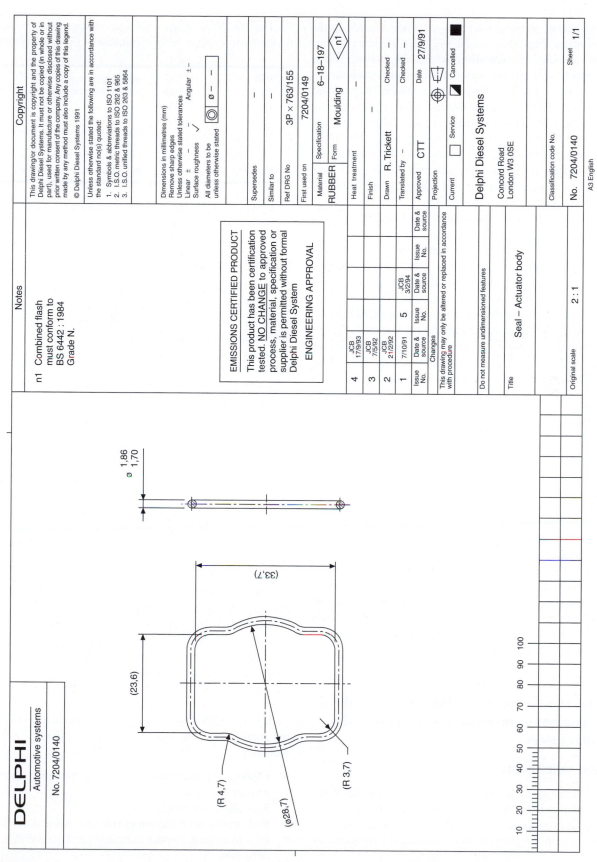

FIGURE 35.7

Engineering progress over the years has resulted in the issue of many Standards and clearly there are literally millions of drawings in circulation, which may contain minor details that have been superseded. It has been the custom in a drawing title block to record a brief note of dates when changes and modifications have occurred. A typical drawing number could be known as, for example: HB 345, for the original issue. Then HB 345A for the first reissue, HB 345B, for the second reissue. A brief note may be added to the drawing relating to the change. Each manufacturer will have their own system for recording full details and also advising current users that a change has occurred.

Examples are shown on some of the following illustrations.

In the case of textbooks, the front page generally states the publishing dates and obviously general standards used are appropriate to that date. Illustrations of typical layouts are reproduced by kind permission of Delphi Diesel Systems.

An advantage of producing a drawing on a CAD system is that the draughtsman can complete the illustration of the component or assembly on the screen. He may reposition some parts if space permits, to avoid congestion. The outline of the drawing sheet will be taken from the database on a separate layer and placed over the drawing to give a pleasing layout.

Draughtsmen producing manual drawings have always derived much personal satisfaction from producing drawings in ink or pencil where the linework and lettering is uniform and neat, and where the illustration completely defines the requirements of the specification. A similar degree of satisfaction can be obtained by producing drawings on CAD equipment. Particular care with the spacings of centre lines, contrasts in linework, crosshatchings, positioning of dimensions, notes, etc., are all small details in their own way, but they collectively contribute to quality, balance and overall clarity. Check the reason for the position of each and every line, ensure that it conforms with the appropriate standard and that it conveys a clear and unambiguous meaning to the reader. Engineering drawings can be considered to be works of art. We both hope you will produce many.

Further standards for design, project and risk management of interest to engineers and manufacturers

Design management

The first priority of an organization is to ensure that it survives and prospers. A business that fails to continuously develop its product range is unlikely to grow in real terms, especially with the increase in open competition across the globe.

This major series of management standards helps organizations plan ahead for products and services into the future thus ensuring continuity of revenue streams.

The BS 7000 series – Design Management Systems is divided into concise parts applicable to different sectors.

BS 7000 – 1	Guide to managing innovation
BS 7000 – 2	Guide to managing the design of manufactured products
BS 7000 – 3	Guide to managing service design
BS 7000 – 4	Guide to managing design in construction
BS 7000 – 6	Guide to managing inclusive design
BS 7000 – 10	Glossary of terms used in design management
BS EN 62402	Obsolescence management. Application guide.

(This standard replaced BS 7000-5)

Project management

BS 6079–1 Guide to Project Management describes a full range of project management procedures, techniques and tools that you can select as appropriate to your project. It gives guidance on the planning and execution of projects and the application of project management techniques. The standard has a broad relevance to projects in many industries including the public sector, both at home and abroad. The principles and procedures outlined are relevant to all sizes of organization.

This standard aims primarily to provide guidance for relative newcomers to project management and to act as an aide-memoire for more experienced practitioners and those who interact with project management teams.

The other parts of BS 6079 are also available.

BS 6079–2:2000 Project Management – Vocabulary

BS 6079–3:2000 Project Management – Guide to the Management of Business Related Project Risk Project Management.

These three standards can be purchased as kit 3.

BS 6079 part 4 is published as a PD 6079-4 Project Management in the Construction Industry BS IEC 62198. Project Risk Management Application Guidelines – This standard provides an internationally approved process for managing project and project-related risks in a systematic and consistent way. It is relevant to decision-makers, including project managers, risk managers and business managers.

BS IEC 62198 provides a general introduction to project risk management, its sub-processes and influencing factors, such as:

- Establishing the context, including confirmation of project objectives
- Risk identification
- Risk assessment, including risk analysis and evaluation
- Risk treatment, impact mitigation and probability reduction
- Review and monitoring
- Communication (including consultation)
- Learning from the project

Guidelines are also provided on the organizational requirements for implementing the process of risk management appropriate to the various phases of a project.

BS 8888 on CD-ROM.

Contains BS 8888 and the full set of approximately 160 cross-referenced documents.

- All the publications on one disk, accessible at the click of a button

- A complete and comprehensive collection of all cross-referenced documents
- A concise route-map to the complex web of ISO standards within this area

Details regarding content and current prices can be obtained from the BSI Information Customers Services, 389 Chiswick High Road. London W4 4AL website: www. bsi-global.com/bsonline. Tel.: +44 20 8996 9001. E-mail: orders@bsi-global.com.

Chapter 36

Drawing solutions

1. Solutions to Fig. 4.17

(a)

(b)

(c)

2. Solutions to Fig. 4.18

(a)

(b)

(c)

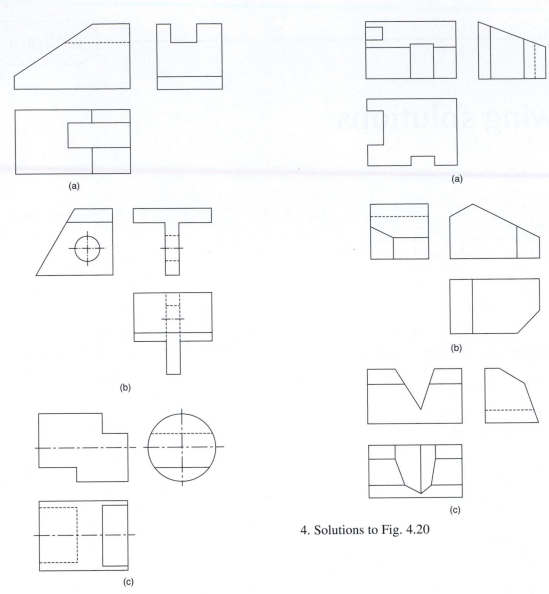

(a)

(b)

(c)

3. Solutions to Fig. 4.19

(a)

(b)

(c)

4. Solutions to Fig. 4.20

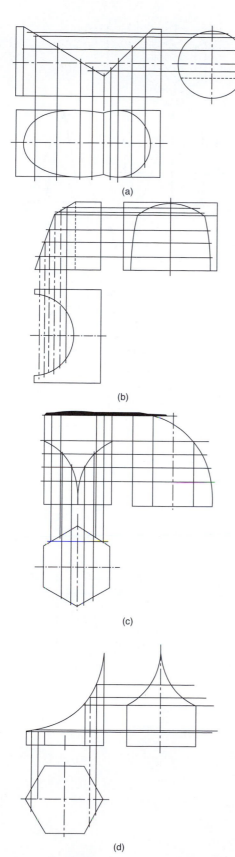

(a)

(b)

(c)

(d)

5. Solutions to Fig. 4.22

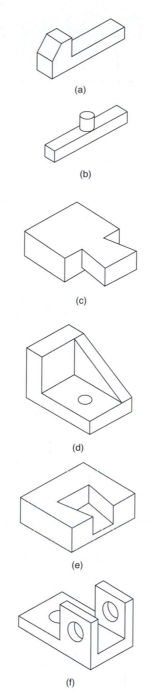

(a)

(b)

(c)

(d)

(e)

(f)

6. Solutions to Fig. 4.23

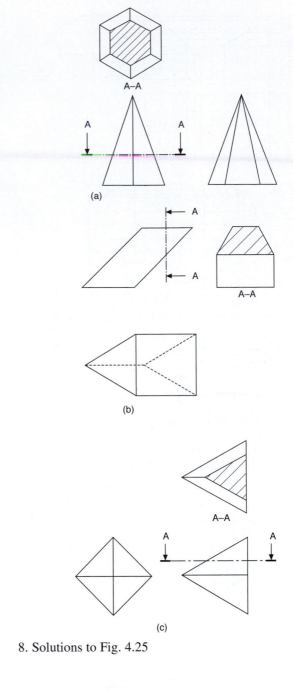

(a)

(b)

(c)

7. Solutions to Fig. 4.24

8. Solutions to Fig. 4.25

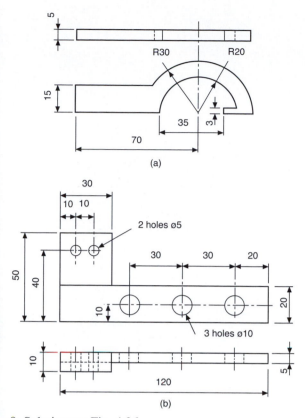

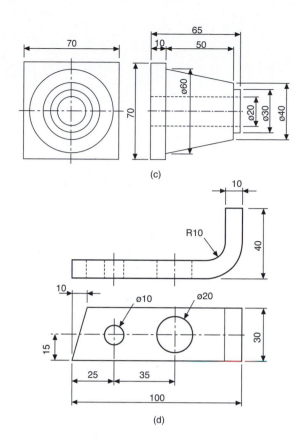

9. Solutions to Fig. 4.26

Index